Changing Energy

Changing Energy

The Transition to a Sustainable Future

John H. Perkins

UNIVERSITY OF CALIFORNIA PRESS

University of California Press, one of the most
distinguished university presses in the United States,
enriches lives around the world by advancing scholarship
in the humanities, social sciences, and natural sciences. Its
activities are supported by the UC Press Foundation and
by philanthropic contributions from individuals and
institutions. For more information, visit www.ucpress.edu.

University of California Press
Oakland, California

Library of Congress Cataloging-in-Publication Data

Names: Perkins, John H., author.
Title: Changing energy : the transition to a sustainable
    future / John H. Perkins.
Description: Oakland, California : University of
    California Press, [2017] | Includes bibliographical
    references and index.
Identifiers: LCCN 2017001098 (print) | LCCN 2017004076
    (ebook) | ISBN 9780520287785 (cloth : alk. paper) |
    ISBN 9780520287792 (pbk. : alk. paper) |
    ISBN 9780520962842 (ebook)
Subjects: LCSH: Energy consumption. | Renewable energy
    sources. | Fossil fuels. | Power resources. | Sustainable
    development.
Classification: LCC HD9502.A2 P465 2017 (print) |
    LCC HD9502.A2 (ebook) | DDC 333.79/4—dc23
LC record available at https://lccn.loc.gov/2017001098

Manufactured in the United States of America

26 25 24 23 22 21 20 19 18 17
10 9 8 7 6 5 4 3 2 1

For

Milo and Linus

And their cousins and peers

Their generation and those that follow stand at risk from unchanged energy.

Contents

Preface

Authors generally explain something about the origins of a book in the preface, but does it matter why someone decided to sit down long enough to grind out a narrative text? I think it does, in at least one sense: authors must have a passion that leads them to write, and readers benefit from knowing what that passion might be.

In my case, the decision to delve deeply into energy and write a book about it solidified with the births of two grandchildren in the first decade of the twenty-first century. As I looked at these marvelous, wiggling babies, I realized they had entered a world that was rapidly changing into something very different from the world that I have spent my life in.

I mused about the fact that my father and mother, both born about one hundred years earlier than my grandchildren, had entered a world in which automobiles and electricity were just beginning to appear, at least in the United States and Europe. For them, after the Great Depression and World War II, life was filled with incredible new machines and rapidly growing uses of energy, but they and their parents and grandparents also remembered the days of horses, wagons, and kerosene lamps.

By the time my sister and I, plus our cousins, arrived from the late 1930s to the 1950s, our family was firmly entrenched in the luxuries of the automobile, electric lights, radios, refrigerators, telephones, and gas-heated homes. Horses were strictly for recreational riding, and kerosene lamps provided a quaintly old-fashioned and rather dim light. Obsolete! And a fire hazard to boot. Moreover, we were never concerned about drinking water from the tap, because sewage water and drinking water didn't mix to threaten our health. For our generation, the cool, new things were jet planes, television, computers, and cell phones. Amazingly, one could just assume that there was enough fuel and electricity to run all these things.

Our son and his cousins came into a world truly at its peak for the abundance of energy and the services it provided, again at least for certain segments of the United States and other highly industrialized countries. On the horizon, however, the first murmurings of future problems had begun to appear. Generating all that electricity with coal and oil polluted the air. Automobiles demanded ever more space for highways and parking and likewise dumped toxic materials into air and water. Gasoline ready-to-buy could suddenly become scarce due to conflicts far away, and critics began to assail the dangers of nuclear power. Maybe energy services had a serious downside that might get worse? Rachel Carson's *Silent Spring* eloquently told a story of how modern technology could come back to bite its users, despite its genuine benefits.

Soon after our son's generation arrived, the new science of climate change gathered enough confidence in its findings to make unnerving predictions of risk; nuclear power plants exploded; health effects from air pollution grew worse; a country that could embrace nuclear power also acquired the skills to make nuclear weapons; and mining for fuels became more difficult and dangerous. By the time my grandchildren and their cousins arrived, it had become ever more obvious that—as much as we might like, indeed need, energy and energy services—the rapidly rising uses of coal, oil, gas, and uranium threatened the genuine benefits they provided.

This musing about the life-altering effects of energy and energy services, all within the short span of five generations of people I have known personally, combined with the risks and threats that had appeared, mostly within my lifetime, led me to focus on energy and energy services as problems of highest priority in the twenty-first century. Was there a way to preserve and expand the benefits of energy services with fuels and technology that have fewer intolerable downsides than coal, oil, gas, and uranium? Would my grandchildren, and their children and their grandchildren, draw on the resources of the earth to have a prosperous, healthy, and stable life? These are the questions that fueled my passion to write this book.

When I started, I thought the book would lay out both the strategic goal for changing energy and an assessment of tactics to reach the goal. As I progressed, however, I came to realize that consensus about the best strategic goal did not exist in discussions about energy. Without consensus on strategy, setting priorities was difficult if not impossible. Without priorities, policy choices remained captured by existing industries. Therefore, I decided to focus on just one thing: making a case for the optimal strategic goal.

The short statement of optimal strategy is easy to formulate: *countries must move as close as possible to 100 percent renewable energy used with high efficiency*. More pointedly, technologies based on natural gas, nuclear power, and carbon capture and sequestration are not part of the goal.

This book is not the first to suggest that 100 percent renewable energy is both possible and desirable as a target, but it seeks to make a comprehensive case for it. I believe that is its main contribution, and without consensus on that goal energy policies will remain muddled and ineffective.

The task of the next book is clear: How can humanity achieve the goal? What tactics will work, and how do successful tactics differ from country to country and person to person? Just as many arguments have surrounded discussions about the right strategic goal, so, too, will they envelop debates about the best tactics.

It is my hope that this book will usefully inform and educate engineers, scientists, political and business leaders, leaders in the labor and religious communities—indeed all citizens—as we grapple with some of the most difficult political, cultural, and moral problems that have arisen in the past three hundred years.

As the author of any book knows, it's not possible to bring one into the world without a great deal of help from others. Although I remain responsible for everything here—especially any mistakes—I had wonderful assistance on many fronts from others. I'm particularly indebted to the advice and suggestions from reviewers of early drafts.

- Dustin Mulvaney and an anonymous reviewer read the entire early draft for the University of California Press and provided excellent suggestions and encouragement, particularly Mulvaney.

- Two other anonymous reviewers for the press read the first complete prospectus for the book and encouraged its writing. I'm very indebted for this early, positive response.

- Kevin Francis, George Irwin, and Barbara Whitten read the first five chapters of an early draft, and their suggestions and critiques led to many changes for the better and added further encouragement. Each of them pointed out that the historical chapters had way too much detail, which obscured the points readers needed to grasp. This critique, plus others, proved invaluable. Both Irwin and Whitten are physicists, an educational background I had only a bit of (first-year physics as an undergraduate and physical chemistry as a graduate student), and I welcomed their abilities to comment in depth on the physics of energy. Their help, however, does not affect my complete and sole responsibility for any remaining errors.

- Mark Gilmore read and provided excellent suggestions for improvement on an early draft of chapter 3 on energy and the modern state; I appreciated his long experience in the banking industry. The economist Peter Dorman also provided excellent advice on this subject.

- Cheri Lucas Jennings and I developed a class at The Evergreen State College called Energy Matters, a title Cheri originated that helped me grasp what was at stake. The title of this book, *Changing Energy*, descends directly from that course.

- Bruce Jennings also critiqued an early draft of chapter 3 and raised a question that I could not answer then and am still thinking about now: would the "modern state" be better designated as the "market state"?

- Cathy French brought my attention to solar power installations that heated water by thermal absorption and electricity by photovoltaic methods.

Words remain indispensable for energy, but pictures and graphs frequently show one facet or another more eloquently and simply. The following individuals provided much assistance in helping me obtain suitable illustrations.

- Andrew Aldin, geologist and photographer
- Jesbin Baidya, Intergovernmental Panel on Climate Change
- Jonty Clark, cartoonist and illustrator
- Wayne Hicks, National Renewable Energy Laboratory
- David K. Hulse, engineer and historian of technology
- Rune Likvern, Resilience
- Kurt Menke, Birds Eye View GIS

- Gareth Peers, Science Photo Library
- Kathleen M. Saul, The Evergreen State College and University of Delaware
- Sophie Schlingemann, Intergovernmental Panel on Climate Change
- Anne M. Stark, Lawrence Livermore National Laboratory
- Vivian Stockman, Ohio Valley Environmental Coalition
- Mark A. Wilson, the College of Wooster

Teaching at the Evergreen State College involved prolonged interactions with colleagues, who expanded my horizons on dealing with challenging issues surrounding technology. Ralph Murphy and Tom Rainey imparted their wisdom on political economy, which with further input from Peter Dorman, Jeanne Hahn, Cheri Lucas Jennings, and Ted Whitesell prompted me on the development of political ecology as an analytical framework. Rob Knapp first introduced me to energy-flow charts (figure 5.2), an invaluable visual representation of energy economies, which has helped me understand the relationships among various primary energy sources. Lin Nelson, José Suarez, and Jude van Buren helped me grasp essential issues in public health. Paul Butler, Larry Eickstaedt, Steve Herman, Pat Labine, and Bob Sluss enlarged my appreciation for ecology, natural history, and geology.

Students, too, contributed in many ways to the development of the materials in this book. The class Energy Matters was given twice, in 2007 and 2009. The approximately sixty-five students who took the class responded with enthusiasm to the subject, convincing me that students knew that questions of energy and climate change were going to have significant effects on their lives. This was not just an academic subject; it was also a learning-to-cope-with-life subject.

Three graduate students strongly affected the development of the ideas expressed here. Tetyana Murza encouraged me to attend the "Chornobyl +20" conference in 2006 in Kyiv, Ukraine. I was grateful for the financial assistance to attend arranged by Michael Mariotte, and it was here that I came to see the Chernobyl catastrophe in its full scope. Murza and I in 2007 developed and co-taught a field study course that took seven Evergreen students to Ukraine for two weeks to study the lingering effects of the disaster. Natalie Kopytko and Kathleen Saul, two graduate students who took that course, subsequently developed their masters' theses on issues surrounding nuclear power, which led to two publications that further enhanced my understanding of the issues. They have subsequently completed PhD work on issues related to energy and climate change.

Outside of Evergreen, it has been my pleasure to learn from and exchange ideas with others also drawn to energy. David E. Blockstein (National Council for Science and the Environment), Catherine H. Middlecamp (University of Wisconsin), and I coauthored a paper on the challenges of energy education. In addition, the three of us joined with four others (Jennifer Rivers Cole, Robert H. Knapp, Kathleen M. Saul, Shirley Vincent) to publish an article on linking climate education with energy education. I spent six months as a senior fellow in residence at the National Council for Science and the Environment, which allowed me to interact with Blockstein, Peter Saundry, and Virginia Brown, each of whom further contributed to my understandings of energy.

David Blockstein deserves special thanks and praise for bringing into existence and nurturing the Council for Research and Educational Leaders (CEREL), a program of the National Council for Science and the Environment. CEREL has successfully organized two National Summits on Energy Education, in January 2015 and in June 2016. These conferences assembled, for the first time, a highly diverse collection of academics seeking to initiate and improve energy education in colleges and universities. I have been inspired by their enthusiasm, and I hope this book may be of assistance in their respective efforts. Personally, I have benefited from the multiple perspectives on energy expressed at these conferences.

This book is about energy, but climate change occupies the pivotal point on why energy economies must change. My understanding of the challenges of dealing with climate change expanded as I collaborated with three classmates from undergraduate days: Robert A. Knox, Richard E. Sparks, and Paul C. Stern. We published a paper in the Policy Forum of *Science* magazine, which argued for better and more comprehensive risk assessments of changing climates, use of findings in decision science, and improved simple models for education about climate change. The effects of that work appear in chapter 6. In addition, Sparks was very helpful in helping me locate articles on damage to wildlife from renewable energy sources.

After retirement from full-time teaching at Evergreen, I joined in the work of the Center for Safe Energy (CSE), a small nonprofit located in Berkeley, California, and dedicated to promoting expert exchanges between the United States and the independent republics of the former

USSR. This work has taken me to Ukraine twice and Kazakhstan once, to work with NGOs in those two countries on issues of energy and climate change. I owe a great deal to the wisdom of Enid Schreibman and Melissa Prager, my two colleagues at CSE, and to the financial support of the Trust for Mutual Understanding for these trips.

Through work with CSE, I have met an amazingly talented and enthusiastic group of folks working on energy and climate change in those two republics. I have learned a great deal especially from Iryna Holovko and Oleg Savitsky (National Ecological Center of Ukraine, Kyiv) and Andriy Martynyuk and Illiya Yeremenko (Ecoclub, Rivne) during these exchanges. Martynyuk was also very helpful in advising for the Evergreen class on Chernobyl in 2007, and he and I co-led a study tour on Chernobyl for university and high school faculty in 2010. Rita Zhenchuk of Ivano-Frankivsk, Ukraine, provided additional help for that trip. The Trust for Mutual Understanding provided financial support for the latter group, for which I'm very grateful.

After my retirement from Evergreen, I enjoyed the support offered to Visiting Scholars at the University of California, Berkeley. I thank Susan Jenkins and Carolyn Merchant for supporting my appointment, which has been of immense value in writing this book. The librarians at the University of California have unfailingly been helpful. Similarly, although I am now geographically remote from Evergreen, I have continued to receive prompt and helpful assistance in tracking down journal articles from Michiko Francis and Nancy Brewer in Interlibrary Loan at Evergreen.

Editors at the University of California Press have continually encouraged me and managed the production processes. Blake Edgar first approached me about the possibility of the press being my publisher and nudged me gently into sending him a prospectus for the book. This led to the contract I signed with the press, and then Blake moved to another position, leaving me in the good hands of Merrik Bush-Pirkle. She was quite helpful in questions I had during manuscript preparation before handing me to Kate Marshall and Bradley Depew. Kate cleared the way to final acceptance of the manuscript before taking maternity leave, and Bradley shepherded the way to final publication, with important substantive suggestions. Sheila Berg and Francisco Reinking made many helpful suggestions on style and substance. In all ways, the staff helped me see the flaws and find ways around them. I have enjoyed working with all at the press.

There are still others who have contributed to this effort, and I apologize if I've forgotten to thank someone I should have. I also want to acknowledge my indebtedness to Wikipedia, an encyclopedia I used to disdain but have begun, with unseemly grumpiness, to appreciate. I still won't use Wikipedia as an authoritative citation source, but at many junctures I found myself using it to find references to primary sources and for quick fact checks, for example, of dates. Whether I have the grace to admit it or not, I admire and thank the sincere and dedicated efforts of many souls who brought Wikipedia into existence and made it a source of information.

I'm also very indebted to my family members, who have supported my writing both substantively and personally. Barbara Bridgman Perkins, a fellow author writing on issues of business structure and technology in health care, has shared the delights and anguish of writing books. She invariably supported my writing efforts and at many times offered timely advice when I seemed to be heading down dead-end paths. Ivan Perkins, an author and lawyer who has expanded my understanding of political power, and Nicole Perkins have continually given friendly encouragement to the process. And it was their children, Milo and Linus, who sent their grandad delving into energy, because their generation is most at risk from climate change. In addition, over the years I have long enjoyed the cheerleader support of Ellen Perkins Ivy Bates. My parents, Eulalia, Henry, and Mary Louise Perkins have long been gone, but their initial support was key. In so many ways, all these people have made my life better; without them, it would have been difficult to even contemplate this book.

Prologue

Energy. The very word carries uplifting overtones. Just compare energy's common synonyms—power, vigor, force, strength, spirit—with its opposites—exhaustion, lethargy, debility, enervation, feebleness. Who wouldn't welcome energy? Our language alone signals that we like it, we want it, we need it!

But exactly what is energy? How does it accomplish the things that make it so appealing? The very fact that you've opened this book means that you want to know more about the subject, even though you undoubtedly already know a great deal. We know, for example, about electric lights and automobiles, and that these things run on energy, even though we usually just refer to it as electricity and gasoline.

The term *energy* seems abstract and a bit mysterious, but we know energy improves life. But is it really the energy? We actually don't want the electricity or the gasoline but the light and mobility they provide, that is, the energy services, not the energy itself.

But we also know that controversy surrounds energy. If the price of energy goes up or if suddenly it's not available, unhappiness erupts. We structure our lives around energy services, and we insist those services remain affordable, safe, and secure. But consider the following examples. Climate is changing dangerously because of carbon dioxide emissions from burning fossil fuels. Air pollution from burning coal makes people sick and kills them. Depletion of easily accessible oil has forced oil exploration into deep ocean waters and inhospitable places like the

Arctic with increased chances of destructive spills. Nuclear power plants have catastrophic accidents.

Following the complaints come proposals to alleviate the problems. For example, use solar and wind energy, use energy more efficiently, use homegrown bioenergy, get out of your car and ride the bus, and change your lightbulbs to LEDs. Whatever the proposal, critics stand ready to defend the status quo: those proposals will make energy too expensive, kill jobs and prosperity, subject people to unreliable energy supply, and imperil national security. Besides, wind turbines ruin the look of the neighborhood and kill birds.

Political leaders have long recognized the importance of energy and energy services and sought resolution to complaints, claims, counter-claims, and proposals for new energy sources. Laws and policies enacted over more than a hundred years ago have, for example, regulated prices, controlled the structure of energy businesses, promoted new energy supplies, mandated pollution controls, regulated energy-mining practices, fought wars abroad to procure energy supplies, provided favorable tax rates and other subsidies to selected energy sources, and provided education to train technicians and engineers in energy technology.

But for every law or policy enacted and enforced, a new bevy of complaints inevitably arise. The law is too lenient. It's full of loopholes. It's good, but it doesn't go far enough. Or, on the other side, it's too strict, a job-killer. Government shouldn't be in the business of making energy choices; let the markets decide. If government chooses energy technology, the choices will not work as well as individuals making up their own minds.

Most people remain uncertain about the best pathways forward. Some people gravitate to the proposition that the energy economies of modern, industrial nations have reached a serious, perhaps crisis stage: climate change, damages to health and environment, insecurity of supply and prices, and depletion of resources. These worriers insist that governments act. Others feel little or no sense of crisis, merely everyday problems that markets can sort out, maybe with a little help from government, but not too much.

No agreement has emerged on the best strategy for action. Energy policy in the United States for over forty years has been best described as an all-of-the-above strategy, that is, a strategy without priorities, other than to guarantee supplies of energy, particularly from fossil fuels. Or better said, U.S. policy is less a strategy than a handbasket full of policies and subsidies to please existing energy suppliers and their lobbyists.

Why is it so hard to agree on a strategy for change, or even the need for a strategy? The answer comes from a simple fact: the energy sources on which the world now relies have become deeply embedded in the structure of nation-states and their economies. Tinkering with energy sources and technology touches a sensitive nerve leading to the economy, political stability, and national security.

This book seeks to increase knowledge about energy. It identifies the First and Second Energy Transitions that occurred many millennia ago and then turns to the Third Energy Transition that began in about 1700 and ended in the 1950s. It explains (a) the genuine benefits conferred by this new energy economy, (b) energy's integration into the foundations of modern states, (c) the origins and spread of energy science and energy technology, (d) the weaknesses of this energy economy that threaten its benefits, and (e) a strategy for directing needed change, the Fourth Energy Transition to energy efficiency and renewable energy. Without a clear strategy and priorities, successful tactics for change will remain invisible.

Connections with Everyday Life

All people live in a culture, those aspects of life so heavily ingrained in everyday behavior and thought that they are assumed, not consciously thought about. Culture is part of what people know as "habit" and "normal," not a puzzle or problem that needs constant attention.

People in modern cultures think nothing of turning on a light switch to dispel darkness or of taking the car to the market to buy a week's worth of groceries. At the store, maybe they see the trucks that delivered vegetables, fruit, and meat from around the world, but probably they don't even see the trucks. They don't see the machinery that enabled 2 percent of the population to raise abundant food for 98 percent, nor do they see or think about the fertilizers applied to the soil to enable high yields, year after year. Maybe they have never even been on a farm to see an orange tree, corn field, or dairy cow. They certainly have never done the work of raising food.

At home, in schools, and at work, people assume that turning on a faucet delivers clean, abundant water for drinking, cooking, bathing, and flushing toilets. Maybe the water came from hundreds of kilometers away. When they walk outside, they don't smell raw sewage; all that stuff flows through buried pipes to the sewage treatment plant.

This modern culture is less than three hundred years old, and it exists only because of energy services. This chapter recounts the major steps that brought modern life into existence and brings the invisible onto center stage for all to see—and perhaps for the first time to think about energy services and how unusual it is for people to assume they are normal.

CHAPTER I

The Invisible Keystone of the Modern World

All animals, including human beings, consume food for energy. Every human acutely recognizes the imperative to eat or perish. This form of energy is not invisible. Similar as we may be to other animals in terms of food, humans uniquely acquired fire, which brought light, warmth, and protection from predators. Of equal importance, fire cooked food, and its advantages separated our evolutionary pathway from that of our other primate cousins.

Wood fires, combined later with beasts of burden and a little water and wind, powered human society for thousands of years. In the 1500s, the enormous energy from coal began to supplant the earlier sources in England. Later, oil, gas, and uranium joined coal as the big-four primary energy sources or fuels. In the late 1800s, a new form of water power, hydroelectricity, joined the big-four fuels, and these five now supply most energy in the world, outside the unique role occupied by food.

Based on these energy sources, people leaped from the agrarian to the modern, industrial world, and the material benefits of the big-four fuels lie beyond dispute and beyond calculation. Despite the keystone centrality of energy to modern human life, most people think little about it. These forms of energy shrink to invisibility, which makes us vulnerable to the problems they pose. Exploring the pathways to fire, food, and subsequently the big four brings the keystone of modern life into focus.

THE FIRST ENERGY TRANSITION:
HOMO EMBRACES FIRE

Evolutionary processes—long before the appearance of primates—established food as the energy foundation for all animals, but humans are different from other animals in their reliance on cooked food. Although many animals, including nonhuman primates, prefer cooked food to raw, only *Homo* fully mastered the use of fire. Darwin speculated that learning to use fire ranked with language as one of the most important traits determining human evolutionary success. Chimpanzees may be able to understand the behavior of fire and thus avoid wildfires without panic,[1] but they don't regularly make use of it. Only humans fully integrated fire into their normal daily behavior.[2]

The use of fire for warmth, light, protection, and cooking, however, does not lie far in the antiquity of evolution. In 2012, microscopic remains of plant material, bones, and minerals in a cave in South Africa showed that regular use of fire was occurring in the cave about one million years ago, and the materials were unlikely to have originated in any way other than regular use of fire by *Homo erectus*, a species that appeared between 1.9 and 1.5 million years ago.[3] Other firm evidence for fire dates to about 780,000 years ago at Gesher Benot Ya'aqov in Israel, before the evolution of *Homo sapiens*.[4]

Archaeological evidence of fire is persuasive that early hominins used it regularly, but anthropological findings suggest that hominins began to use fire about the time that *Homo habilis* disappeared and *Homo erectus* appeared. Significant reductions in the size of teeth and the volume of the gut suggested *habilis* maybe and *erectus* for sure relied on cooked food. It is easier to digest, and organisms extract more energy from it than they do from raw food. In addition, reliance on cooked food requires considerably smaller amounts of time devoted to eating and chewing.[5]

Homo erectus possessed distinct traits consistent with survival by the use of fire in addition to its smaller gut and teeth. This hominin had lost the ability to move about on all four limbs and to climb trees adroitly. It slept on the ground, and to avoid predators it may have used fire for protection as well as warmth and light. The finding of regular use of fire by *Homo erectus* in South Africa one million years ago supports these inferences.

If *Homo erectus*, an evolutionary predecessor of *Homo sapiens*, had mastered fire, then in all likelihood use of fire was an integral part of human life from before the time that modern humans evolved. Now only *Homo sapiens* regularly and mandatorily uses fire, and no people live without it. If this reasoning is correct, then mastery of fire became "natural," and traits supporting the mastery of fire lie in the human genome. Only *Homo*, the primate genus that completely embraced fire, colonized the entire globe in ever increasing numbers. Embrace of fire was evolutionarily very successful, and, as some have quipped, perhaps *Homo sapiens* should be named *Homo incendius*.[6]

THE SECOND ENERGY TRANSITION: *HOMO SAPIENS* LEARNS TO FARM

Until about 10,000 years ago, *Homo erectus* and then *Homo sapiens* survived and expanded to all continents except Antarctica. Populations grew slowly and, based on changing climates, sometimes contracted. Human life relied on a foundation of food to run bodies and fire to heat, light, cook, and protect against predators. Survival of the species required no further advance in the mastery of energy, but a few scattered settlements built a new energy economy by domesticating plants and animals for agriculture, a change that vastly increased the availability of food and thus energy supplies.[7] Farming and animal husbandry may have originated with improvement of climate after the last ice age, and it enabled settled living as opposed to nomadism, hunting, and gathering.[8] Settled living in turn enabled the rise of cities, written languages, social divisions, and vastly faster development of new or more refined materials like ceramics, metal tools, and jewelry.

Anthropologists named this change the Neolithic Transition, but this book uses the term *Second Energy Transition*. No comparable name demarcates hominins before and after fire, but here it's called the First Energy Transition. Embrace of fire and agriculture underlay a lifestyle that persisted in nearly all human cultures from about 10,000 years ago to 1600. By that time, some hunting-gathering cultures survived using only gathered food and fire, but most people derived most of their food energy from domesticated plants and animals and "extra" energy from wood fires. Some people supplemented food and fire with windmills and waterwheels to harvest small amounts of energy from wind and falling water.

This was the agrarian economy in which most people tilled the soil and a much smaller proportion served as merchants, artisans, scholars, priests, soldiers, government servants, and rulers. Civilizations rose and fell in Asia, Europe, Africa, and the Americas, and these various cultures steadily increased both technical prowess and academic learning. A hallmark of all agrarian economies, however, was that they drew energy supplies solely from the yearly input of solar energy. Photosynthesis made "biomass," which provided food, feed for animals, fiber for clothing, and woody materials for fire, tools, and shelter. Wind and falling water came indirectly from the heat of the sun.

The historian Alfred Crosby named *Homo sapiens* "children of the sun."[9] They were much more energy-rich than they had been as hunters and gatherers, but their material wealth was constrained by the annual input of solar energy harvested by plants, windmills, and waterwheels. Greater amounts of stable food energy fueled population growth that could not have occurred based on the food supplies available from hunting and gathering.

In the minds of classical economists like Adam Smith, David Ricardo, and Thomas Malthus, the creation of wealth depended on three elements: labor, capital, and land. Land, however, really represented energy, because photosynthesis for food, feed, and fiber depended on the amount of land controlled.[10]

Classical economists, especially Malthus, were highly pessimistic about the improvement of material living conditions above subsistence levels. For Malthus, a small minority, through provident behavior, might aspire to a more comfortable material standard of living, but the vast majority of humanity must live with much less. As Malthus famously said, the geometric potential for population to increase would always in the end outpace the ability of land to provide more food and other goods. If population levels dropped, then the bulk of humanity might temporarily have a richer life, but the proclivity to reproduce would in the end bring population levels back up to the maximum that land could support. At that point, mortality would balance fertility, and inevitably, Malthus argued, most people would lead an impoverished life of bare subsistence.

THE THIRD ENERGY TRANSITION: *HOMO SAPIENS* CREATES THE MODERN WORLD

People living in "developed" countries think of themselves as "modern," based on democracy, nation-states, individualism, economic systems to organize capital investments for growth, science, industry built with new technology, and the idea of progress. Sometimes modernity distinguishes itself from predecessors with negatives: not feudal, not an absolute monarchy, not agrarian, not rural, and not superstitious. In a modern society, most people live in cities and do not farm, the biggest contrast with agrarian societies.

A modern person's material life has far more "stuff" and "conveniences" than even royalty and the wealthiest premodern societies commanded. What medieval monarch in Europe, for example, could enjoy a hot shower with clean water by turning a valve, a ride to another continent in a comfortable jet, painless surgery to heal an injured joint, and instantaneous communication with his far-flung armies?

Material abundance characterized the "modern world" as much as did the standard components: nation-states, democracy, large business organizations, and scientific enlightenment. A philosopher living in Britain, France, or the United States in 1800 could point to great changes in politics, new scientific knowledge, and new ways of organizing economic activity, all in a nation-state that transcended individual leaders and governments.

Yet the vast majority of people in these three countries remained mostly rural and lived very much like their ancestors of 1,000 or even 6,000 years earlier. They farmed with human and livestock muscle power. If they traveled at all, it was on foot, horseback, or wind-driven ship. Their housing and water supply had changed but little. At night, the world darkened except for the feeble light of candles. They had a few more iron, bronze, or brass tools and ornaments. Maybe their clothes included textiles woven in the newly mechanized mills of Lancashire, but probably they wore homemade clothes. A person from 2000 suddenly launched backward to 1800 would be hard pressed to feel that he or she was still in the modern world, even if democracy, freedom from royal tyranny, and scientific knowledge animated public conversations.

The transition from premodern to modern life, in short, rested heavily on material shifts in living circumstances. Without the huge shifts in material life, most of which occurred after 1800, life in the 2000s would have continued to look amazingly like that of over 200 years ago, which in turn looked not all that different from 8,000 years ago. Mastery of energy sources and technology created the Third Energy Transition with major consequences, but all too often the centrality of energy remains underappreciated and ignored.

The economic historian E. A. Wrigley, in his studies of the English industrial revolution, rectified the oversights about energy. He had a vastly richer set of concepts from the physical and biological sciences on which to draw compared to Smith, Ricardo, and Malthus. After the mid-1800s, the physical concept of energy, defined as the ability to do work, became a fundamental part of science, and scientists could measure it quite precisely in units like joules, kilowatt-hours, and calories.

Wrigley drew from biology and ecology to embrace the concept of ecosystems with energy flows and material cycling. British ecologists in the 1920s and 1930s had borrowed from economic thinking to integrate ideas of producers, consumers, ecosystems, efficiency, and energy flow into biology.[11] Wrigley returned the favor by bringing the refined concepts of ecologists back into economics.

He noted that agrarian civilization rested on "organic energy" supplied entirely by the annual flux of solar energy into the biosphere.[12] People harvested this energy directly as food and feed produced by photosynthesis and indirectly from livestock that fed on plants. Firewood plus other plant and animal products supplied fire for light and heat, which had many uses. People also harvested smaller amounts from wind and water power, both driven by solar energy.

Increasing use of coal in place of firewood started in England in the 1500s and ultimately underwrote a new energy economy and vastly expanded the industrial revolution. These events moved first and fastest in England and were virtually complete by 1850.[13] Wrigley named the new regime the "mineral energy economy," which eclipsed the older agrarian "organic energy economy."[14] Agriculture and animal husbandry didn't cease, of course; they remained the primary source of food and feed for almost all of an increasingly large human population. Firewood remained important in economies not yet industrialized.

Wrigley reconceptualized the industrial revolution, which for him rested on the immense supplies of energy that coal provided compared to that supplied by the organic energy economy. It's not that other factors and changes weren't also important as causes or consequences of the Industrial Revolution. To ignore the liberation of human life from the constraint of the annual flux of solar energy, however, was to miss the main point.

Wrigley was one of a long string of historians who attempted to make sense of the industrial revolution, which was so easily visible after 1850. For example, Arnold Toynbee, in his 1884 essay, *The Industrial Revolution*, generally received the most credit for the term, and he celebrated the increased abilities to make things for an easier life. But he lamented the unevenness with which the benefits were shared among different classes of people. Political reform, argued Toynbee, must spread the benefits more evenly.[15]

Karl Marx also postulated that human beings had reached a new stage of development in which dearth of material goods should no longer plague human life. Like Toynbee, Marx argued for more equal sharing, but he argued that this would require a revolution driven by the working class to sweep away the capitalist class.[16] William Stanley Jevons took quite a different tack from Toynbee and Marx: he worried about the future supplies of coal and the possibilities for continued expansion of the British economy. For Jevons, the issue was how to keep the good times rolling in the face of projected future rises in coal prices.[17]

Economists and historians focused on the multiple dimensions of the industrial revolution.[18] How did labor and capital form in factories? How did growth of national and per capita income cause or change in the industrial revolution? When and why did labor move out of agriculture into cities and factory work? What inventions of new machines drove the productivity of labor upward? What role did coal play? Why and when did the changes in England spread to other regions and countries? What consequences followed?

All of these perspectives are valuable, but they reflect the invisibility of energy that characterized the 1900s. Yes, energy involved ideas, costs, politics, social impacts, and technology, but in the 1900s scholarship too often took energy for granted. Wrigley, in contrast, focused on energy as a sine qua non in the modern world. In the 2000s, climate change, health effects, geopolitical tensions, and the difficulties of procuring fossil fuels demand a focus on the pivotal role played by energy.

The Third Energy Transition developed between about 1600 and the 1950s. It began with the sustained increase in the use of coal in England at the start of the 1500s. Increasing use of coal continued in the 1700s through 1900s, supplemented with petroleum, natural gas, and hydropower. The last fuel of the Third Energy Transition came in the 1940s and 1950s when three countries started to use the heat of uranium fission, first for explosives and then to make electricity. Controlled fission made uranium, plutonium, and thorium into actual or potential fuels to produce heat.

Like the fossil fuels (coal, petroleum, gas), uranium is a mineral fuel, mined from the earth. Heat from all four of these mineral fuels frees humanity from the constraint of annual fluxes of solar energy to the earth. Each of the four fuels is also "energy dense"; that is, each can provide high amounts of heat per kilogram of mass compared to, for example, firewood, solar, and wind energy (for more details, see appendix 2). In addition, like the other fuels, uranium creates benefits as well as problems (chapter 7). It thus shares many important characteristics with coal, petroleum, and gas, which makes it fit easily into an assessment of the Third Energy Transition.

Compared to the unknown but considerably longer time it took to make the Second Energy Transition (the Neolithic development of agriculture), the Third was amazingly fast. Wrigley puts its beginnings in 1600. Its reality didn't become clear until the mid-1800s in England, a period of about 250 years, or about ten generations. After the mid- to late 1800s, numerous philosophers tried to assess the big change—where it came from, how it progressed in England and many other countries, and whether its benefits could be transferred.

The Third Energy Transition includes the industrial revolution but moves considerably beyond it. Energy industries developed new fuels, vastly expanded the rate of energy production, spread globally (although not evenly), and found a bewildering array of applications not imaginable in 1600 or even in 1800. In the half century since the 1950s, no new "mineral fuels," as Wrigley calls them, have joined the fossil and nuclear fuels, although work continues on the hope of adding nuclear fusion. But development of new energy services shows no abatement.

If only benefits—and no problems—flowed from the Third Energy Transition, then the rest of the story might consist only of triumphs. For better or worse, however, that's not how things worked out. The radical transformation of human society by the Third Energy Transition has not yet ended, but its side effects now threaten to overwhelm the benefits (see chapters 6 and 7). Quite possibly the problems will even destroy the civilization built on the Second and Third Energy Transitions.

If citizens, consumers, and leaders were fully informed about the roles of energy, then the downsides of "mineral energy" might not constitute such a serious problem. Unfortunately, low energy literacy dominates social and political conversations, and some either fail to see or simply deny any serious problems. Others acknowledge the problems, but solutions commensurate with the magnitudes of the threats continue to elude nations around the world.

We'll delve into the downsides and solutions later (chapters 5–11), but first I want to explore the Third Energy Transition in more detail. How and when did it occur? What actually changed? What effects followed? Questions like these illuminate energy's role as a keystone of modern life, the modern state, and modern science.

Various steps defined the Third Energy Transition. Step 1 was the embrace of coal as a replacement for firewood. This step enabled the first escape from the constraints of the organic energy economy. Steps 2 and 3 each started with new inventions and their spread. Step 2 used heat to produce motion with the "atmospheric engine" of Thomas Newcomen and, later, the "steam engine" of James Watt. Step 3 changed motion to mobility by adapting steam engines to power ships and locomotives. Steps 1 through 3 alone profoundly altered Britain between about 1600 and 1830, and the technologies on which they depended began spreading around the world, a process still not complete.

STEP 1—FROM WOOD TO COAL FOR HEAT

At one level, coal replacing wood involved no significant change, because the use of coal for heat already had a long history by the 1500s. People in Roman Britain made systematic use of coal for heating and forging iron, and trade routes took it to areas like the Fens on the southeastern coast in exchange for grain and pottery. Earlier uses of coal dated to the Bronze Age, about four thousand years ago. When the Romans left Britain, however, coal use declined.[19]

After 1500, firewood became quite scarce in some areas of England, which led people in those areas increasingly to turn to coal for heat.[20] Access to firewood and coal depended on transport costs; coastal routes, rivers, and eventually canals made coal cheaper than firewood if the latter had to be hauled overland.[21] In England, shortage of firewood developed earlier than in continental Europe, because the long, fast-flowing rivers of the continent allowed reliance on firewood imported from distant mountainous areas. England, in contrast, did not have large areas of mountainous woodlands, but its rivers and coastline easily supported long-distance hauling of coal.[22]

The switch from firewood to coal reflected changes in Britain's environment, especially changes driven by human population size. At the Norman Conquest in 1066, Britain's population was probably about one million.[23] By 1300, it increased as much as threefold, leading to increased clearing of land for agriculture. Woodlands covered 15 percent of Britain at the time of the conquest but only 10 percent by the mid-1300s.[24] About one-third of England's people died in the plague years beginning in 1349, which diminished pressure on woodlands and generally led to better living conditions for those who survived.[25] The population recovered to over 4 million by 1600 and over 5 million by 1700. Increased clearing of woodland for agriculture and increased use of firewood again shrank Britain's supply of wood fuel.[26]

London became a big user of coal by the early 1300s, because heavy clearing removed woodlands ever farther from the city and coal was easily imported from the north of England via coastal shipping routes. Coal use steadily increased again after 1560 and by 1700 comprised 50 percent of the total energy consumption in England and Wales.[27] By 1700, woodlands had again shrunk, to less than 10 percent of Britain's area and could not match the heat produced by the over 2 million tons of coal then produced. The switch to coal was not entirely popular, and critics bemoaned the way coal smoke fouled the air.[28] Nevertheless, its heat proved so valuable that people tolerated the nuisance.

So long as coal was simply burned for heat, the embrace of coal looked very much like the long-standing use of firewood. By the early 1600s, coal had replaced firewood and charcoal as a source of heat in brewing and distilling; making bricks, tiles, glass, pottery, nails, cutlery, and hardware; producing salt, sugar, and soap; and smelting and casting brass. In many cases, coal mining formed part of the operations of making a commercial product, especially salt in Scotland, bricks in Lancashire, and copper smelting in Wales.[29]

Inventions, however, opened the door for coal to take on even more new tasks previously not possible. Undoubtedly one of the most important inventions came from Abraham Darby's use of coal to make coke for smelting iron for casting, probably in 1708–9. Prior to Darby's work, iron smelting required charcoal from wood. Either coal or wood could heat iron for forging, but the chemical processes involved in smelting iron worked better with charcoal. Darby's success probably stemmed from the properties of the coal he used, and he continued to have problems with some batches of iron made from coke.[30]

Energy in England and Wales changed drastically between 1560 and 1850. Draft animals and firewood figured prominently and coal very little early in the period. By later in the period, draft animals were somewhat more important, firewood essentially disappeared, and coal took first place as a source of heat. The total energy available expanded more than twenty-eight times in the period 1560–1850, and the amount of energy available per person expanded by more than five times.[31] In sharp contrast, the energy use per person in Italy in the period 1861–70 matched that of England in 1561–70. In other words, Italy lagged behind England in changing its energy use by at least three hundred years (figure 1.1)

As markets for coal expanded after the mid-1500s, landowners sought to join or increase the production of coal. Colliers dug mines

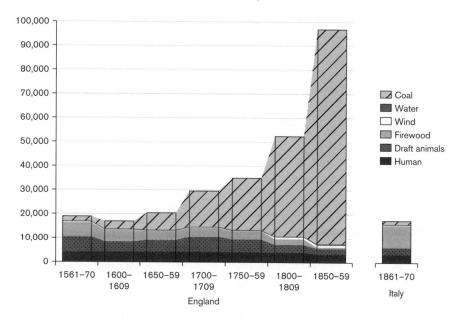

FIGURE 1.1. Annual energy consumption per head (megajoules) in England and Wales 1561–70 to 1850–59 and in Italy 1861–70.

SOURCE: E.A. Wrigley, *Energy and the English Industrial Revolution* (Cambridge: Cambridge University Press, 2010), figure 4.1, p. 95, used by permission of the author.

increasingly deeper into the ground in search of the fuel. Digging deeper unfortunately uncovered more than coal, because water flooded the mines, and mining therefore entailed getting the water out. Pumps run by men, draft animals, and water power at first sufficed to clear the water, but these methods were expensive and had limits.[32] Thus miners and engineers began to search for an alternative way to remove the water from mines. This search led directly to Step 2, the ability to use coal to pump water.

STEP 2—FROM HEAT TO MOTION

The first invention addressed a long-standing need: to move things without using the muscle power of people or other animals. Wind and water power had successfully satisfied this need in limited circumstances long before coal became an important fuel in Britain. Water power drove the hammers of fulling mills for making wool textiles and of forging shops that turned pig iron into wrought iron. Wind drove sailing vessels and windmills for pumping water. Both water power and wind, however, were in limited supply, not always available when needed, and geographically constrained.

Breakthroughs in moving things came first with water. This was no coincidence, because water is heavy as well as vital. Hauling it vertically or horizontally demands considerable energy or work. From time immemorial, people have had to secure reliable access to good water, and they have always sought to minimize the work needed.

During the 1500s and 1600s, various inventors in Europe sought ways to use steam to move water. They combined the use of expanding steam to move a piston in a cylinder with the use of steam on condensation in a cylinder to create a partial vacuum.[33] Although most devices did not work very well, Thomas Savery (ca. 1650–1715), from Devon in England received a patent for his machine.

Savery's machine condensed steam to make a vacuum and suck water into a container, thus putting air pressure to useful work. It then drove water out of the container by using steam pressure. This was the first practical pump run by atmospheric and steam pressure, but atmospheric pressure could not raise water more than about 20 feet, and steam pressure to empty the container lifted not much more than that. His patent, "Raising water by the impellent force of fire," ran from 1698 to 1733 and significantly affected the device invented by Thomas Newcomen.[34]

Newcomen (ca. 1663–1729), also from Devon, became an ironmonger in Dartmouth in about 1685. Ironmongers engaged in retail sales of iron goods, manufacturing of iron tools and devices, and regional trade in these goods. Newcomen probably had blacksmithing skills, and his partner, John Calley, had experience working with iron, brass, copper, tin, and lead. In addition, Newcomen had talents in design and engineering, and Calley was a plumber and glazier. Regional trade routes took Newcomen to Cornwall and Devon on business trips, so he had firsthand familiarity with the problem of water in mines.

Historical archives are too spotty to document when Newcomen first conceived his engine or the details of the arrangements he made with Savery to protect his engine under Savery's patent. Nevertheless, Newcomen's engine successfully pumped water from mines using the heat of burning coal. Miners widely adopted it during the 1700s, and some of his engines continued working well into the 1800s. Fire under a boiler made steam, the steam was injected into a cylinder with a piston that rose due to the downward pull of a counterweight and steam pressure, cold water condensed the steam and created a vacuum, and atmospheric pressure pushed the piston down, which in turn pulled a beam down and actuated a pump to bring water out of the mine (figure 1.2). The first engines were probably placed in use around 1710, but the engine built at Dudley Castle in 1712 worked very well and established the reputations of Newcomen and Calley.[35]

It is unlikely that Newcomen himself thought he was launching a "Third Energy Transition." Instead, it's more reasonable to think that he saw a problem—water that had to be removed to work the mine—and figured out a way to solve it. Many of the mines produced coal, so fuel supplies were readily available and cheap. Scarcity of firewood may or may not have affected Newcomen's thinking. Nevertheless, Newcomen's device, a new energy service, launched a deluge of inventions and innovations over the next two and a half centuries, all of which depended on the ability to make heat move things.

In 1733, Newcomen's engine lost the protection of Savery's patent, which opened the door for many more people to further improve the design.[36] In 1759, Watt, a Scottish instrument maker and surveyor, thought about the potential for steam to create movement and propel a vehicle. Watt also experimented with a device, the digester, or what would now be called a pressure cooker. Again he was seeking insights into making steam produce motion.[37]

The real breakthroughs began in 1763, when Watt tinkered with a model Newcomen engine at the University of Glasgow. He was perplexed at the large amount of steam required and the difficulty of making the model work for any significant time. Watt found that most of the steam in the Newcomen engine served only to reheat the cylinder after it had been cooled to condense the steam and make a vacuum. Watt's experiments also showed that steam itself possessed a great deal of heat, and Watt saw steam as the product of a chemical reaction between heat and water,[38] a view no longer held. In 1765, Watt realized that condensing the steam in a separate container, not the cylinder with the piston, would enable the cylinder to stay hot and thus require less steam and less fuel.[39]

Watt moved the condensation of steam to make a vacuum from the cylinder to a separate condensing vessel. Watt also separated the cylinder and piston from the atmosphere and relied on the expansive force of steam alone to push the piston once a vacuum had been formed on the other side of the piston. His first patent in 1769 makes clear that the claim to novelty stemmed from the economy of fuel use.[40]

Watt's changes also made the engine a "steam engine," one that worked by the expansive force of steam, not an "atmospheric engine"

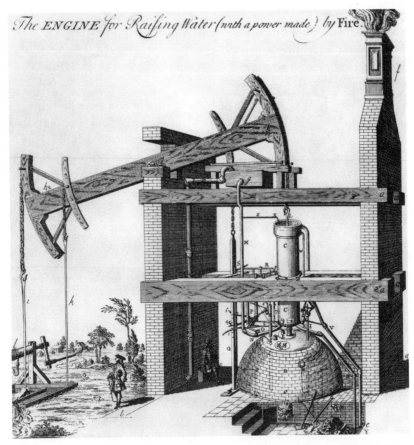

The ENGINE *for Raising Water (with a power made) by* Fire.

FIGURE 1.2. Newcomen engine used to pump water out of coal mines in England. The Newcomen atmospheric engine was the first to have a "walking beam" pivoted arm (left to upper center) to transfer power between the piston (at right end of arm) and the pump rod (at left end of arm). The boiler (bottom right) released steam into the cylinder containing the piston, forcing the piston up and the rod down. As the valve between the boiler and piston cylinder closed another opened that sprayed a small quantity of cold water (from the tank labeled *g*). This caused the steam in the cylinder to condense, creating a partial vacuum. The weight of the atmosphere forced the piston down and the rod up. Opening of the valve between the boiler and cylinder restarted the process. Walls of the building housing the engine not shown.

SOURCE: Science Photo Library Ltd., London. Used by permission.

like Newcomen's. Both, however, turned heat into motion by making steam, replacing air with steam in a cylinder, and then rapidly condensing steam to make a vacuum and thus making a piston move.

The fame and fortune that came to Watt for his inventions was more than sufficient to earn him a place in the pantheon of inventors, but the first markets for the new engines remained exactly the same as those for the Newcomen engine: mines flooded with water. Had the use of steam engines gone no further than pumping water out of mines, the scope of the Third Energy Transition would have remained modest at best. Watt himself and many subsequent inventors, however, went far beyond the vexing problems of miners.

Watt's contributions after 1769 lay in three major arenas, and collectively these developments set the stage for the vast array of changes of the Third Energy Transition. First, he continued improving his steam engine in ways that made it even more efficient in terms of fuel use and eminently more suitable for applications beyond pumping water from mines. His subsequent patents involved the capacity for rotary motion in addition to up-and-down motion of a mine pump, more even delivery of power through the piston's cycle of motion, and better control of the magnitude of an engine's power (figure 1.3).[41]

Second, Watt and his manufacturing partner, Matthew Boulton (1728–1809), developed an industrial complex in Birmingham, England, that efficiently made and installed engines in many locations, earned money, and used public policy to serve the proprietors' interests.[42] Starting in 1776, Boulton and Watt successfully sold and installed steam engines. Most of their first customers were metal mines in Cornwall, where coal was expensive. Their business model rested on fuel economy: they took royalties amounting to one-third the value of the coal saved compared to that consumed by a Newcomen engine. Later sales came from all over Britain and abroad. Of equal importance, Boulton and Watt successfully obtained an extension of the 1769 patent to 1800. The original patent would have expired in 1783, and the two partners knew that it was not worth the capital investment needed in Birmingham without patent protection.[43]

The patents of the 1780s drove their business success. Almost 2,200 steam engines were operating in Britain during the 1700s. Of this total, over 1,400 (about 64 percent) began operation between 1780 and 1800.. Watt-type engines (manufactured by Boulton and Watt) comprised 478 of the total (22 percent). An additional 63 Watt-type engines made in violation of the patent brought the total to 541 (25 percent). Many of

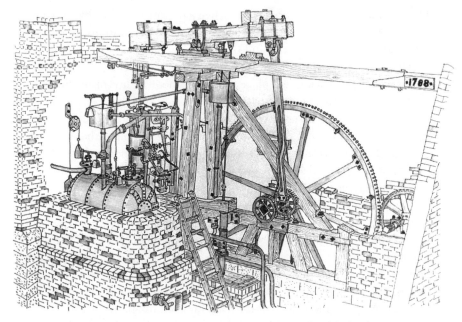

FIGURE 1.3. Watt steam engine. In 1788 James Watt designed and built a beam engine the sole purpose of which was to provide the rotary drive for the lapping and polishing machines at Matthew Boulton's Soho Manufactory at Handsworth in Birmingham. This engine is possibly the most famous rotary beam steam engine in the world, and it is now preserved in the Science Museum, South Kensington, London. The Lap Engine is one of the first engines in the world to have its power output rated in horsepower. The drawing shows how the engine would have appeared in 1788 when it was first assembled at the Soho Manufactory. The flywheel is 16 feet in diameter. This type of engine was used to drive machinery that had previously been driven by horses; customers asked James Watt, "How many horses will one of your engines replace?" A simple calculation established that the engine would do the work of ten horses and became Boulton and Watt's standard 10-horsepower engine; in 1788 this engine cost £800.

SOURCE: Drawing and text by David K Hulse (www.davidhulse.co.uk), used by permission.

the new installations produced rotary motion rather than the up-and-down motion of a water pump and thus rivaled water power for the first time. Rotary engines came into use in manufacturing, such as in textile plants. Boulton and Watt didn't have a monopoly on the market, but they were the biggest company, and no rivals came close to their sales.[44]

Watt's engine, along with Newcomen's, founded a new economy based on mineral fuels. Newcomen atmospheric engines, in fact, totaled 1,022 (47 percent) of the total installed engines in the 1700s, despite the vast increase in the more efficient and more versatile Watt-type engines in the last part of the period.[45] Firewood, water power, draft animals, and human muscle power all continued as important sources of energy into the 1800s, but the steady growth of skills in using coal and steam surpassed other sources for moving things.[46]

The steam engine launched a powerful positive feedback loop, first in England, where water pumps at mines made coal cheaper and more plentiful, which made it easier and cheaper to make iron, which made it cheaper to make steam engines, and thus begin the loop again. This positive feedback loop catapulted Britain to freedom from the constraints of the organic energy economy. The wealth of England, based on coal, iron, and steam engines, powered military and political power across the globe, a new global economy, and the growth of the British Empire, which far surpassed all previous imperial ambitions. To be sure, many other factors contributed to the changes England underwent, but without the ability to turn heat into motion, it is unlikely that a small island could have achieved such political and economic power.

England's triumph of being the first to turn heat into motion was not, however, a secret that could be kept. As soon as it became clear that turning heat into motion was possible, clever engineers in Europe, the United States, Japan, and elsewhere began to imitate the British. Exporting the devices helped transfer the skills needed to build and run heat engines. No country that embraced the energy cornucopia of mineral energy has ever turned its back on it. Material wealth, money, and power captivated societies, and a steady stream of new applications of heat-into-motion begun in the 1800s and has not abated.

Retrospectively, it's possible now to see that Step 2 stimulated the development of the science of heat in the 1800s. Heat could produce motion, and motion could produce heat. In the process, energy was neither created nor destroyed; it just changed form. As simple as this concept seems now, such was certainly not the case in 1700. Chapter 2 delves into these developments.

STEP 3—FROM MOTION TO MOBILITY

Easy mobility permeates every nook and cranny of modern life: commerce, urban design, the food system, education, health care, entertainment, recreation, and more. Lack of mobility separated premodern from modern lifestyles. The first efforts to achieve mobility from coal's heat predate Newcomen's engine, but success came over a century later.[47]

For millennia, people had moved themselves and their goods over land and water with muscles as well as wind. Travel overland was difficult, slow, expensive, and often dangerous. Roads, with a few exceptions, were wretched, especially in wet weather. Water travel was often less expensive, but it, too, suffered delays when winds or tides were not favorable, and danger always accompanied the voyage.

Consider just a few examples. The journey from London to Liverpool, a distance of 352 kilometers (km) (219 miles) served by regularly scheduled wagons, took ten days in summer and twelve days in winter in the 1500s and 1600s; that from Paris to Marseilles (774 km, or 481 miles) in 1672 took a month.[48] In 1606, an Italian traveler near Strasburg reported roads with mud, stones, holes, and water up to the horses' stomachs. The French monarchy was improving the roads in the early 1600s, but the improvement was not "bad roads to good, but from very bad to a state of uncertainty."[49] Roads in North America were no better in the 1700s and 1800s.[50]

Not all roads were bad. Rome, China, and Japan had built highly serviceable roads in the premodern period;[51] but these were the exception, and travel, such as it was, moved no faster than a horse. Water travel was little better. The trip from Dover, England, to Calais, France, for example, could be as little as four hours, but unfavorable winds and tides could make it two weeks or a trip that ended somewhere else.[52]

Early inventors recognized the potential to use steam for mobility, but there were impediments. Achieving mobility was more difficult than pumping water from a mine. The first atmospheric and steam engines were stationary, of great size and weight, and limited to up-and-down motion. Steam engines on boats and locomotives, in contrast, had to be movable and of smaller size and weight and required rotary motion. Heat and vibrations from the engine had to avoid setting the vehicle on fire and tearing it apart. Smaller engines would benefit from higher steam pressure, but that posed the danger of explosions. Higher pressures also dispensed with the need for a condenser and its supply of cold water,[53] a considerable saving on weight and size. Boats and locomotives had to minimize friction with water and land and maintain reliable propulsion and steering over swells and hills, respectively.

Given these challenges, it's amazing how quickly solutions were found. Watt's development of a pure steam engine provided a key advance by making the idea of higher steam pressures useful. Oliver Evans in the United States (1755–1819) and Richard Trevithick in England (1771–1833) successfully demonstrated the feasibility of higher pressures and of the utility of small engines to propel wagons and boats. Evans began thinking seriously about steam engines for powering land vehicles in

1777, and he developed a successful, high-pressure steam engine that propelled a large amphibious vehicle in 1805.[54] Trevithick successfully operated a land vehicle with his high-pressure engine in 1801.[55]

Use of high-pressure engines did not occur, however, in the first successful steamboats. Controversy still swirls over who exactly invented the steamboat, but the priority nod goes to James Rumsey from present-day West Virginia. In 1787, Rumsey sufficiently integrated all the necessary components of a working steamship, and his low-pressure steam engine propelled his boat along the Potomac River near Sheperdstown at 3 to 4 miles per hour. Twenty years later, Robert Fulton achieved commercial success on the Hudson River in New York using a low-pressure Boulton and Watt steam engine.[56]

Adapting the steam engine to boats avoided two challenges of adapting them for land transport. First, boats were larger than land vehicles, and they accommodated the size and weight of steam engines relatively easily. Buoyancy provided by the water helped support the weight of a steam engine. Second, existing rivers, lakes, and the ocean provided a ready-made highway for boats. Land vehicles, in contrast, were entirely dependent on construction of some sort of roadway. In turn, construction of roadways was so expensive that the size and weight of a land vehicle faced immediate constraints, which in turn constrained the size of a practical engine.

Canals made water into a "roadway," and extensive canal development in England began in the 1500s, over one hundred years before Newcomen's steam engine.[57] Comparable enthusiasm arose in the United States shortly after the Revolutionary War. In the early 1800s, canals seriously competed with schemes to develop land-based, steam-powered transport in both England and the United States.

Trevithick in England and Evans in the United States had first successfully demonstrated propulsion of land vehicles by steam, but only in the 1810s did practical ventures begin. Once again coal was the motivation, but this time it was to move the coal itself, not the water in the mines.

The north of England had good coal deposits, but some lay miles from water transport, a barrier for hauling the heavy mineral to markets. Railways made of wooden rails, later sometimes covered with iron or entirely of cast iron, had been used since the 1600s to ease the work of horses hauling coal wagons. In the early 1810s, however, the wars against Napoleonic France had driven up agricultural prices,[58] a significant blow to those who depended on horses to haul their product. John Blenkinsop (1783–1831), manager of Middleton Colliery near Leeds,

Yorkshire, confronted these increased costs of hauling coal by inventing a new method.

Blenkinsop produced a steam-driven locomotive that pulled up to thirty coal wagons three and a half miles at a little over 3 miles per hour. The device used a two-cylinder steam engine and a cogged drive-wheel that engaged a rail with cogs.[59] This colliery continued to use the device into the mid-1830s. Although Blenkinsop didn't develop his invention further, he showed for the first time the practicality of hauling very heavy loads overland at a speed never previously attained, without human or animal muscle power. His successful invention, although technically obsolete within twenty years, inspired a cascade of refinements and new inventions that made overland mobility something taken for granted.

Compared to others who followed, especially George Stephenson, Blenkinsop has attracted relatively little attention, perhaps because his vision for steam locomotion seemed to go no further than hauling coal to Leeds. In contrast Stephenson's vision encompassed the transport of passengers as well as freight for much longer distances. A self-taught engineer who grew up in the coal fields of Northumberland, Stephenson learned to operate, maintain, and eventually design and build steam engines for various jobs around coal mines.

After 1807, he became familiar with Blenkinsop's work, and by 1813, he believed that he could build a better engine. His first, the Blucher, showed in 1814 that it could draw a weight of 30 tons up a slight incline at 4 miles per hour. As an important simplification, its drive wheels had no cogs like Blenkinsop's machine. By 1815, he had made several important improvements that increased the power of the engine substantially and had a much better connection between the engine and the wheels.[60]

Today Stephenson's fame rests on his stunning success as chief engineer for the first two railroad lines that moved something other than coal. The Stockton and Darlington Railway, established in 1821, began hauling goods and passengers in 1825. When regular operations began, steam locomotives hauled coal and horses hauled passengers. Stephenson became chief engineer for the even larger venture, the Liverpool and Manchester Railway, which opened in 1830 and was powered by his engines. One year earlier the line had held a contest to test whether steam locomotives really were the best choice for propulsion, and Stephenson's Rocket locomotive had won handily (figure 1.4). Rapid expansion of railway building all over Britain and elsewhere quickly followed.[61]

FIGURE 1.4. Rainhill Trials, which *Rocket* wins, October 1829. These trials for early steam locomotives were run at Rainhill, near Liverpool, UK, to determine which would be used on the newly completed railway from Liverpool to Manchester. Ten trains entered, but only five competed. The two shown here are the *Rocket* (right, by George Stephenson) and the *Sanspareil* (left, by Timothy Hackworth). The *Sanspareil* completed eight of the required ten trips before suffering engine failure. It was later used on the railway, along with the winner, the *Rocket*, the only locomotive to complete the trials. Artwork from the third volume (first period of 1889) of the French popular science weekly *La Science Illustrée*.

SOURCE: Science Photo Library, London. Used by permission.

In the United States, proponents of a rail connection between Baltimore and the Ohio River valley organized a new company in 1827. They studied Stephenson's work closely, reaffirmed his decision that steam locomotives provided the best propulsion, and began running trains over the partially completed route in 1830. The most difficult part of the route, over the Allegheny Mountains to Wheeling, opened in 1853, and train travel from Baltimore to Saint Louis began in 1857.[62]

Steam locomotives moved people at speeds previously unimaginable. By 1900, steam was moving huge ships across the seas and trains across continents. After oil joined coal and powered internal combustion engines, cars and trucks moved people and goods along roadways and airplanes through the air. Within another 70 years, heat-to-motion machines took people to the moon and sent spaceships careening toward other planets and even the stars.

In the scant period of just over 250 years, the embrace of coal turned heat into motion. Human mobility underwent a transformation from slow, rare, dangerous, and expensive to rapid, common, safe, and cheap. Similarly, heat-to-motion machines in factories changed manufactured items from rare and expensive to common and cheap. These transitions occupied about 1/1,000th (0.001) of the time *Homo sapiens* had been on earth and separated premodern from modern humans.

People who have lived their entire lives with easy mobility—as is now common throughout the industrialized world—take it for granted, which makes it invisible. Consider just three dimensions in which mobility underlies how modern people live: commuting, family location, and the movement of food.

As agriculture became mechanized in the 1900s (another example of using heat to make motion), people increasingly moved to cities in search of work. In the 1700s and before, cities tended to be small and traversed easily on foot, on horseback, or in a carriage. As cities grew, the difficulty of travel constrained the spread of cities and promoted congestion. Mechanized transport started appearing in the last half of the 1800s, and urban-suburban areas covered huge expanses of land. By the end of the 1900s, people in energy-rich countries often traveled 5 to 25 miles or more to work, to school, to worship, and to shop. People know they need means of transport, but few think of this as turning heat into motion into mobility. Many consider the energy required necessary and even an entitlement.

Not only did daily life involve movement over large distances; families moved farther apart without feeling cut off from loved ones. Before 1700, people generally lived their lives within 100 miles of where they were born, and families of necessity were physically close. For the few that moved far away, it could be tantamount to saying goodby forever. Travel was too slow, too expensive, and too dangerous to warrant even one visit to parents or siblings. In contrast, it is now common in energy-rich countries for grandparents, parents, children, and grandchildren to live hundreds or even thousands of miles apart, yet not feel estranged. At holidays and other special occasions, reunions bring far-flung family members together, all courtesy of rapid mobility.

Before the 1800s, people ate food raised nearby. Vegetables and fruit were seasonal, and the meat, dairy products, eggs, and cereals came from nearby farms or from your own farm. Those near the coasts or inland waters might have fish, but they were probably caught nearby. Diets were more monotonous, embellished with few spices and exotic products. For long periods of cold weather, only preserved fruits and vegetables were at hand.

Mobility transformed agricultural production, fisheries, and diets. A calf born on the prairies of Texas might move to Iowa for fattening and end up as a hamburger in England. Bananas, mangos, and other tropical fruits appeared in Boston markets year-round. Coffee and tea moved thousands of miles before appearing for sale from a favorite barista. A fish caught off the coast of New Zealand could easily reach a dinner plate in Buenos Aires. Relatively few walked to a neighborhood grocery store, instead preferring to drive to a regional megastore with fresher goods at cheaper prices. Consumers, however, seldom connected their food source to heat and mobility.

Today people construct their lives around cheap and easy mobility. If that mobility for any reason became unavailable or too expensive, a human catastrophe would follow, as neither cities nor rural areas could function to keep each other alive. Economically, socially, and physically, countries with rich energy services rebuilt themselves—to create the modern world—in ways that would collapse without energy services. It's not that people are "addicted" to energy as if it were a recreational drug; instead, modern life both assumes and requires it.

CONCLUSION

Control of fire separated humans from all other species, and agriculture stabilized and expanded human food supply. These two changes alone had profound effects on human ecology. The pivot toward coal in England, however, remade the world in ways vastly beyond that of fire and agriculture.

New machines put the heat from combustion to new kinds of work, which vastly expanded commerce and the ability to move its products and people great distances. Modern life today still relies on heat to create motion and mobility, a topic to which I return later. But this was just the beginning of the Third Energy Transition.

The next two chapters enlarge the story by showing how energy emerged as a foundation of modern science and as a component of the fully modern state. Thomas Newcomen's unique new use of coal to pump water from mines blossomed far beyond what anybody at the time could possibly have imagined.

Connections with Everyday Life

Consider the following stories about two events in everyday life.

A student is writing a report on her experiments in making biofuels from corn stover (the stalks and cobs of corn plants). Her working concepts of the chemistry involved rest on the notion that corn plants used the electromagnetic energy of sunlight to make lignocellulose, molecules composed of electrically charged particles. Her experiments rearrange those molecules by breaking and reforming chemical bonds that were electrical interactions. When the biofuel burns, it will release energy as the bonds once again break and re-form. Electric lights, electrically heated vats, and electric pumps assist her work. Computer-driven instruments collect her data and store them as magnetically oriented particles arranged by electricity on the hard drive of her laptop computer. As she writes, her computer turns electric signals into letters and numbers on a screen. When finished, she sends electrical signals to the printer. As she worked in the laboratory, she heard her favorite music, sometimes broadcast through the air as electromagnetic radiation and sometimes transmitted digitally through the internet. The electricity powering her many devices come from distant sources.

A dairy farmer turns on an electric switch, and a machine with rotating components generates a vacuum that pulls milk from the udders of his cows. Milk flows rapidly and cleanly directly into a refrigerated tank kept cold by an electric motor. Once full, a truck powered by a gasoline engine fired by an electric spark collects the milk for delivery. Electrical devices displaced labor by allowing the dairyman to do the work previously done by many more. The electricity comes from a generating station run by his farm cooperative and fueled by natural gas.

Both stories portray common events in modern life, and both show that energy is deeply embedded in different dimensions of everyday events. But how did the concept of energy arise? What is energy? This chapter highlights the invention of the concept of energy and its central role in the science and technology of modern life.

CHAPTER 2

Energy and Energy Services

When Newcomen developed his steam-powered, atmospheric engine, societies already had considerable knowledge about the practical uses of fire, heat, light, waterwheels, windmills, and wind-powered sailing ships. In other words, societies were already using a wide variety of what we now call energy services. These services, however, originated as practical arts rather than from a philosophical or scientific concept of energy. In fact, in the early 1700s the word *energy* meant forceful and effective speech, thought, and actions. It had no precise, mathematical, philosophical meaning. Moreover, no one could see the various practical arts collectively as energy services; they were simply things skilled artisans could do, learned by trial and error.

Between 1600 and 1900, a span of just three hundred years, philosophers, now called scientists, learned to express energy in mathematical terms and with models and metaphors. Energy now plays a central role in all the modern natural sciences (physics, chemistry, earth sciences, biology) and in all the engineering disciplines. Energy also organizes the idea of energy services as the class of practical arts that fully define modern life and the modern state.

This chapter traces inquiries about electricity and heat as philosophers and practical artisans worked to understand and exploit them. The concept of energy ultimately synthesized these two lines of practical and philosophical inquiry, from which we draw the scientific meaning of "energy." Energy services in turn are the practical arts requiring energy sources. We begin with a preview of key ideas that emerged from this synthesis.

ENERGY, A PREVIEW OF KEY IDEAS

Energy is not something you pick up and examine; instead its presence is inferred and its amount is computed. In many ways, energy is nonintuitive and a "difficult idea,"[1] but it's easy to understand its essence based on five key ideas.

Force and motion. All societies have known that things, both living and dead, move and change. Historically, major thinking on motion and change through the late 1600s drew on Aristotle (384–322 BCE) and Thomas Aquinas (1225–74). Aristotle's notion, simplified, emphasized that part of the cause of motion and change lay within the material that moved or changed.[2]

Galileo Galilei (1564–1642) and especially Isaac Newton (1643–1727) postulated the tendency for a body to remain exactly as it was unless compelled to change by a force. In other words, the cause of motion lay outside the body moved. A body, or mass, had inertia and would remain stationary or keep moving in the same direction and at the same speed unless accelerated by a force. Forces caused acceleration (a change in speed, direction, or both). No net force, no change in motion; no change in motion, no net force.[3]

People had known since time immemorial that an object held in the air fell to earth if dropped, but Galileo, Newton, and others changed this common wisdom into a philosophical event. The word *gravity* entered the English language in the 1500s and in the late 1600s became the scientific term for an attractive force between two bodies.[4] Newton inferred the existence of and quantified gravitational force by calculating its effect of making the moon orbit the earth. He didn't explain how the gravitational attraction worked physically, but he expressed the relationship in mathematical equations.

Forces arise from attractions and repulsions. Three distinct kinds of forces exist: gravitational, electromagnetic, and the nuclear force holding protons and neutrons together in atomic nuclei.[5] More recent work recognizes that nuclear forces consist of two distinct types, weak and strong.[6] Electromagnetic forces consist of attractions and repulsions. The physical nature of attractions and repulsions remains elusive, but scientists and engineers can express their magnitudes and behaviors with mathematics.

Energy is the ability to do work. This key idea united science and the practical arts. Both focused on the motion of things and the ability of forces to change that motion. Where Newton wanted to know how gravitational force determined the movement of the moon, Newcomen used movement of a piston in a cylinder, caused by the weight of the atmosphere, to pump water out of a mine. Watt and others changed the force moving the piston entirely to expanding steam, but Newcomen's and Watt's engines both performed work in the everyday meaning of the term. They moved something people wanted to move. Science absorbed the concept of work—making a mass increase its speed, that is, accelerate, over a distance—and expressed that if work was done on a system, then the energy of the system changed by an identical amount. Work and energy are measured with the same units, and engineers use the laws of thermodynamics to extract the maximum amount of work from an energy source.

Energy behavior expressed by the laws of thermodynamics. The existence of distinct types of forces came from two different lines of practical and scientific work on electricity and heat. For many years, these two types of investigations had no theoretical connections, and the practical artisans using electricity and heat, respectively, remained distinct specialties.

The concept of energy arose first in theories about heat and steam engines, but experiments demonstrating a "mechanical equivalent of heat" synthesized forces and movements from heat and from electricity into energy. Energy occurred in different forms—heat and a moving body—governed by two overarching laws, the laws of thermodynamics. These laws occupy center stage in energy science, used universally by scientists and engineers. Occasional efforts to preach against them read, to adherents, as the mutterings of a confused soul.[7]

The two laws have appeared in many forms. For example, the first law focuses on conservation: energy can be transformed from one form to another, but the total amount remains unchanged. The total amount of energy is conserved during processes of change or performance of work. The second law holds that during a transformation not all energy can do useful work; some appears as heat that performed no work or as an increase in disorder of the system.

Energy unites science, technology, and political economy. Political and commercial leaders saw the connections between science, technology, work, and political-economic power.[8] Energy science both derived from and promoted technological developments. In retrospect, scientific laws governed and constrained the First, Second, and Third Energy Transitions, even though those transitions happened before energy science explained them. These laws will also affect and constrain the Fourth Transition, but they enable a better understanding of how energy, through technology, performs useful work, that is, energy services. This narrative is a brief sketch; other sources provide more detailed histories.[9]

ELECTRICITY AND ENERGY

The science of electricity began with William Gilbert's book, *De magnete* (1600). He noted that lodestones acted as magnets, which attracted iron materials. Rubbing amber caused it to attract lightweight items such as hair or lint. This latter sort of attraction had been known at least since observations in ancient Greece, and Gilbert's brief treatment of electrical attractions carefully separated them from magnetism.[10] In addition, Gilbert's work preceded that of Newton, and Gilbert did not attach the concept of force to electric or magnetic attractions or repulsions.

From 1700 to 1800, numerous natural philosophers learned that electrical effects included attractions, repulsions, light, sparks, and crackling.[11] One of the factors keeping electricity in the arena of arcane effects was its subtlety. The effects were interesting but small, hard to observe and reproduce, affected by many factors, and of no apparent practical use. A serendipitous invention in 1745–46 in Leyden, Netherlands, (the Leyden jar) suddenly made electricity more interesting.

With the Leyden jar, for the first time a philosopher could easily and reliably make large electric effects, which were no longer subtle sparks, light, or little attractions or repulsions of feathers and lint. One of the discoverers of the jar received a horrific and unpleasant shock: "I would like to tell you about a new but terrible experiment, which I advise you never to try yourself, nor would I, who have experienced it and survived by the grace of God, do it again for all the kingdom of France."[12]

A Leyden jar turned out to be a rather simple device, but the explanation of how it worked took time. Benjamin Franklin, an American printer and later revolutionary, explained the Leyden jar by assuming that materials had some sort of subtle fluid. Where others had seen rubbing to make the electric effect as an excitation of the fluid, Franklin instead said rubbing just moved the fluid around.[13] He used the word *charge* to indicate a surplus of this subtle electric fluid,[14] and his application of moving an electric charge from clouds to the ground harmlessly—through a lightning rod—became the first of many practical applications of electrical knowledge.

The evolution of words to describe electric attractions reflected the challenges of naming physical phenomena. "Electrics" were substances that, when rubbed, attracted bits of chaff. *Amber* in Greek (electron, or ἤλεκτρον) and Latin (*ēlectrum*) morphed in the 1600s into the English adjective *electric:* an "electric body" attracted other lightweight things. And then the wonderful flexibility of English turned the adjective *electric* into a noun, *electricity*, meaning "a something" with the power to attract light bodies. The equivalent is saying, "Look at the blank effect," and then claiming that blankicity caused the effect. Electricity was simply "a something" with many unusual effects.

Franklin, like Gilbert, did not use the concept of forces, and he wasn't bothered by the fact that his own descriptions and explanations of electrical events did not explain why the events happened.

> Nor is it of much importance to us, to know the manner in which nature executes her laws; 'tis enough if we know the laws themselves. 'Tis of real use to know, that china left in the air unsupported will fall and break; but how it comes to fall, and why it breaks, are matters of speculation. 'Tis a pleasure indeed to know them, but we can preserve our china without it.[15]

After Franklin, quantitative measurements suggested electric attractions and repulsions behaved mathematically like Newton's gravitational attractions; that is, they constituted a force.[16] The physical causes of electric attractions and repulsions had no ready explanation, nor did their appearance in living tissue.[17] Electric effects simply meant electricity existed.

Gilbert had used friction to create the electric effect, also the dominant source of electric effects for the Leyden jar, but by 1800 and after various investigators found several new ways of generating electric effects. In the late 1700s, Luigi Galvani in Italy touched metal to a dissected frog's leg, which made the muscles of the leg twitch. Galvani believed that he had discovered a new form of electricity, "animal electricity," generated in the animal tissue itself.[18]

Alessandro Volta, also in Italy, doubted Galvani's identification of a new and distinct type of electricity. He showed that the muscle contractions in frogs could be induced by touching the frog's leg and back with a metal arc made of two different metals. Volta went on to construct an "artificial electric organ" by placing zinc discs and silver coins together, separated by cloth moistened with salt water. This "organ" or "voltaic pile," or "battery" in today's language, could make electricity constantly and in larger amounts than could be made and stored in a Leyden jar.[19]

Volta's new way of generating electrical effects launched many follow-up experiments. English scientists found that the electricity from a voltaic pile created bubbles of hydrogen at one of the wires connecting the water to the battery. At the other wire, oxygen combined with the wire to change it chemically. In other words, Volta's battery supplied electricity by means of chemical reactions, confirmed by many subsequent experiments, especially those of Humphrey Davy and Michael Faraday.[20] Davy's experiments with a voltaic pile clearly linked electric effects to chemical change and the composition of matter,[21] but for energy services Michael Faraday, hired originally as Davy's assistant, did the most important work.

Faraday (1791–1867), born into England's lower classes, had little in the way of formal education but was a genius at self-instruction (figure 2.1). Before joining Davy, Faraday had built his own Leyden jar and voltaic pile, and gained wide exposure to many areas of science. He had noticed that a voltaic pile made from zinc and copper left metallic copper on the zinc electrode and zinc oxide on the copper electrode. Like Davy, Faraday saw early on the close connections between electricity and chemical reactions.

Between the 1820s and 1850s, Faraday fleshed out Davy's proposition that electric effects had something to do with matter and went far beyond to build a unified science of electricity.[22] On the way, he also demonstrated in 1831 that a moving or changing magnetic field could induce an electric current in a metal conductor. The concept of force field had far-reaching effects for energy science, and the discovery of electromagnetic induction of electricity had equally profound effects on the practical arts and modern states. I present below just a few highlights of this pathway of discovery.

Hans Christian Oersted (1777–1851) of Denmark opened a new pathway in 1820 when he showed that electricity from a voltaic pile caused a compass needle to change its direction. A reversal of the current flow deflected the needle in the opposite direction. Moreover, the magnetic force created by the current formed a circle around the wire.[23] Such a simple observation, such profound consequences.

FIGURE 2.1. Michael Faraday (1791–1867), English chemist and physicist, who devised the first electric motor along with equipment such as transformers and dynamos. Faraday had no formal education but gave his first Royal Institution lecture when a scheduled speaker failed to appear. It was a great success, and his lectures would popularize science. He discovered the organic molecule benzene and was later made professor of chemistry at the Institution in 1833. His other notable works centered on electromagnetic induction and the laws of electrolysis.

SOURCE: Science Photo Library, London. Used by permission.

Gilbert in 1600 had carefully separated electric from magnetic effects, but Oersted's finding suggested a relationship. Immanuel Kant (1724–1804), a Prussian philosopher, had prepared Oersted to put magnetism and electricity back together, because for Kant, material reality lay in the combined effects of attractive and repulsive forces, all of which should be interconvertible. Oersted had studied Kant's philosophy, and by 1813 Oersted the Kantian suggested that electrical forces should be convertible to magnetic forces.[24] His 1820 experiment, however, startled electricians, some of whom reacted with disbelief and others of whom immediately rushed to study it.[25]

Experiments in France confirmed that a current in a wire made the wire a magnet and that two wires parallel to each other attracted each other if the current in both wires ran in the same direction. If a single wire was made into a coil and a current run through it, the wires adjacent to one another were attracted magnetically.[26] Electricity and magnetism became essentially one phenomenon, electromagnetism.[27]

Davy and Faraday repeated these experiments, and Faraday confirmed that the attractive and repulsive forces were circular. In 1821, Faraday made a current-carrying wire rotate around a magnet and a magnet rotate around a current-carrying wire.[28] This experiment laid the theoretical groundwork for an electric motor: electric attractions and repulsions could force motion, just as could heat in a steam engine.

In the 1820s, Faraday sought the reverse of Oersted's experiment: if an electric current could deflect a magnet, could a magnet make an electric current flow? Faraday's question reflected his own Kantian bent that all natural forces should be interconvertible. In 1831, Faraday used the powerful magnetic forces of the newly discovered electromagnet: electricity through a wire wound around a bar of iron turned the iron into a magnet that could be turned on and off by disconnecting the wire from the voltaic pile. In the 1830s, the iron industry employed electromagnets to separate iron ore from surrounding rock,[29] another practical art based on electricity.

Faraday found that an electromagnet induced an electric current in a conductor as he turned it on and off.[30] Within a year, French artisans had built a new machine, the magneto, in which a waterwheel or steam engine moved a magnet and conducting wires relative to one another to produce electric currents. Also in the 1830s, improvements in the voltaic battery increased the amounts of current available for practical tasks, such as electrometallurgy and electroplating to deposit and shape metals, make metallic molds, and produce printing plates and thus lower the costs of printed materials.[31] Together, the magneto and better batteries transformed the electrical practical arts; and Faraday's discovery of electromagnetic induction of a current utterly transformed economies, lifestyles, and nation-states (see chapter 3).

Faraday didn't turn induction of electric currents into practical devices, but his scientific work produced models and explanations of the action of electric forces of attraction and repulsion that had a comparable transformative effect on the development of energy science. Guided by Kant's and other philosophical speculations on the convertibility of forces, Faraday experimented with electrical and magnetic effects over a twenty-year period, from the 1830s to the 1850s, and his work attracted William Thomson (1824–1907, often referred to by his honorary title, Lord Kelvin), who expressed Faraday's results with mathematical models. Faraday saw a force field created by electrical and magnetic attractions and repulsions in lines of force in the space around the body. He at first thought these lines of force were more a metaphor than real, but ultimately they became real for him. He believed only God knew why the lines of force existed and how they worked.[32]

Electric telegraphy played an important part in Faraday's thoughts, because it brought him into contact with electrical effects on a far grander scale than any laboratory could offer. In the mid-1840s, telegraphers sent electric pulses along a wire for many miles to activate electromagnetic devices and transmit a signal in code. Problems arose when oceans separated the sender and receiver or when the wires went underground. Here the wires had to be insulated. A natural rubbery plastic from trees, gutta-percha, insulated well and made underwater telegraphy possible.[33]

A signal sent over a submerged or underground insulated wire did not travel nearly instantaneously, unlike signals sent over insulated or uninsulated overhead wires. The longer the distance of these undersea or underground wires, the more retardation delayed the signal and made it less distinct. Telegraph wires from England to France or Belgium showed the effect but not to the point of rendering them ineffective. Retardation between England and North America, however, could destroy the signal.[34] Faraday's interpretation of retardation became prime evidence that electric attractions and repulsions acted as force fields, passing forces from one molecule to another to induce "strain" on them in the gutta-percha.[35]

Faraday also found that a polarized ray of light could be rotated by a magnet, the "Faraday effect." This was the first solid evidence connecting light and electromagnetism. Again, Faraday's interpretation centered on magnetic forces creating a space around the magnet with lines of force in it, which could rotate the polarized light.

Just as Thomson had translated Faraday's ideas about retardation into mathematics, James Clerk Maxwell (1831–79) translated Faraday's lines of force into general mathematical models and suggested that light might be waves comparable to those that produced electric and magnetic forces.[36] By 1888, German investigations confirmed the experimental induction of electromagnetic waves, as predicted by Maxwell. Radio broadcasting put these waves to practical work,[37] and light, too, was a similar kind of wave with a shorter wavelength.

Revisions of Maxwell's equations put them into the form now learned by every physics major and electrical engineer.[38] In simple English, the four equations said (a) "[a] changing Magnetic Field gives rise to a changing Electric Field" and (b) "a changing Electric Field gives rise to a changing Magnetic Field."[39] Faraday's idea of a field, a place influenced by a force, is now at the heart of modern physics.

Heat, Steam Engines, and Energy

Electromagnetic investigations, except for light and chemical reactions, generally focused on attractions and repulsions (forces) that made tangible objects move. A second line of development focused on heat and gases expanding and contracting in cylinders to make a piston move. Newcomen and Watt engines made things move (water in a mine), but the study of gases and engines did not directly examine forces that made things move. The piston moved in the cylinder, but the artisans had applied the ability of expanding steam to move objects well before science could fully explain the motion.

Two questions plagued the development of an explanatory science. First, what made up a gas, atoms and molecules? But what were they, and did they really exist? Second, what was heat? Burning coal gave heat, but was heat a substance in its own right, or did it consist of the movement of atoms and molecules? Full consensus on the reality of atoms and molecules came only in the 1900s, but the concept of heat as the movement of atoms and molecules gained considerable strength during the 1800s. Here I sketch the more important points.

Everyday observations indicated some things were warm and others cool, but what caused warmth? One hypothesis proposed that the cause of heat lay in the movement of atoms and attracted leading philosophers of the 1600s, such as Francis Bacon, Robert Boyle, Isaac Newton, and John Locke. Locke, for example, speculated, "Heat is a very brisk agitation of the insensible parts of the object, which produces in us that sensation, from whence we denominate the object hot; so what in our sensation is heat, in the object is nothing but motion."[40]

Unfortunately, no experimental evidence conclusively demonstrated either the existence of atoms or that motion of atoms caused heat. Despite lack of evidence, this dynamic hypothesis of *heat* remained attractive during the 1700s and into the 1800s, along with another hypothesis about a postulated material, caloric. Caloric played an important role in the theoretical explanation of the operation of a steam engine, but shortly thereafter the dynamic theory of heat replaced it.

Two French engineers, Lazare Carnot (1753–1823) and Sadi Carnot (1796–1832), father and son, built a theory of engines that ultimately led to the concept of energy. Lazare Carnot led the way in 1783 with his *Essay on Machines in General*, which was expanded by his son, Sadi (figure 2.2), in his 1824 essay, *Réflexions sur la puissance motrice du feu* (Reflections on the Motive Power of Fire).[41]

Sadi Carnot's imagination took him where no one had gone before. He imagined a perfect engine that goes through a full cycle of operation. The gas expands to make the piston move, and then in a reverse operation the piston and gas moves back to its original state, ready to begin another cycle. The actions of this Carnot cycle occur in infinitely small, reversible steps.

Sadi's "perfect" engine wasn't a real engine, but he had abstracted the processes of a steam engine and brought steam engines into the purview of engine theory. Engineers today still use his methods to assess the vast panoply of engines in use today. Sadi Carnot died of cholera in 1832 at the age of thirty-six, and his work remained relatively unknown until it was revived by Emile Clapeyron in 1834.[42] Sadi Carnot never saw that he laid the groundwork for modern thermodynamics, the science of energy.

Sadi Carnot focused attention on heat, not steam. People born after 1950 may never have seen a steam engine in operation, but it is impressive with its belching smoke, clanging metal, and hissing steam. It was easy to be distracted by the steam, smoke, and size, perhaps even to suppose that steam itself ran the engine. Heat, by contrast, was immaterial and subtle. The name "steam engine" makes more intuitive sense than heat engine, but Carnot redirected attention.

In Carnot's perfect engine, steam was merely the carrier of heat, and in theory any material that expanded when warmed could provide the motive power of the engine. He discussed at some length the possibilities of using air or the vapors of alcohol instead of steam, and observed that—in theory—even a solid or liquid could carry the heat. Carnot dismissed solids and liquids, however, as essentially impossible in practice, because their expansions and contractions were so minute compared to gases. He concluded that air would probably carry heat better, despite successes with steam.

FIGURE 2.2. Nicolas Leonard Sadi Carnot (1796–1832), French physicist and founder of the science of thermodynamics. Carnot was interested in the amount of work that could be obtained from a heat engine. In Carnot's time the efficiency of the steam engine was only about 5 percent, meaning that 95 percent of the heat energy of the burning fuel was wasted. He was able to demonstrate theoretically that the maximum possible efficency depended on the difference in temperature between the hottest (T) and coolest (t) parts of the engine: Efficiency = (T-t)/t. This important result later led to Clausius's formulation of the second law of thermodynamics.

SOURCE: Science Photo Library, London. Used by permission.

In a new vein, Carnot postulated a maximum efficiency that a heat engine could achieve. No engine could ever be more efficient than his ideal heat engine. If such a machine were claimed, it would mean a perpetual motion machine had been found, which Carnot argued had not been found and never would be. In addition, the maximum efficiency possible depended solely on the fall of heat, which for Carnot was caloric, from hot places to cooler places. Fall in temperature was proportional to fall in heat, and today the maximum theoretical efficiency of a heat engine is found with the simple equation: Efficiency = $(T_{hot} - T_{cool})/T_{hot}$, where T is temperature in degrees Kelvin.[43] I'll return shortly to the Kelvin scale of temperature.

Most remarkable, at least in retrospect, Sadi Carnot saw a relationship between amounts of heat created by combustion and the motive power (now called work) created by an engine. He did not emphasize this finding, but he lamented the shortfall of power from practical engines compared to his predictions for an ideal engine. The best engines at tin and copper mines in Cornwall, England, produced only 5 percent of the power theoretically available.[44]

Clearly Sadi drew on his father's work, and quite likely he drew analogies from the power of falling water in picturing the fall of caloric. It's also possible that the idea of a cycle may have been drawn from Volta's battery, in which electricity, another subtle fluid, required a circuit or "cycle" in order for the electric effects to occur.[45]

One final point about Carnot's essay: as all authors do, he began by framing the importance of his topic. Significantly, the essay's first three pages indicate Carnot's intense interest in the positive effects of the steam engine and suggest his envy of England's political power derived from these devices.

> [The steam engine] has already developed mines, propelled ships, and dredged rivers and harbors. It forges iron, saws wood, grinds grain, spins and weaves stuffs, and transports the heaviest loads. . . . Iron and fire, as every one knows, are the mainstays of the mechanical arts. Perhaps there is not in all England a single industry whose existence is not dependent on these agents. . . . If England were to-day to lose its steam-engines it would lose also its coal and iron, and this loss would dry up all its sources of wealth and destroy its prosperity; it would annihilate this colossal power. The destruction of its navy, which it considers its strongest support, would be, perhaps, less fatal.[46]

The French Revolution had affected the lives of both Lazare and Sadi Carnot profoundly, and both served revolutionary France. Is it reasonable to suggest that Sadi Carnot regretted the final defeat of Napoleon's forces at Waterloo in 1815, led by the English, which forced his father into exile? Sadi Carnot remained in the French army after Napoleon's defeat, but does his essay, published nine years after Waterloo, suggest his desire for France to have the same geopolitical power the English had achieved? Had he been interested only in heat and engines, he could have confined his introduction to scientific and philosophical ideas. He didn't, suggesting he understood the argument made in chapter 3: political-economic and geo-political power played an integral part in the story about the Third Energy Transition.

ELECTRICITY, HEAT, AND ENERGY: THE SYNTHESIS

Electric telegraphs and steam engines developed before science could explain how they functioned, but those technologies had relied on ear-lier scientific discoveries about the behavior of electrical attractions and repulsions and the behavior of gases when pressure and volume changed. In the quarter century between the 1840s and the early 1870s, studies of heat and electricity ultimately crowned energy and its conservation the foundation of all natural philosophy.

The new synthesis started with claims of a fixed and measurable mechanical equivalent of heat, signifying the convertibility of heat to mechanical action and vice versa. Some years later this conversion became a basis for the doctrine that energy can't be created or destroyed, but it can be converted to different forms and to work.

In Germany, Julius Robert Mayer (1814–78) first calculated the mechanical equivalent of heat in 1842, and his ideas rested on the con-version of heat to work in both physical and physiological systems, but he did not have a concept of energy.[47] In England, James Prescott Joule (1818–89) noted that Mayer may have had priority in stating the prin-ciple of the mechanical equivalent of heat, but Joule claimed, based on extensive experimental evidence, to have established the correctness of the equivalency (figure 2.3).

Joule began his research on a different problem: given the attractions and repulsions of magnetic forces, is it possible to make an electric motor? Had this line of inquiry succeeded, we might today remember Joule as the one who beat James Watt at his own game. Lofty goals, however, do not preordain future events. By 1841, Joule sadly con-cluded that the potential for electric motors could not match the exist-ing performance of steam engines.

FIGURE 2.3. James Prescott Joule (1818–89), English physicist, the discoverer of the mechanical equivalent of heat. Joule was born at Salford, the second son of a wealthy brewer. Having suffered a spinal injury early in life, he was allowed to study rather than work in his father's company. By and large, Joule was self-taught, and he became devoted to accurate measurements of heat. In 1847 he published his first paper, showing that a given quantity of work always produced a particular quantity of heat. He was also the first to propose the principle of conservation of energy and that the temperature of a gas drops as the gas expands. The SI unit of work (joule) is named after him.

With the consumption of one pound of zinc in a battery, his device could lift 331,400 pounds one foot, while one pound of coal in the best steam engines in Cornwall could lift 1,500,000 pounds one foot. Not only could the heat of coal provide more lifting power, but the price of zinc and other materials in the battery was much higher than that of coal. Electric motors worked, but they flopped miserably on economic grounds.[48] This situation changed later when cheaper ways of generating electricity appeared (see chapter 3).

Despite his disappointment about not replacing steam engines, Joule's subsequent research turned him from a would-be inventor to one of the most noted natural philosophers of the 1800s. Like Faraday, who spent the 1830s laying the basic groundwork for electricity, Joule in the 1840s conducted an exacting set of experiments to understand the relationships between heat and mechanical work.

In virtually every combination imaginable, he measured the evolution of heat, in tiny fractions of degrees, from mechanical work, electrical currents, electric batteries, chemical reactions, and expanding gases. In multiple experiments, he found that the amount of heat needed to raise one pound of water by one degree Fahrenheit was the equivalent of mechanical work that raised between 587 and 1,040 pounds one foot of height, with an average of 838 foot-pounds (i.e., number of pounds in weight multiplied by height through which the weight fell). He also measured the equivalency by comparing the heat needed to warm water as well as oil and mercury and found the results were essentially identical.[49]

For example, one experiment involved a set of falling weights hooked through a pulley system to a paddle wheel in water. As the weights fell, they turned the paddle wheel; Joule measured the temperature of the water before and after rotation of the paddle wheel and found it went up; that is, the friction of the wheel against the water created heat. Joule calculated the mechanical work done in foot-pounds and estimated the mechanical equivalent of heat at 772.24 foot-pounds for a 1° Fahrenheit rise in temperature of one pound of water.[50]

Joule believed his estimates of the mechanical equivalent of heat said something important about heat, but until 1847 he failed to attract much interest from other scientists. Finally, at meetings in Oxford, his work interested William Thomson (Lord Kelvin) (figure 2.4). Thomson, professor of natural philosophy at the University of Glasgow, had just returned from a year's work with Henri Victor Regnault (1810–78) in Paris, measuring the thermal behavior of gases, a research program

FIGURE 2.4. William Thomson (Lord Kelvin) (1824–1907),
physicist and mathematician. Thomson was born in Belfast, son
of a mathematics professor. He moved with his family to Glasgow
and at an early age showed an ability with mathematics and
physical science. He did well in his studies at Cambridge
University, studied in Paris for a year, and was then appointed a
professor at the University of Glasgow. Thomson found Joule's
experiments interesting and important, a boost to Joule, and
contributed heavily to the development of the laws of
thermodynamics. He was honored by a peerage in 1892. (From
Jennifer Coopersmith, *Energy: A Subtle Concept* [New York:
Oxford University Press, 2010], 284–85.)

SOURCE: Science Photo Library, London. Used by permission.

sponsored by the French government to increase the efficiency of steam engines.[51] During his year in Paris, Thomson had learned of Carnot's and Clapeyron's work on heat engines.

While impressed by Joule's findings, Thomson saw an immediate conflict. For Carnot, heat went from a high temperature source to a low temperature source while making an engine work, and the amount of heat at the end was the same as at the start, because heat was indestructible. Joule's paddle wheel, however, measured mechanical work turning into heat. Thomson balked at full acceptance of Joule's work, because, Thomson said, Joule had not converted heat into work, even though he had indirectly done so in experiments with electric motors and mechanical work. For Joule, the action of an electric motor was a proxy measure of heat.[52]

Between 1847 and the early 1860s, Thomson fully reconciled with Joule's ideas about conversion of heat, and, along with Rudolf Clausius (1822–88) of Germany (figure 2.5), founded thermodynamics, the bedrock of energy science. Thomson and Clausius did not collaborate directly, but each wrote papers that clearly built on the other's work.

The Thomson-Clausius interplay began in 1848 with Thomson constructing an absolute temperature scale. At the time, measuring heat involved thermometers that measured temperature by expansions of air. Any two numbers on the scale were supposed to have equal changes of volume as the thermometer went from the first to the second number. Recent investigations by Regnault, however, indicated that the "equal volumes" between any sets of two numbers were only approximately related to equal heat changes between the sets. Thomson wanted a temperature scale that reflected an equal amount of heat changed for each degree. He argued that Carnot's theory that falling heat makes work (of a heat engine) provided a basis for an absolute temperature scale. Thomson's new scale, later called the Kelvin scale, was based on the idea that the amount of heat lost in dropping one degree (now written 1 K) was the same no matter the starting, higher temperature.[53]

Thomson's first attempt to set an absolute scale for temperature still relied on Carnot's postulate that the ideal heat engine produced work but that no heat was lost in the Carnot cycle. He was still stuck with the idea that heat was caloric, a substance that could not be created or destroyed. Yet he also understood Joule's experiment, in which falling weights produced heat by the friction of the moving paddle wheel. Thomson could not resolve the conflict, but Clausius moved things along with the help of Thomson's absolute scale of temperature.

FIGURE 2.5. Rudolf Clausius (1822–88), German theoretical physicist. Clausius was the first to propose the electrical dissociation of molecules in solution into ions, an idea not fully accepted until decades later. His fundamental contributions were in thermodynamics, especially the concept of entropy. He stated that within a closed system, the ratio of the heat content to the absolute temperature may increase but will never decrease and that the system may become more disordered but never more ordered. This is now known as the second law of thermodynamics. Clausius was awarded the Royal Society's Copley Medal in 1879.

SOURCE: Science Photo Library, London. Used by permission.

Clausius in 1850 resolved the conflict by declaring Carnot incorrect on one important point: in the Carnot cycle, not all the heat remained; some changed into work to move the piston. At the start of the cycle, the gas in the engine had a temperature and volume, called U. During the cycle, the gas expanded at constant temperature, doing work; the gas then continued expanding, but the temperature dropped as the piston continued moving and more work appeared; then the ideal cycle reversed as work compressed the gas at the constant lower temperature; and finally the compression raised the temperature with more work back to the original temperature and volume.

Clausius said that U had changed by an amount Q + W, where Q was the heat transferred from the higher to lower temperature and W was the work done by the engine. Thomson shortly thereafter gave U the name "energy," but it was Clausius, relying on Joule's insights, who saw that Carnot hadn't understood everything. The relationship $\Delta U = Q + W$ is one way of stating the first law of thermodynamics: energy isn't destroyed or created but is transformed from one form (e.g., heat) to another (in this case work, an expanding gas makes a piston move and the engine do mechanical work like pump water from a mine) (ΔU, or delta-U, means change in U.)[54]

Now came the tricky part of the theory of energy. Carnot saw *transfer* of heat, and Joule saw *transformation* between mechanical work and heat. Clausius reasoned that both could occur in a Carnot cycle. He also knew that real engines involved transfer of heat from the hotter to the cooler temperatures, not the other way around. Clausius also synthesized the mathematical analysis of Clapeyron and Thomson and the data on heat effects on gases from Regnault to deduce that the efficiency of an ideal engine varied as $1/T$, where T was the absolute temperature on Thomson's temperature scale.

By 1851, Thomson had absorbed the work of Clausius as well as Joule, settled the conflict between Joule and Carnot in Joule's favor, and embraced the dynamic theory of heat advocated by Joule. Caloric as a substance disappeared, and heat was the movement of atoms and molecules. With these presuppositions, Thomson said (a) Q was always proportional to T (again using his absolute scale) and (b) around every ideal Carnot cycle the sum of all values of Q/T had to be zero. Heat transferred out of the engine was positive, and heat transferred in during the ideal compression was negative. Thomas wrote the relationship for an ideal engine as $\Sigma (Q/T) = 0$, where Σ means sum of all values of Q divided by the T of the Q.[55]

Clausius, building on Thomson's $\Sigma (Q/T) = 0$, built a new conceptual scheme for the changes of heat in an ideal Carnot cycle and a real cycle of an actual engine. He distinguished between heat that did work and heat that merely transferred from the hot source to the cooler sink. Clausius realized that in real engines no ideal Carnot cycle existed. In other words, in a real engine some heat would come out of the temperature change unaccompanied by work.

This "wasted" heat he called entropy. The definition of entropy was $S = \int dQ/T$, in which S designated entropy, and $\int dQ/T$ was equivalent to Thomson's $\Sigma \, (Q/T)$ put into the language of calculus, in which dQ/T represented infinitesimally small changes in Q divided by the temperature at which the change in heat occurred. Clausius reasoned that some heat was always wasted, and thus the value of $\int dQ/T$ had to be positive, that is, greater than 0, in a real engine. Thus Clausius restated Thomson's equation as $\int dQ/T \geq 0$; for any real engine, the value was greater than 0, but for an ideal engine, the value was 0. The symbol ≥ 0 means "greater than or equal to 0." Some entropy, or wasted heat, always appeared in any real process. Thomson had given a preliminary form to the second law of thermodynamics, $\Sigma \, (Q/T) = 0$, for an ideal engine, and Clausius provided the more formal statement, $S = \int dQ/T \geq 0$.[56]

Mayer, Joule, Thomson, and Clausius developed the conceptual and mathematical foundations of energy science, summarized by the two laws of thermodynamics. The first law described this new concept, energy, by asserting that any change in energy potentially contained heat plus work. The second law asserted that in any real change of energy some heat would leave the engine or process without doing work, and no transformation could be 100 percent efficient.

Newcomen and Watt also played an essential role in the origins of energy science, because their engines that moved things prompted development of the idea that work by the engine (accelerating a body through a distance) derived from energy (heat). Heat in turn was the pressure or force of moving atoms and molecules that moved a piston in a cylinder.

Energy and its transformations appeared first in 1859 in a textbook by a colleague of Thomson's at the University of Glasgow, where Watt had developed his steam engine.[57] Thomson and Peter Guthrie Tait at the University of Edinburgh, prepared other classroom materials as early as 1863, and their book, *Treatise on Natural Philosophy*, appeared in 1867. As they noted in the preface, "One object which we have constantly kept in view is the grand principle of the Conservation of Energy. According to modern experimental results, especially those of Joule, Energy is as real and indestructible as Matter."[58]

Both Clausius and Thomson believed that the concept of heat transformations with increases in entropy applied far beyond the world of practical steam engines. Energy, not force, was the conceptual glue of all natural processes. Entropy, a measure of heat lost in every process and no longer available for work, meant that the cosmos was headed for a future state of uniform temperatures, from which no work could be extracted. No work meant no processes of anything, including life, could occur—a very glum view, the ultimate heat death. Scottish Calvinist theology, embraced by Thomson and Tait, sought ardently to use all energy available as efficiently as possible. Energy gone to waste was a sin. They believed the ultimate pessimism expressed in the second law, but in the meantime they wanted to let no temperature change happen without extracting maximum work from it.[59]

SUMMARY OF THE IMPACTS OF THE SYNTHESIS

Ideas about energy in the late 1860s synthesized over two hundred years of work in the science of mechanics (movements of masses on earth and in the heavens), electrical science (attractions, repulsions, and chemical reactions), and developments in technology (steam engines, electric lighting, electroplating, telegraphy, and electromagnetic induction). Energy and entropy offered unprecedented insights into both natural processes of change and the rapidly increasing human-caused changes of industrial technologies.

Two central ideas captured the nature of energy and entropy. First, the laws of thermodynamics originated from models of mechanical work: a steam engine running on the heat of burning coal. Movement of the piston eventually led to a fall of temperature of the steam as heat transformed into work. Energy/work was transformed heat. At the same time, not all heat transformed into work. Some simply left the engine as heat at a lower temperature, and this heat was entropy. Most important, every real engine produced both work and entropy, meaning that in every process at least some heat does no useful work. Inevitable increases in entropy constituted a constraint on the work that a fuel like coal could produce.

As a corollary, energy theory was more consistent with the dynamic theory of heat rather than heat as a substance, caloric. In other words, what we feel and measure as heat comes from the motion of atoms and molecules; atoms and molecules in warmer materials move faster than those in cooler materials. In the steam engine, both energy and entropy resulted from the movement of water molecules in the steam. This inference strongly supported atomic theory, and later direct experimental evidence confirmed it.[60]

Second, Faraday's fields of force moved into the realm of energy. Magnetic and electrical attractions and repulsions meant that an electromagnetic field derived from and surrounded matter, and the forces or energy in this field could do work, such as run an electric motor. Joule's comprehensive studies on interconversions of electrical effects, heat, and mechanical work gave every reason to see work done by electrical attractions and repulsions in the same way as work done by a steam engine.[61]

Other changes also followed from the findings of energy science. One seems trivial in retrospect but in fact solidified important links between energy science and the modern state. Many of the technical problems encountered building long-distance, underwater telegraph lines stemmed from a lack of standard units to measure electrical events. Engineers working for the entrepreneurs building the telegraph lines worked with scientists such as Thomson and Maxwell to build a system of units for measuring events, building devices that worked, and maintaining the underwater lines.

Between the late 1850s and 1880s, collaboration among engineers and scientists in Britain, Germany, France, the United States, Japan, and elsewhere resulted in agreement on an absolute system of units based on the work done by energy, not the forces of attraction of gravity and of attraction and repulsion of electricity (see appendix 1). "Absolute" meant basing the units on measures considered invariable: mass (kilogram), distance (meter), and time (second). Mechanical work is the ability to accelerate a mass over a distance (i.e., to change the velocity of the mass or change its direction).

Thomson and Maxwell were particularly keen that work done made the best basis for units, because they saw that commerce and the state valued work. German proposals to base units on measuring the strength of forces did not advance. British experiences with underwater telegraphy and steam engines had heightened their sensitivities to the commercial importance of work accomplished, but most of the German scientists involved had not worked on those practical problems. German scientists who had experience with practical electrical events had favored a "relative" unit, based on a measuring instrument with carefully specified properties.[62]

Universally accepted units helped energy science enter commerce as a product ready-to-buy, but scientists seeking to understand the physical world never ceased working on discrepancies and loose ends. As a result, the Newtonian world of invariant space and time, filled with atoms and molecules moving in a void and altered by forces of attraction and repulsion, gave way after 1900 to a vastly different model of the universe. The engines and electrical technologies of the 1800s, however, behaved in ways consistent with the Newtonian view of space and time, so they did not immediately pressure science for new explanations.

One set of experimental observations on electricity, however, had not yet become a technology at the time of the energy synthesis. In 1839, Alexandre Edmond Becquerel (1820–91) of France had found that sunlight striking certain materials created an electric effect. This was a fourth way to make electrics following the earlier discoveries of electricity from rubbing (Gilbert, 1600), chemical reactions (Volta, 1800), and movement of metals in magnetic fields (Faraday, 1831). Practical use of transforming light into electricity, however, did not arise for over a century due to multiple factors.

The ability to make light a practical source of electrical energy used a set of new ideas from work by Max Planck and Albert Einstein between 1900 and 1905. They interpreted experiments on light and other electromagnetic radiation to mean that energy behaved as if it were in packages: quanta or photons, not waves as suggested by earlier experiments. The energy in each quantum, however, depended on the wavelength, a paradox: quanta or photons of shorter wavelengths had more energy than those of longer wavelengths.

Quantum mechanics blossomed when new models for atoms pictured electrons in orbits around nuclei made up of protons and neutrons. Electrons could rise to new orbits of higher energy or escape the atom by absorbing a photon; an atom emitted a photon if an electron dropped to an orbit of lower energy. This model of the universe was anything but intuitive, unlike the Newtonian world, which seems like common sense. Nevertheless, quantum mechanics ultimately, after the 1950s, modeled photovoltaic cells that transformed light energy into electrical energy (see chapter 8) and light-emitting diodes (LEDs) that transformed electrical energy into light.

The concept of energy as quanta was only one of numerous discoveries in the years around 1900. A second finding between 1896 and 1903 by Henri Becquerel (1852–1908; son of Alexandre Edmond) found that uranium emitted a new type of invisible radiation.[63] Ultimately, this discovery—radioactivity—broadened the concept of matter and supported discoveries of subatomic particles: protons, electrons, and neutrons. The ability of a neutron to fission atoms of uranium under certain conditions released heat, an energy source useful in weapons and in generating electricity. This turned uranium (and thorium) into fuels, the last mineral fuels of the Third Energy Transition.

Beyond quantum mechanics and radioactivity, physical science, dominated by the concept of energy, found experimental evidence and theories that replaced the Newtonian world with the relativistic world of Einstein. Faraday's field concept expanded to absorb a plethora of subatomic particles and two new forces—weak and strong forces—to add to the attractions of gravity and the attractions and repulsions of electromagnetism.[64] Those discoveries have yet to enter the practical world, however, and they lie beyond the scope of this book.

CONCLUSION

The three hundred years from 1600 to 1900 brought new devices, ideas, commerce, and political institutions to human life. Energy services transformed modern life, and energy science was as much a product of these changes as a source. Energy science provided descriptions, explanations, predictions, and control of energy processes; engineering marched hand in hand to build the energy services of the Third Energy Transition, based on the heat of burning coal, oil, gas, and the fission of uranium (see chapter 4). Energy and energy services became a fundamental part of the modern state, to which we turn next.

Connections with Everyday Life

What exactly is a "modern state"? People who live in modern states have a general notion of their nature, and some everyday rituals capture the essence. In the United States, for example, the Pledge of Allegiance equates the flag with the state and summarizes its obligations.

> I pledge allegiance to the flag of the United States of America, and to the republic for which it stands, one nation under God, indivisible, with liberty and justice for all.

This simple, thirty-one-word oath grounds the pledge to a flag (usually a piece of cloth) representing a republic that is subservient to God, that cannot be divided, and that promises every person freedom and fair treatment. The pledge is not to a leader, nor does it refer to any particular administration that happens to be in power. The "state" transcends individual political leaders and governments and comprises an important part of modernity.

What does the nature and shape of a modern state have to do with energy? Why should a book on energy plunge into the political history of the past eight hundred years? At the simplest level, the answers lie in the wealth and military power that energy enables. More completely, energy is one of several building blocks on which modern states rest and without which they would collapse. This chapter outlines the relationships and argues that advocates of sustainable energy must recognize the connections, because the passions about reform of energy are generally linked to passions about modern states.

CHAPTER 3

Energy and the Modern Nation-State

> Modern states have always considered the energy sector as
> an exceptional case within the general panorama of
> industrial activities. Its importance in the development of
> their economies, in their currency balance, in the distribu-
> tion of wealth within and among them, and in their status
> in international relations, was considered so great that
> states could not be expected *to let markets alone achieve*
> *goals of general interest that . . . would not otherwise have*
> *been followed spontaneously.*[1]

Modern industrial states consume enormous amounts of mineral energy
(coal, oil, gas, uranium), and their engineers and scientists use energy
science to produce energy services of many kinds. Before 1600, however, neither modern states nor energy science existed. Energy services
rested mostly on firewood for heat and light during the First Energy
Transition and agriculture for food and feed during the Second Energy
Transition, supplemented by small amounts of wind and water power.
In England coal had made inroads, supplying heat like firewood, but the
vast majority of people worked in agriculture.

After 1600, over the course of about three hundred years, the modern state, energy science, and vast new arrays of energy services
appeared. How did modern states and energy services/science interact
during this period? Is it just a coincidence that the two sets of changes
occurred at the same time, or did new sources of energy and new energy
services interact in synergistic ways to affect the control of wealth and
political power? How did wealth and power relate to the work done by
energy? To what extent did these interactions affect the course of the

Third Energy Transition? Will similar interactions between energy science, energy services, and the modern state also affect the Fourth Transition? Will the Fourth Transition itself create new interactions between energy science/services and the nation-state?

These are challenging questions that lie at the heart of energy issues. Too often only science and technology frame discussions about energy and energy policy. This chapter argues that energy technology, energy science, and modern states coevolved and that it's difficult to grasp the essence of the Third Energy Transition without incorporating the coevolution. In the movement to a new energy economy—the Fourth Transition—we dare not ignore these issues, because the interactions affect the distribution of benefits and risks from energy, and they can support, encourage, and promote, just as they can also stymie, block, or warp.

THE IDEA OF MODERN STATES

The exact nature of a modern state eludes precise definition. The Declaration of Independence of the United States (1776) captures part of the essence succinctly: "The United Colonies . . . as Free and Independent States, . . . have full Power to levy War, conclude Peace, contract Alliances, establish Commerce, and to do all other Acts and things which Independent States may of right do."

Put starkly, American revolutionaries saw four attributes worth noting explicitly: going to war, making peace, making alliances with other states, and setting the rules for trading and business. A "state" was real, it had duties and rights, and it could act for or on behalf of the people who lived within its boundaries. As developed below, Parliament established and constrained the king's power in Britain, and American revolutionaries argued that King George and Parliament had reverted to tyranny in the colonies.

The political scientist David Held pictures the state as an impersonal, legal constitutional order that can administer or control territory. For the West, the idea of "the state" began with Rome and the Roman Empire, but the concept was little used until the 1500s. After that time in England many questions arose about the nature and limits of political authority exercised by the king. By 1700 the "state" had become the object of acute analysis by political philosophers.[2]

Intense questions about the state and the powers of rulers coincided with developments in new philosophies about nature, and some philosophers, such as Francis Bacon and John Locke, played important roles in both lines of inquiry. Also simultaneously, major shifts occurred in demographics, commerce, trade, and money and finance in England and other parts of Europe. Changes in politics, economics, and science, respectively, were mutually reinforcing..

It is easiest to follow the trail of comingled changes by looking at two periods: before Newcomen's engine and after. Changes in energy science and technology came mostly but not entirely after 1712; changes in politics, commerce, trade, and money continued apace in both periods.

BEFORE 1712: THE EMBRYONIC MODERN STATE AND SCIENCE

When Thomas Newcomen launched his first successful engine in 1712, England already had many trappings of a modern state: increased legal rights for individuals, increased effectiveness of the "rule of law," Parliament supreme over the monarch, expanding cities, increased industrial production, increased trade, patents, significant uses of coal, scientific societies, and a bank solidifying the use of credit to create money. A person from today's industrialized world would find England of 1712 quaintly old-fashioned at best. But the England of 1712 had changed mightily compared to that of five hundred years earlier and to most of its neighbors.

What caused England's transformation? Key events between 1066 and 1712 point to many factors. England of course had a history before 1066, but that year—the year of the conquest of England by William of Normandy—established the context for profound changes.

Before the Conquest, aristocratic wealth flowed from the military, political, and judicial subjugation of the common people, and the Normans mostly replaced the existing ruling structure with their own aristocrats. Ownership of land depended on services rendered. The nobility gained control over land and people—the source of almost all wealth—by rendering military and administrative services to William, and commoners who worked land rendered labor, agricultural produce, and cash to their lords.[3]

Towns provided manufactured goods such as metals, ceramics, glass, beer, and other products. Townspeople, however, were distinctly a minority, reaching only 3 to 5 percent of the total population by 1500.[4] England was typical of northern Europe after the eleventh century: rigid social hierarchy based on who owned, controlled, and labored on the land, with virtually all economic activity centered on agriculture.[5]

FIGURE 3.1. The barons and bishops constrain King John with the Magna Carta.
SOURCE: Jonty Clark, used by permission.

England before and after the conquest was a feudal-agrarian state, not a modern or industrial state. Its energy supply was organic, food, feed, and firewood, not mineral.

By 1700, however, England was well on its way to becoming the first modern state. What happened? The story has many twists and turns, but five interwoven factors stand out.

First, in 1215, disgruntled aristocrats, church officials, town leaders, and merchants forced King John (r. 1199–1216) to agree to a treaty that specified liberties of the upper class that could not to be violated by the monarch (figure 3.1).[6] Some dealt with freedom of commerce and some with procedures for enforcement of laws. The Magna Carta, or Great Charter, seeded the growth in England of the most important idea underlying the modern state, the rule of law. In the 1600s, it fueled a rallying cry to tame the English monarch. Rule of law forced the king to abide by rules, and a modern state requires the rule of law.

Second, a horrific plague decimated the populations of Europe between about 1347 and 1353. Between 25 and 33 percent of the people perished from the Black Death.[7] Plague reached England in 1348, and 30 to 40 percent mortality may be an underestimate for this area

THE 'BLACK DEATH'
ENTERED ENGLAND IN 1348
THROUGH THIS PORT.

IT KILLED 30-50%
OF THE COUNTRY'S
TOTAL POPULATION

FIGURE 3.2. Plaque commemorating the Black Death, Weymouth, England.
SOURCE: Photo by Mark A. Wilson, The College of Wooster, used by permission.

(figure 3.2). The historian John Hatcher places the mortality at 60 percent.[8] The disease continued to suppress the population for 150 years. For those who endured the period, it was a horror, but it unleashed social change that altered England spectacularly.

The deaths of so many created a surplus of land and a shortage of labor. The slavery-like conditions of feudal servitude crumbled as commoners found better wages and working conditions from aristocrats or left farming for the towns. Some land went untilled, but food supplies remained adequate, except in some years when unfavorable weather prevailed. Demand increased for meat, shoes, metal utensils, cloth, and other goods. Shortage of labor in the countryside and the promise of profits from livestock prompted landowners to enclose fields and raise more animals, for more meat and shoes.[9]

"English society in 1500 was not as it had been in 1350," noted the historian Jim Bolton. Plague in the 1300s and 1400s played an essential role in these changes, and they produced a variety of winners and losers. People at the time would not have recognized they were living in a transition from feudalism to capitalism, but that was the upshot of the Black Death for England.[10] The rule of law began to expand first— unsuccessfully—to preserve feudal servitude by commoners[11] and later to slowly expand the freedoms of the Magna Carta to everyone, not just aristocrats. England, a rural backwater compared to other areas of

Europe in the 1300s, began steady urbanization with rapidly growing industry that put it in the forefront of Europe by the 1600s.[12]

Third, as noted in chapter 1, England began a serious love affair with coal, beyond heating of buildings, especially after 1500. The plague's decimation of the population slowed the transition to coal by allowing reforestation and more use of firewood between 1350 and 1500, but deforestation resumed after 1500. Increased use of coal, even before Newcomen's engine, powered English manufacturing. In the 1560s, coal provided 10.6 percent of England's energy use, which was more than the 7 percent coal provided in Italy three hundred years later. By 1710 coal provided about one-half of England's energy needs, and the proportion reached over 90 percent by 1860 (see figure 1.1). If human and animal energy are excluded, in the 1560s, 74 percent of energy came from firewood and 24 percent from coal. By the 1860s, firewood provided 0.1 percent and coal 98 percent. In both periods, wind and water provided the remainder.[13]

The uses of coal changed and grew between the 1200s and 1600, and English trade and manufacturing changed in parallel ways. In earlier periods, coal had mostly two uses: burning lime to make cement and heating iron for blacksmiths to make specific objects. Firewood, either as open fires or charcoal, provided other sources of heat, including smelting of iron ore to make metallic iron for blacksmiths. In such an economy, the major trading items from England were wool and woolen cloth, and these exports proved quite lucrative from the time of the plague to the early 1500s. The late medieval English economy brought in more gold and silver bullion than left for the purchase of imports. In modern terms, England enjoyed a balance of payment surplus, and no reason existed for a shortage of money in the form of coins.[14]

Recovery of the population and increased uses of coal in the 1500s, however, began to make the English economy more elaborate and productive. By 1600, coal had replaced charcoal for making beer and ale, distilled spirits, bricks, tiles, pottery, salt, sugar, soap, glass, nails, hardware, cutlery, and the smelting and casting of brass.[15] In the last few decades of the 1500s, the total energy consumed in England rose by 20 percent, almost all of which came from increased uses of coal.[16] But the population also increased during this period, and wages fell, in contrast to the rising wages after the Black Death and the population decline.[17]

At the start of the 1600s, in spite of a coal-powered expansion of its economy, England again faced social instability, dislocation, unemployment, and poverty. To this array of problems was added a chronic lack of money in circulation to facilitate the growing trade between towns and rural areas. Population growth, expanded economic activity, and taxes to be paid in money had combined to render the supply of coins grossly insufficient. In the 1500s, merchants developed credit mechanisms to enable trade, but the credit instruments—pieces of paper—did not circulate as money. Credit was a private arrangement between two parties.[18]

Solving the lack of money depended on a powerful shift in ideas about the purposes and nature of the political economy and money. To make a long story short, the fourth factor enabling the modern state was the rise of a new institution, the Bank of England, and new ideas about the nature of wealth, property, the use of credit money (see Glossary), and religious freedom, all of which both shaped and resulted from the radical changes in England during the revolution of 1688–89.

The most visible changes occurred in government and legal affairs, all triggered by the failure of the Stuart kings, especially Charles I and James II, to respect the rule of law launched in the Magna Carta. Four parliamentary actions and two civil wars solidified England as a constitutional monarchy ruled by Parliament by the start of the 1700s.[19] The Petition of Right (1628) begged Charles I to curtail arbitrary power and recognize the Magna Carta. Resistance by Charles provoked the English Civil War and ultimately his execution by the victorious Parliament. Reversal of Parliament's victory reestablished the monarchy under Charles II. Like Charles I, however, Charles II's actions incited his subjects, and the Habeas Corpus Act was passed in 1679, curtailing indefinite imprisonment without trial.

Turmoil about governance continued to build after the death of Charles II in 1685. His brother James II succeeded to the throne and ignited passions once again in the Glorious Revolution of 1688. James fled England, and Parliament conferred the crown on James's daughter Mary and her husband, William. Parliament's assertion of power to name the monarch symbolized the rise of rule of law. Parliament further defined its powers and the liberties of all people in the Bill of Rights (1689) and the Act of Settlement (1701). Many people—all women and many men—could not vote, so England was not yet a real democracy, but the king could no longer evade the law with impunity.

This tumultuous change involved struggle between Whigs (drawn largely from the manufacturers and merchants who had blossomed in the previous century) and Tories (the long-standing, politically powerful, aristocratic, royalist land owners) for political primacy. The political agenda of the Whigs during the Glorious Revolution included formation of the

FIGURE 3.3. Bank of England, London.

SOURCE: Diliff, *Wikimedia*, Creative Commons license http://creativecommons.org/licenses
/by-sa/3.0/.

Bank of England to promote more credit money circulating in commerce. This, they believed, correctly, would promote manufactures and trade, which the Whigs argued were property, wealth, and the ability to make war. Tories, in contrast, generally opposed the Bank of England, believing that only land underlay real wealth and the ability to make war.[20]

The Bank of England, launched in 1694 (figure 3.3), set in motion the growing willingness to use paper money in place of coins for commercial transactions. But why would people accept pieces of paper when buying and selling goods and services? The answer lay, for example, in the trust held by a craftsman selling a pair of shoes or a brewer selling a pint of ale that they in turn could use the piece of paper to buy their bread or pay their taxes. Generating trust rested heavily on the Bank's relations with the new government.

King William, legitimated by Parliament to replace James II, used the Bank of England to finance war with France, considered by the Whigs England's major threat. William's government issued bonds to creditors, promising to repay them from tax receipts based on taxes approved by Parliament, a form of security that created trust that the paper currency issued by the Bank had a known, stable value. The Bank also became the center for all the government's financial transactions, further enhancing its prestige and solidity.[21]

England was not the first country to embrace credit money, as numerous Italian banks and the Bank of Amsterdam had already used versions of credit money. England, however, moved the institution of credit money faster and further than others. People today take credit money for granted, as long as they maintain trust in the central bank issuing paper currency. Governments and central banks can suffer loss of trust and consequent loss of value in credit currency, and macroeconomics is the academic discipline that explains the theory and functioning of the world's central banks that control and manage credit currency.[22]

In the United States, for example, paper currency bears the statement, "Federal Reserve Note," and, importantly, "This note is legal tender for all debts, public and private." Similarly, in the United Kingdom, a £5 note bears the statement, "I promise to pay the bearer on demand the sum of five pounds," signed by the Chief Cashier for the Governors and Company of the Bank of England. Do these pieces of paper have any intrinsic value? No, but the fact that merchants and the tax collector will honor them gives them great authority.

For the study of energy, the significance of the Bank of England lies in the power of credit money to facilitate the expansion of commerce powered by coal. No longer hobbled by a shortage of silver and gold coins, the economy could have as much money as it needed to match its size. Credit money facilitated trade and thus gave incentive to increase the amounts of goods made, and coal provided inexpensive heat to make more goods.

Creation of credit money is linked to the fifth and final factor pushing England toward a modern state: science, or natural philosophy as it was more commonly known in the 1600s. Again, to make a long story short, England played a leadership role in the development of the modern sciences, particularly in the 1600s. The historian Carolyn Merchant has assessed the growth of science as "the death of nature," a perspective that reformulated the world as dead, spiritless matter rather than an organic, feminine, living entity.[23] Francis Bacon, a true polymath of the late 1500s and early 1600s, argued that science should tame wild nature and make her do man's bidding. Bacon's *New Atlantis* (1627) stimulated many empirical and experimental studies and underlay the establishment of the Royal Society in 1660. Promotion of interchange among philosophers undoubtedly led to more intensive levels of investigation and the spread of scientific and technological knowledge. The Royal Society, and its *Philosophical Transactions*, was one of the first if not the first such organization in the world, and its founding charter from Charles II signified that the modern state took science seriously (figure 3.4).

The historian Carl Wennerlind argues that the creation of credit money interacted with the emergence of modern science. Alchemical

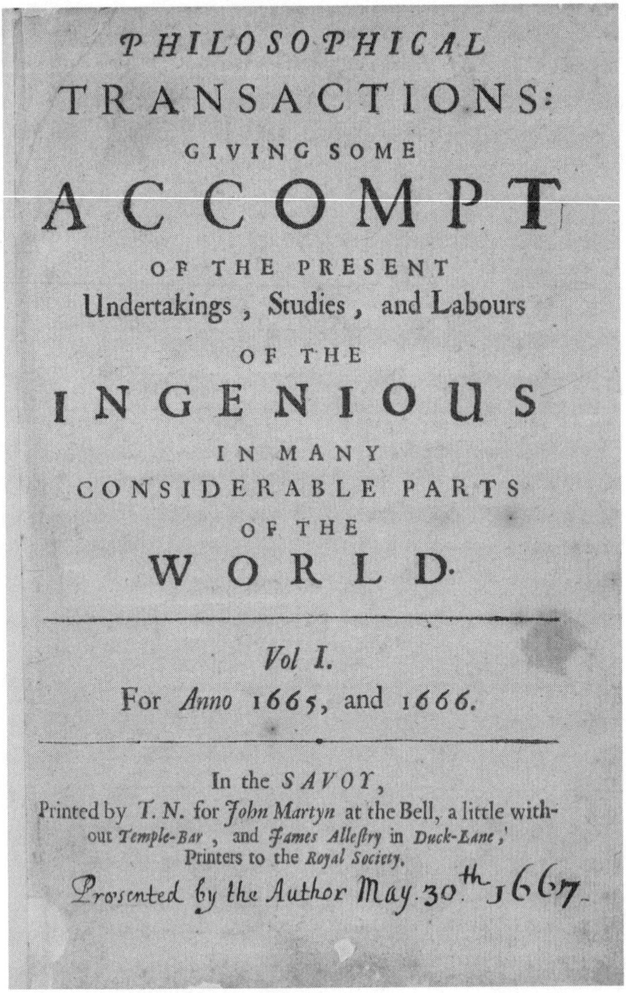

PHILOSOPHICAL
TRANSACTIONS:
GIVING SOME
ACCOMPT
OF THE PRESENT
Undertakings, Studies, and Labours
OF THE
INGENIOUS
IN MANY
CONSIDERABLE PARTS
OF THE
WORLD.

Vol I.
For *Anno* 1665, and 1666.

In the *SAVOY*,
Printed by *T. N.* for *John Martyn* at the Bell, a little without *Temple-Bar*, and *James Allestry* in *Duck-Lane*, Printers to the *Royal Society*.

Presented by the Author May. 30th 1667.

FIGURE 3.4. Frontispiece of the first volume of *Philosophical Transactions of the Royal Society*, 1666.

SOURCE: Henry Oldenburg, Creative Commons Attribution 4.0, http://creativecommons.org/licenses/by/4.0/legalcode.

ideas, widespread in 1600s England but now discarded, saw the world as organic, alive, growing, and always striving to reach its ultimate endpoint. When combined with Bacon's new empirical science, nature's bounty was was seen as neither meager nor confined. The alchemical tradition saw money as a means to facilitate exchange and store value between trades. With it, life could become better for everyone. Money limited to a finite number of gold and silver coins constrained and thwarted human potential. Money in the form of circulating, trustworthy credit certificates came from these essentially alchemical ideas, along with modern science,[24] even though modern science moved away from the idea of a living nature.

These five factors—the beginnings and growth of rule of law embodied in the Magna Carta, increased social mobility and the collapse of feudal servitude, the benefits of coal, credit money and banking, and modern science—melded in mutually reinforcing ways. Political and legal changes rested on social change, the rise of coal, and a new way of creating money. People escaping feudal servitude and fleeing to towns provided a labor force that enabled an increase in the manufacturing of goods. Coal provided the energy to make more goods at lower costs compared to firewood. More people with higher wages bought these goods and encouraged additional manufacturing. With more money, commerce grew. Embryonic scientific and technical organizations began systematically to investigate nature and find new ways to control it. England did not invent capitalism, but English merchants and manufacturers successfully joined the quest to profit endlessly from new investment. Between 1600 and 1700, England's embrace of coal, rule of law, credit money, and science brought it from a feudal backwater to the prototype of the modern state.

The rising power and influence of the urban manufacturers and merchants lay behind the rise of England's constitutional monarchy. Entrepreneurs could not thrive under the arbitrary and capricious actions and taxes of an unrestrained monarch, because their investments and properties were not safe. The new economy—powered increasingly by coal, credit money, and new technology—provided the social and political base to defang the king. We know little about the financial arrangements underlying Newcomen's invention of his engine, but Newcomen benefited from the protection of a patent issued to Thomas Savery. In this way the rule of law that enhanced commerce embraced the first device to move coal into an entirely new realm and vastly increase its supply and use.

AFTER 1712: ENERGY COMPLETES THE MODERN STATE

England had emerged as a modern state when Newcomen introduced his atmospheric engine, but by 1812, England had changed in even more ways, some of these changes directly or indirectly attributable to the development and deployment of steam engines. Within another hundred years, these changes had moved far beyond with developments of more engines and, especially, electricity. Moreover, England had very few counterparts in the early 1800s, but by the early 1900s, other countries had mimicked the English. Belgium, France, Germany, Sweden, the United States, and Japan had also become or started to become modern industrial states.

To follow the paths of these changes to modernity in detail would take us far afield, but a few examples, drawn from England and the United States, illustrate the patterns. Specifically, changes in the production of coal, iron, and textiles, accompanied by other changes in transport, made early 1800s England very different from England of 1712, and very different from all other countries at the time. Subsequent changes in the late 1800s with electricity, initiated mostly in the United States but rapidly spreading, amplified the effects. By the mid-twentieth century, a country could not consider itself a modern state unless its economy and culture rested on intensive uses of fuels, heat engines, and electricity for manufacturing, agriculture, lighting, heating, transport, and communications.

Coal

Coal use in England declined to almost nothing after the Romans left in the 400s, but it had become a significant fuel again by the 1300s as woodlands shrank and firewood became scarce and expensive. London began to use substantial amounts in the 1200s. Nevertheless, significant mining of coal did not pick up until the 1500s. The most accessible coal lay at or just beneath the surface, but by the 1600s mines were reaching a depth of 100 feet. Miners had to drain water continuously to keep the mines open.[25] Mining was hardly sophisticated, but it produced an energy source that— uniquely in Europe—allowed the English capital city to grow far beyond the limits it would have faced had it continued to rely on firewood alone.

By 1700 many mines had exhausted the easily reached surface deposits of coal and had burrowed deeper into the ground. Inevitably, more water intruded. Removal of water from coal mines before Newcomen's engine rested on gravity drainage channels, wind and water power, and horses. By 1700, however, these water removal methods would have left Britain's coal industry stymied and subject to decline. Mines deeper than 150 feet were subject to troublesome water intrusion, and only Newcomen's device allowed the industry to continue and to vastly expand its outputs. The steady spread of the engine's use during the 1700s opened a new chapter in obtaining prodigious amounts of coal to burn for heat and operate machines. British coal production soared from 3 million tons per year in 1700 to nearly 31 million in 1830. Prices remained competitive and fairly stable.[26] Sufficient coal at stable prices fueled profound change during this period.

Iron

Britain had made iron and iron products long before 1712, and by the time Newcomen's engine came into use the industry worked on a capitalistic basis: ironmasters hired wage labor, supplied the raw materials (ore and charcoal), and closely supervised the work at the master's facility.[27]

Locations had to be near woodlands, which provided the source for charcoal. British pig iron was useful mostly for cast iron products, not the stronger and more versatile products, for which pig iron was made into wrought iron and steel. Charcoal-made pig iron was brittle, and for the finer products the country depended on imports of higher-quality Swedish wrought iron.

Shrinkage of the woodlands prompted a search for smelting iron using coke, made by heating coal in the absence of air. The coal lost many volatile compounds and became like charcoal, nearly pure carbon. Abraham Darby, an ironmaster, successfully used coke instead of charcoal in 1709, and by the end of the 1700s, coke had replaced charcoal. This in addition to other inventions in the iron-making process made British iron as good as Swedish iron.

Coke did not burn as fast or hot as charcoal, so coke-fired smelters had to use bellows to blast air to encourage faster, hotter conditions. Water power ran the bellows before 1775, but in that year Boulton and Watt tied a steam engine to the bellows, thus freeing the iron blast furnace from dependence on water.

From the late 1700s through the 1800s, the British iron industry rose steadily to dominate the world's production. In 1800, Britain produced 19 percent of the world's pig iron, but in 1820 it was 40 percent and in 1840, 52 percent. During this vast expansion, other innovations enabled iron masters to make use of low-grade iron ores, often found in the same mines as coal. Integrated iron works produced everything from coal and ore to finished products in Staffordshire, Yorkshire, and South Wales.

Britain's prowess in iron making by the end of the 1700s, at the outbreak of the Napoleonic Wars, enabled the country to fashion guns and cannons for the navy and army that defeated Napoleon in 1815. In the 1830s, Britain began massive construction of railroads, and locomotives (powered by steam engines) and rails demanded large amounts of coal and iron (see chapter 1). By 1871, iron manufacture absorbed 25 percent of the steam power in factories and occupied 40 percent of the labor force. Once British railroad building was complete, British products found export markets to build railroads elsewhere. Rail transport, like credit money, made commerce easier for all products, which in turn provided the incentive to dig more coal, make more iron, and manufacture more of everything else.

During this time of expansion, the efficiency of using resources to make iron constantly improved. In the 1770s and 1780s, it required up to 10 tons of coal to make one ton of pig iron. By 1840, only 3.5 tons were needed. British prices undercut Swedish prices by the early 1800s. British expertise in making iron of high quality and low cost transferred into other industries that used the iron to make other precision machines. As the historian Phyllis Deane put it, "The iron industry played a role that . . . provided . . . the commodity on which . . . modern industry was to depend."[28] Modern life would be unrecognizable without the vast expansion of iron, other metals, and many other materials that occurred as part of the Third Energy Transition.

Textiles

Woven textiles for clothing, bed covers, wall hangings, and other items have been standard fare for thousands of years.[29] Before the 1800s in England, weavers used yarn made from wool, linen, silk, cotton, and other materials. The rich had more and higher-quality textiles compared to the poor, but everyone had some. Some cotton found its way into these textiles, but wool supplied England's most common textile. When cotton was used, it was generally in combination with linen, probably for undergarments. Cotton yarn was generally not strong enough to be the warp for weaving, so strong linen yarn made the warp and cotton the weft.[30]

In the mid-1700s, England's production of cotton-linen cloth amounted to £600,000 per year, with £200,000 of that exported. In contrast, wool textiles accounted for about £5.5 million per year. Cotton clearly lagged behind wool as the fiber of choice early in the Third Energy Transition. Production of textiles, wool or cotton, occurred in homes, generally as an adjunct to agricultural work. During times of the year when rural people had little to do, children carded the fibers, women spun, and men wove. No factories concentrated the work, and human muscles powered the processes. One weaver could keep three or four spinners occupied, so spinning the raw fibers into thread or yarn created a labor barrier to producing more textiles. Clothing remained expensive, and even the rich had little of it.

This situation changed radically after 1764, the year James Hargreaves developed the first labor-saving device for spinning, the spinning jenny. This early machine remained human-powered, but it enabled one person to spin eight yarns at a time rather than just one. Useful for both wool and cotton fibers, this device began the revolution in labor requirements for textile manufacturing: one spinner could keep two weavers supplied rather than one weaver keeping three to four spinners busy.

Between the 1770s and 1850s, a combination of other factors combined to completely revamp textile manufacturing worldwide, make cotton the dominant fabric, make Britain the major textile producer, and make textiles and clothing much less expensive. Invention of the water frame by Richard Arkwright in 1769 and the spinning mule by Samuel Crompton in 1779 vastly expanded the effect of Hargreaves's jenny. One spinning machine, powered first by water and later by steam engines, could spin over 1,300 threads at a time, continuously. Ultimately weaving, too, became mechanized. In the 1700s, a hand spinner required 50,000 hours to spin 100 pounds of raw cotton. After 1825, 135 hours sufficed, a 370-fold increase in labor productivity. Invention and deployment of the cotton gin in the United States reduced the price of raw cotton fiber.[31]

In 1780, a piece of muslin cloth cost 116 shillings, but fifty years later only 28 shillings. Britain's cotton textile industry undermined and outcompeted India's, the former source of the world's high-quality cotton textiles. Britain's exports of cotton roared from £355,000 in 1780 to £5,854,000 in 1800. Wages increased somewhat, but the lion's share of profits went to Britain's new class of industrial entrepreneurs.[32]

Human Labor, Slavery, and Empire in the Third Energy Transition

What did all the changes between about 1200 and the present mean? For Europe, historians refer to the period between about 400 and 1600 as the Middle Ages. The seventeenth century was the beginning of the "modern" era, but the 1600s and 1700s were only "early" modern. Only after the 1800s do common conventions begin to speak of "modern."

Important features of modernity began to appear even before 1600, however, and by 1700 England had institutions that later formed the core of what became modern life (rule of law, organized scientific inquiry blessed by government charter, commerce supported by credit currency). And still, as noted above, a person from 2015 England, France, the United States, Japan, and a host of other countries suddenly deposited in 1700 England would not have thought they had arrived at a "modern" place. The major difference? Energy from fuels and the heat engines that power truly modern life.

Did energy and heat engines make life better? By the measures of plentiful material goods and liberation from most physically demanding labor, the answer is overwhelmingly yes. With energy services, human populations grew faster, more infants survived childhood, food supplies became more plentiful, clean water became available from taps, clothing and shoes became less expensive and of higher quality, homes were heated in winter and often cooled in summer, mobility became faster, safer, and cheaper, labor became more productive, fewer farmers could feed more people—the list of benefits goes on and on.

We all harbor complaints about modern life, but at their core these grumblings do not seek the abolition of modern states, energy science, or energy technology. In democracies, a government that fails to provide power from energy risks becoming a government no longer with political power. Even countries with severely constrained citizens' rights dare not neglect energy once energy services become established. Government departments in charge of energy and the industries selling energy services are intrinsic to the operation of modern nation-states. These energy services thrive best under widespread democratic franchise, rule of law, well-regulated credit money and commerce, and a robust penetration of scientific and technological skills among citizens.

Despite the benefits and the entrenchment of energy in modern states, the Third Energy Transition came with more than a bucketful of benefits and happiness for everyone. The human capacity to be insensitive, uncaring, exploitative, violent, brutal, cruel, warlike, and murderous did not disappear just because people began to use coal to heat their homes, make their iron and beer, and run their steam engines. New technology did not automatically pave the way to a better life for all, and energy could power evil as well as good.

Consider just two examples. Newcomen's engine, the device that shifted the Third Energy Transition into a new realm, served its owners well in the sense that they could pump water less expensively and more effectively from their coal mines. What about the miners? Their work basically remained the same but with one important difference. Before the engine they worked with pick-axes, buckets, and carts to break coal from the surface of the seam and haul it to the near surface. After the engine's deployment? They worked with pick-axes, buckets, and carts to break coal from the surface of the seam and haul it to the surface but from a great deal farther underground. Their wages did not increase, and the work became much more dangerous as the mines penetrated deeper into the earth.

At the same time, customers benefited from lower prices. In addition, continued steady supplies of coal prevented an "energy famine," because deforestation and population growth precluded a return to firewood.

Second, consider the launching place of the industrial revolution: cotton textile production in Lancashire in northwestern England and especially in what became the cities of Manchester and Liverpool. For hundreds of years before 1700, Lancashire remained a poor part of England, especially its rural areas. For hundreds of years, Lancashire had a great deal of woolen textile production. This rainy, hilly area was not England's best area for wheat, but grass and sheep thrived. Textile production utilized the wool and comprised a significant part of Lancashire's economy in the 1600s and 1700s.

The situation changed radically over the 1700s and into the 1800s, in both the rural and the rapidly growing urban areas, for a number of reasons. In the process, many of the rural people of Lancashire became workers in the urban textile factories. What happened? First, "enclosure" of rural lands created private ownership and management of formerly common areas. Some families had previously eked out a poor but independent living on their own small plots plus pasturing a cow on the commons.[33] Perhaps they also did spinning and weaving to make ends meet. Disappearance of the commons foreclosed independent employment, more people came to rely on spinning and weaving, and many eventually moved to centers like Manchester to work in the factories.

Coal mines had long produced fuel for the area, and canal construction made transport of large amounts of coal into Manchester very inexpensive. Manchester thus became an area well supplied with fuel to power steam engines. At the same time, good water transport developed between Manchester and the port of Liverpool. Liverpool in turn became the port of entry for cotton, particularly from the American South after the United States gained independence from Britain. Manchester thus sat at the geographic junction of flows of coal and cotton, and labor came from the surrounding area. The Liverpool-Manchester network of trade strengthened further when Britain's first railroad—powered by coal—for freight and passengers connected the two cities (see chapter 1). Iron production also increased in Lancashire.

Now what about labor: did workers benefit from industrialization of cotton production? Some may have found the cotton mill jobs a good way out of a rural life with little to offer. Others, particularly those driven to wage labor by enclosures of common areas, may have found the cotton mills a step down in their standard of living. In either case, the hours were long and the pay was low. Essentially no one has celebrated the conditions of factory work in Manchester's early mills. Workers many generations later—if they had the vote and could unionize for better working conditions—benefited from the path beaten by those who left for the cities. At best, therefore, the benefits of mechanized cotton textile production were decidedly mixed. Most people today would not dream of seeking a better life through farming (exceptions exist, of course), but the first waves of rural folks in Lancashire turned factory worker might have preferred the agrarian life possible before enclosures.

The development of Lancashire, Liverpool, and Manchester offers a fuller understanding of the dark side of the Third Energy Transition: empire and slavery. Britain, a country clothed by wool and linen, with a few silks for the wealthy, gained a taste for fine cottons starting in the 1700s, after the British East India Company began importing fine calicoes and other fabrics from India. Cotton spinning and weaving began on a small scale, but early British cotton textiles primarily found their way into "fustians," undergarments made of linen and cotton. Only in the 1700s did cotton textiles woven in Britain start to appear in larger quantities. Yet cotton had advantages: it didn't mold or suffer moth attacks like wool, and its lightness in warm weather was welcome.

Unlike wool, however, cotton was not a natural for Britain, because cotton can't grow well in its northerly, cool climate. Imports of raw cotton had to come into Britain from somewhere, and its price had to be low enough to compete with imported Indian cotton fabrics. British merchants established their base for cotton textile production with a dual strategy: empire and slavery. Liverpool gained enormous wealth in the 1700s and into the 1800s as a major center for the Atlantic slave trade.

British slave ships took manufactured goods to Africa to trade for slaves. They took the slaves to the Americas and sold them to plantation owners, sugar at first, cotton later. Plantation products then returned to England to begin the cycle again. Eli Whitney (1765–1825) developed a new device in 1793, the cotton gin, which revolutionized cotton production in the United States, which in turn undergirded the making of cotton textiles in Lancashire.

In 1790, the United States produced only 1.5 million pounds of cotton and exported only 12,000 pounds, all to Britain. After Whitney's invention, and concurrent with mechanization of cotton textile production, U.S. cotton production and exports to Britain soared at 20 percent to over 30 percent per decade, a great majority of which went to Lancashire. Although most of the capturing of slaves in Africa ended in the early 1800s, slavery in the United States lasted until after the Civil War. By 1860, at the start of the war, nearly 4 million black people labored as slaves, most in the South on cotton plantations; the United States produced 2.3 billion pounds of cotton that year, about 58 percent of which went to Britain and comprised 89 percent of Britain's cotton supply. As the historian Ronald Bailey put it bluntly, however, "too many people treat the history of cotton textile industrialization as if the source of cotton was the moon, and not mainly . . . the system of slave labor in the southern United States."[34]

Let us return now to the question of whether labor benefited from the introduction of mechanized cotton production powered largely by coal and steam engines. If slaves producing cotton constituted part of the workforce for mechanized cotton production, it's impossible to argue that this part of the Third Energy Transition benefited the enslaved labor force.

But wait. One might ask, Was slavery in the United States *necessary* for the successes of the factories of Lancashire? Maybe, maybe not. This is another "what if" question of history that is impossible to answer. Several facts, however, remain clear. Perhaps most important is that Britain considered supporting the independence of the Southern states in order to safeguard its cotton imports and one of its most important exports: cotton textiles.[35] Britain had already abolished slavery in its territories before the U.S. Civil War, but it seriously contemplated collaboration with the Confederate States of America and its economy based on slave labor to raise cotton.

And to where did the exports go? In the case of Britain, to the empire, as well as to any other country to which they could sell. India's cotton industry declined, because it could not compete with Lancashire. The coercive, undemocratic violence of imperialism differed somewhat from slavery, but Britain ran the British Empire for the benefit of Britain, not its colonial possessions. Lancashire's textile industry, based jointly on coal and on American slavery, severely damaged the economy of India. As with slavery, imperialism showed the dark side of the Third Energy Transition.

What, then, can we conclude about the Third Energy Transition and the fully modern state? First, it remade human ecology. People no longer labored on farms in an agrarian economy; they lived in cities and worked in manufacturing and service industries, with only a few left on the farms. Second, some people made enormous amounts of money, and the capital formed became an essential part of economic and political power. No government turned its back on this source of power. Third, benefits did not flow evenly, especially in the early days of the transition. Fourth and finally, whether it was in some sense *necessary* or not, the transition was partially fueled by slave labor and imperialism as well as coal. It's as if the transition to the mineral energy of coal also intrinsically included the violence of slavery and empire.

One final comment about energy and the modern state as it developed in Lancashire and then elsewhere. The Third Energy Transition rested heavily on the elements of modern life that had appeared before 1700. The rule of law figured heavily in both enclosure and slavery, which built the labor force. It also showed up in patents and security of property of inventors and mill owners, respectively. Science and technology lay at the heart of the whole affair, as did credit and credit money created by modern banks. Both slaves and industrial equipment served as collateral for plantation owners–enslavers and factory owners, respectively.

The events surrounding the industrial revolution in cotton textiles in Lancashire provide one overarching lesson about energy transitions: some may benefit while some suffer devastating losses. The benefits of energy transitions are by no means automatic and self-evident. The application of this lesson to the Fourth Transition is discussed in later chapters.

Electricity: The Mark of the Fully Modern State

Mechanical work and mobility, both powered by steam engines, clearly differentiated the early modern state from what we can designate as the "adolescent" modern state. Mechanical work and mobility alone, however, didn't complete the transition to fully modern states. That final maturation rested on electricity, the crown jewel of the Third Energy Transition.

Chapter 2 explored the development of electrical science as an integral component of energy science, and energy services based on electricity blossomed after 1800. Electroplating and electrometallurgy developed in concert with new types of more efficient batteries in the 1830s through 1850s, and they vastly increased the quality and amount of printed materials and the making of metal components of machines.

It was the electric telegraph, however, that revolutionized business, politics, and warfare. For the first time, information could flow vast distances over land and across oceans in mere seconds.[36] Britain combined its preeminence in steam engines, iron and textiles, and undersea telegraphy to form the commercial-political-military complex of the British Empire. We return shortly to more about the empire.

These first applications of electricity clearly stimulated continued interest and enhanced skills in using electricity. But lightbulbs and electric motors transformed electricity into a commodity sold in massive amounts, first in the United States and Europe and eventually in nearly the entire world.

Pioneers in electrical generation followed two major tracks: turning the heat of burning coal into electricity and transforming the kinetic energy of falling water into electricity. Later, natural gas, oil, and uranium followed coal-fired and hydroelectric plants, but the links between modern states and energy solidified with electrification. I illustrate these two pathways as they developed in the United States.

Heat into Electricity

First, consider the use of burning coal to make electrical energy. Thomas Alva Edison (1847–1931) of the United States, usually cited for his many inventions that included the incandescent lightbulb, probably is more important for having been the first person to create a central generation-distribution system capable of feeding massive amounts of electrical energy into everyday life and commerce.

Edison was not the only or the first person to make significant light with electricity. Over seventy years earlier, street lighting systems using arc lights had served several cities. Others also produced workable incandescent lightbulbs. Edison began concentrated work on developing an incandescent lightbulb in 1878. He knew that a successful bulb required matching the electricity generated to that required by the bulb.

Between 1878 and 1882, Edison succeeded in building a system with the right generators and bulbs, obtained financing for his first project in New York City, protected his inventions with patents (essential for financing), installed dynamos run by coal-fired steam engines, trained workers to install transmission wires to multiple customers, and navigated the politics of the city. His central generating station lit the first bulbs in the offices of his financial partners on Wall Street on September

4, 1882. Within a year over 11,000 bulbs were installed and over 8,500 were in use. Customers received high-quality light for indoor use (something arc lights could not do), and the bulb ultimately outcompeted its rivals, gas, kerosene, and whale oil. All this work used the concepts and methods of electrical and energy science and was conducted at an "innovation factory" at Menlo Park, New Jersey.[37]

Significantly, the competition from manufactured gas, mostly produced from coal, meant that Edison's lighting system was not a guaranteed triumph. Edison's company had to overcome customers' fears of and mystification by electricity, convince customers that they could safely manage electricity in their houses, build a vision in customers that life would be better with electricity, and attend to aesthetic concerns about the appearance of electric lights in homes. Edison and male engineers may have wanted most to work on supplying power and equipment, but ultimate success of this new product also depended on its acceptance by women into their homes. The advent of electrified life was anything but automatic.[38]

Even before success in New York, Edison and his financial partners began developing plans to transfer the concept of central power generation to other American cities and to Europe. Edison companies launched operations in London and Berlin in 1882–83. British law at the time was not receptive to the Edison model, so the London excursion died, but that in Berlin ultimately succeeded.[39]

By 1900, however, Edison's triumph as the first to introduce electrification for lighting based on central power stations had faded into the past. Edison's system involved direct current (DC) transmitted short distances in heavily populated cities for lighting. Electric motors in streetcars could use DC, but electricity from batteries proved too expensive and long distances created problems for powering them with direct current.[40] And then there was the problem that many potential customers did not live in areas as densely crowded as New York City.

Alternating current (AC) transmission remained effective over long distances, and George Westinghouse (1846–1914) and Nikola Tesla (1856–1943) made huge advances with their invention and promotion of AC long-distance transmission and motors that could use AC. These developments made feasible electrification of wide areas with low population density but served by central power plants. Moreover, the uses of electricity expanded tremendously by adding motors to lighting, and for some purposes electric heating became attractive. During the 1890s, the ascendency of AC systems relegated DC systems to a secondary role.[41]

Transforming Falling Water into Electricity

Virtually simultaneously with the opening of the coal-fired power plant in New York City, two pioneers began generating hydroelectricity. The awesome power of water flowing over Niagara Falls had long excited entrepreneurs searching for energy sources. And in fact, the first small-scale electricity generation started in 1881, the year Edison's plant went on line in New York City.[42]

On a larger scale, Henry J. Rogers of Appleton, Wisconsin, began operating the Vulcan Street Plant to produce hydroelectricity. Edison began producing power on September 4, 1882, and Vulcan first successfully produced electricity on September 30, using the dynamos and electric bulbs developed by Edison. Rogers used the electricity to light his and others' factories and homes and in the process launched hydropower as the most important of the renewable energy sources.[43]

Other installations followed in various locations in Europe, the United States, and Canada. Hydropower, however, truly began to reshape the modern world after development in 1895 of the "Niagara system." In a first for electricity generation and for hydropower specifically, the falling water at Niagara Falls generated large quantities of electricity for industry, lighting, and mass transit, some right at the Falls, some in nearby Buffalo, New York.[44]

Hydroelectric generation began as a private enterprise in the United States (see chapter 7), and in the first years of electrification it contributed a bit more than one-third of the total power generated, 36 percent in 1902. By 1957, hydropower had dropped to 19 percent of the total as fossil fuels generated the vast majority of electricity.[45]

Although fossil fuels, joined later by uranium, continued to outpace hydropower as the source of electricity, the shape of the hydropower industry shifted radically to the public sector after 1903. Intense lobbying by agricultural and commercial interests in the western states created what is now the Bureau of Reclamation within the federal government. The bureau, joined later by the U.S. Army Corps of Engineers, the Tennessee Valley Authority, and other agencies, constructed large dams on rivers for irrigation, and hydroelectricity quickly established itself as a valuable by-product of water projects.[46]

The scale of the federal hydropower projects leaped to a qualitatively new level in 1928 when President Calvin Coolidge signed a bill authorizing construction of a dam, finished in 1935 and now known as Hoover Dam, on the Colorado River (figure 3.5). By 1961, the capacity of

FIGURE 3.5. Hoover Dam, blocking the Colorado River between Arizona and Nevada. The dam is 726 feet from the bottom foundation to the crest, and it held back the Colorado River to form Lake Mead, which covers 248 square miles and holds about 29 million acre-feet of water. The dam's construction lasted from 1931 to 1935. The power plant, located in the front of the dam at the bottom of Black Canyon, now has a capacity of 2,080 MW, and it generates about 4 GWh per year. Electric power goes to Arizona, Nevada, and Southern California.

SOURCE: Photo, US Department of Energy, found at www.flickr.com/search/?text=hoover%20dam&license=8, November 3, 2016. Information about the dam found at www.usbr.gov/lc/hooverdam/faqs/faqs.html, November 3, 2016.

its power plant reached 1,345 megawatts (MW).[47] This was the first of the West's huge hydropower projects built between the 1930s and 1970s, and federal agencies now own 49 percent of the hydropower capacity of the United States, with over half of that capacity in Washington, Oregon, and California. Washington, Idaho, and Oregon each get over 60 percent of their electricity from hydropower.[48]

The amounts of electricity generated dramatically demonstrates the rise of electricity. In the United States, statistics date to 1902. Electricity produced on a national scale is measured in terawatt-hours (TWh; "tera" [T] means trillion or 10^{12}). In 1902, central generating stations in the United States generated 5.97 TWh. This grew to 56.6 TWh in 1920 and 716 TWh in 1957.[49] This increase of 120 times in fifty-five years was for a product unknown twenty years before the first record. In 2013, the total was 3,899 TWh, a more than fivefold increase since 1957 and 653-fold since 1902.[50]

A casual look today around homes, schools, commercial businesses, factories, streets, automobiles, and many other places proves that whatever barriers lay before Edison and subsequent electrifiers disappeared in most parts of the world in the years after 1882. People accustomed to electricity shriek with dismay when their lights go out, but more profoundly a prolonged loss of electric power would threaten the delivery of clean water as well as food preparation and preservation. Loss of clean water and food would threaten health and life, not just convenience. In addition, electric motors power everything from dentists' drills to train locomotives to naval warships. Electricity became the crown jewel of the Third Energy Transition, and its luster will be even brighter in the Fourth (see chapter 8).

CONCLUSION

Several changes, beginning in England and spreading to other countries, distinguished premodern from modern societies. Coal, used on a scale previously unprecedented, played a key role in these transformations, and steam engines plus electricity propelled modern people into lifestyles previously unimaginable. Later, oil, gas, and uranium joined coal as primary sources of energy to provide energy services. We turn to those primary sources in chapter 4.

Connections with Everyday Life

Most people, if they think of energy at all, think about gasoline for the car and electricity for the house. They may also think about natural gas or oil for heating their homes and workplaces. Understanding energy, and its benefits and problems, however, requires acquaintance with all the primary energy sources, or fuels.

The world now uses a total of nine primary energy sources: coal, petroleum, natural gas, uranium (and some plutonium manufactured from uranium), hydropower, solar energy, wind, biomass, and geothermal energy. Of these, most people have directly bought or used only petroleum (refined as gasoline), natural gas, and biomass (firewood, charcoal, or ethanol mixed with gasoline). A few people have direct experience with coal, but almost no one has experience with uranium; instead they buy electricity made from these sources.

A growing number of people have direct experience with solar energy, because they installed photovoltaic panels or solar hot water heaters on their roofs. Also a few people in rural areas, often far from and unconnected to electric grids, have small wind turbines for electricity. But most people using solar power, wind, hydropower, or geothermal energy purchase it only after the primary energy has been converted to electricity.

Except for oil, natural gas, and biomass, therefore, most people are unfamiliar with primary energy sources. In this chapter, I introduce the four mineral fuels that supply 86 percent of the world's energy (as of 2014): coal, oil, gas, and uranium. I also introduce the "virtual fuel," energy efficiency, which is widely familiar in forms such as more economical automobiles and insulation for houses. Efficiency, of course, is not a real fuel, but for hundreds of years it has played far more of a foundational role in the uses of energy than is generally recognized.

CHAPTER 4

Primary Fuels and Energy Efficiency

Energy played key roles in building both the modern state and modern science, but new institutions and knowledge emerged while the energy consumed remained small. By the 1950s, energy use across the world had expanded greatly, and it expanded even further after that time. Four mineral fuels now power human life: coal, oil, gas, and uranium provided 86 percent of the world's primary energy in 2014. The big-four fuels appeared on the world stage at different times, and the amounts of each used per person has shifted over time (figure 4.1).[1]

This chapter sketches these major fuels, which, along with hydropower, built the Third Energy Transition. It also provides an overview of energy efficiency, another major player and key component of the Fourth Energy Transition (see chapter 8).

BIOGRAPHIES OF THE MAJOR FUELS

The biographies below rely on two simple ideas, presented here as background. First, the big-four fuels all enter commerce through mining. After mining recovers the raw materials, they are refined and transported to places of use. Mining operations depend on capital investments to develop a resource field and build refining plants. Companies gain the confidence to mine based on estimates of reserves, a measure of the quantity and quality of resources they expect to find, and the costs of recovery compared to selling prices of refined fuel. The

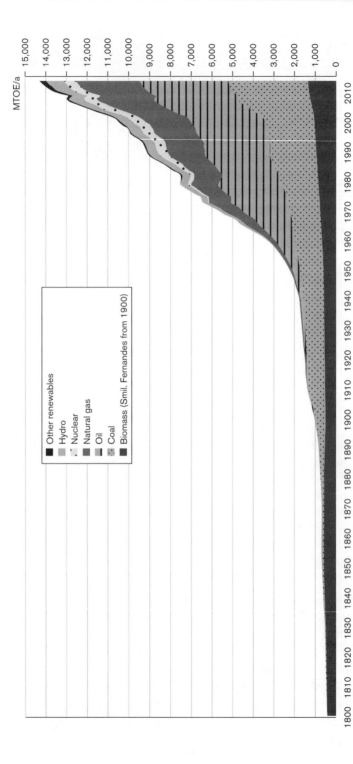

FIGURE 4.1. Changing patterns of world energy use, 1800–2014.

SOURCE: Rune Likvern, used by permission, https://fractionalflow.com/category/world-energy-consumption/ May 27, 2016.

attractiveness of a deposit thus depends on geology, extraction technology, and costs.

Geologists may find an area with the right deposits, but the decision to mine depends on the acceptability of the extraction techniques and their costs. An ore deposit may look unattractive with one extraction technology, but a different extraction method may make it feasible. Similarly, if market prices for refined fuel rise, companies will mine raw fuel that is more difficult to extract. Estimates of geological reserves' presented below assume current technology and costs. Regulations that restrict an existing extraction technology, new technology, or altered market conditions and costs may change the estimates of reserves in the earth.

Second, the three fossil fuels, coal, oil, and gas, deliver heat by burning or combustion. Uranium delivers heat by quite a different method, the fission of uranium atoms. Combustion and fission both change the fuel to produce heat and waste products. Fission of an atom of uranium releases far more heat than combustion of one molecule of fossil fuel. Appendix 2 explains combustion and fission in more detail.

COAL

The British embrace of coal launched the Third Energy Transition, and until the mid-1900s this fuel supplied most of the energy consumed. Today coal remains significant in the energy world, both as a source of heat and as a chemical reactant in the smelting of iron. Not only did coal power the English industrial revolution; it also enabled the transition of many other countries from agrarian to modern industrial economies, most recently China.

Despite coal's importance, for most people it lies hidden in obscure trading and transport routes. As economists would say, the economic demand for coal is a "derived demand." People want electricity and iron, for example, and the makers of those two commodities create the economic demand for coal. For the most part, coal goes directly from mine to place of burning, usually an electric power plant or iron and steel plant. In the highly industrialized world, individuals seldom even see it, unless they live near a power plant or a rail line hauling it. Coal's current invisibility is in sharp contrast to its historical past, when mining coal required huge labor forces. Whole towns clustered around coal mines, and nearly every family worked in coal production, often for generations.

Physical nature. Mineral coal formed from plant debris buried in sediments in all geologic periods since the Devonian, from 408 million to 360 million years ago.[2] Even today coal may be forming, but commercially useful deposits all date to earlier than two million years ago.

Microbial decay processes plus heat and pressure slowly changed dead plants, and a few animals, into a rocklike substance, often with other minerals intruding into the mass. Over time, and depending on the conditions in the sediment, the decayed material formed, in chronological order, peat, lignite, subbituminous, bituminous coal, and anthracite. All these burn readily and find uses as fuels in different parts of the world. The energy density of the different grades of coal varies substantially: lignite has less than 46 percent carbon and 5,500 to 8,300 British thermal units (Btu) per pound, but anthracite has 86 to 98 percent carbon and 13,500 to 16,600 Btu per pound.[3]

Carbon is the dominant chemical element, and much of the carbon forms rings of six carbon atoms. Peat has the lowest percent of carbon, the highest percent of hydrogen and oxygen, and the lowest release of heat during combustion. Anthracite is almost pure carbon, has relatively little hydrogen and oxygen, and provides the most heat per unit of mass.

Physical distribution. Coal deposits lie unevenly scattered across the world. Although many countries have some coal, Eurasia, North America, and Australia have the richest deposits. Relatively little occurs in South America or most of Africa; in the latter, South Africa holds major reserves. The United States has the largest deposits, followed by the Russian Federation and China. The top ten countries have over 90 percent of the world's coal (table 4.1). The United States, Russia, and China alone have over half of the reserves. Ironically, given Britain's leading historical role in using coal, the United Kingdom does not hold significant reserves compared to the world's total: 251 million short tons in 2011, ranking it forty-fifth.

Production levels per year approximately reflect the deposits in the respective countries. Each of the top ten holders of reserves appear in the list of top ten producers, except Ukraine, which was fourteenth in production. Poland was ninth in production but sixteenth in reserves. China led the world by producing 3.9 billion short tons in 2011. The United States, at second place, produced 1.1 billion short tons.[4]

Production levels indicate the economic influence of a country's coal industry but say little about the long-term prospects. As a whole,

TABLE 4.1. TOP TEN COUNTRIES WITH HIGHEST COAL RESERVES, END OF 2015

Area	Rank	Reserves (million metric tonnes)	Percent of World Reserves	Cumulative Percent
World		891,531	100	
United States	1	237,295	26.6	26.6
Russia	2	157,010	17.6	44.2
China	3	114,500	12.8	57.1
Australia	4	76,400	8.6	65.6
India	5	60,600	6.8	72.4
Germany	6	40,548	4.5	77.0
Ukraine	7	33,873	3.8	80.8
Kazakhstan	8	33,600	3.8	84.6
South Africa	9	30,156	3.4	87.9
Indonesia	10	28,017	3.1	91.1

SOURCE: *BP Statistical Review of World Energy June 2016*, 65th ed. (London: BP, 2016), 30. Percent and cumulative percent calculated by author.

projections of the world's reserves and consumption levels indicate that known reserves will last 118 years at current consumption rates. A total of thirty-six countries have more than 118 years of consumption, including eight of the top ten producers. India (93 years) and China (32 years) are below the world level of 118 years.[5]

Consumption as a percent of domestic production indicates countries that import more than they produce or have a surplus for export. A total of 46 countries import coal with no domestic production; that is, they depend solely on imports for their needs. Many importers without domestic production buy relatively small amounts, but two, Japan and Taiwan, together import about 275 million tons per year. Japan takes the larger share, about 200 million tons per year. An additional 23 countries consume more than 110 percent of their production. This group includes the United Kingdom, which consumed 225 percent compared to its domestic production in 2011. Of the countries in the top ten of reserves, 6 consume less than they produce (Indonesia, Australia, Russia, South Africa, Kazakhstan, and the United States). Two, China and Ukraine, have consumption about equal to production; and two, Germany and India, have consumption greater than 110 percent of production (2011 data).[6]

Coal industry. John Hatcher has concluded that a "coal industry" did not exist in Britain before 1700, even though Britain had begun to seriously embrace the use of coal over one hundred years earlier and created the first coal industry. With 25,000 tons per year production as the criterion of a "large" coal operation, most mines fell far below this threshold. Only in northeastern England, in Northumberland and Durham, did approximately twenty mines qualify as large. Despite the predominantly small-scale size of the operations, coal production in all of Great Britain (England, Scotland, and Wales) rose twelvefold between the 1560s and about 1700, from 227,000 to 2.6 million tons per year. The ability to exploit coal for income depended on landownership, so coal mining generally started as an effort by a rural landowner to augment agricultural income. Coal's weight and bulk meant most mines lay close to water transport. London in 1700, with 11 percent of England's population, bought over 70 percent of northeastern coal and about 18 percent of all coal.[7]

The coal industry blossomed first in Britain after 1700. Production expanded over tenfold between 1700 and 1830. This period depended on constantly increasing energy to power the cotton mills of Lancashire (see chapter 2), the launch of railroads (see chapter 1), and the myriad other uses of coal.[8] Steam engines and the many other uses for coal spread to other countries, especially after 1800. In the United States, for example, abundant firewood and water power kept coal production small before 1820, but it grew rapidly in the years afterward. Improved water transport for coal to cities like Philadelphia made coal mines in northeastern Pennsylvania economically sound. Coal production rose rapidly after 1820.[9]

Coal mining began as a collection of isolated collieries before 1700, but during the 1800s and the 1900s it became a very large industry in Britain, the United States, and other countries. At the surface of the mine, coal had essentially no economic value, and in 1800 in Britain half or more of the labor making it valuable lay in transporting it to customers.[10]

Between 1830 and 1913, the coal mining industry grew into one of Britain's largest in terms of people employed: from 109,000 to 1,095,000, or from 5 percent to nearly 12 percent of all workers in mining, manufacturing, and building. Both coal production and mining employment increased tenfold in this eighty-three-year period, far more than manufacturing and building. Well over 90 percent of workers were men. The hewers, who extracted the coal from the seams, undermined the seam, collapsed it, loaded for hauling to the surface, and repeated the process, endlessly, in cramped, dusty, dark, and dangerous quarters. Extraction practices changed little during this period. If workers survived, they generally moved to lighter work at age 55 and retired at 65 to 70. They often lived apart from other communities and occupations, and they transformed Britain into the most powerful nation of the nineteenth century.[11]

In the early 1900s, coal mining companies, some producing coal alone and others integrated with iron and steel production, figured prominently in the top 100 largest companies of the United Kingdom. Employment levels in large firms in the United States and Germany, the two countries most similar to Britain in terms of economic development, also included coal, iron, and steel companies.[12]

In Britain and in the United States, employment in coal mining reached a peak during the period 1900–1930. In Britain, 1.2 million people worked in coal production in 1920, the peak year, while in the United States 1.4 million worked in "mineral extraction," mostly coal, in 1917, the peak year. Employment in this work declined in both countries after the peak times, probably as a result of competition with oil, increasing mechanization, and a switch to surface mining.[13]

The spectacular decline of coal production in Britain, especially at the end of the 1900s, decimated the very industry that had launched the Third Energy Transition. The number of deep mines dropped from its peak of 3,236 in 1893–1902 to 13 in 2008. Surface mining began in 1943 and surpassed the production of deep mines by 2004. Britain began to import coal in 1971, and imports surpassed domestic production in 2001. In 2008, only 6,000 people still worked to produce 18 million metric tons—a precipitous drop from Britain's peak production of 292 million metric tons in 1913 achieved by 1.1 million workers.[14]

The immense workforces producing coal lived near coal mining towns in both Britain and the United States. Mining accidents, particularly explosions and fires underground, traumatized whole towns, and strikes affected everyone, both in the mining towns and far beyond. Decline of coal production left impoverished villages in places such as Wales and Kentucky, places of great poverty in a landscape devastated by mining.

Timothy Mitchell has argued that labor agitation in the coal mining industry played a critical role in labor reforms and the development of democracy in the United Kingdom and the United States. He identified "carbon democracy" as a by-product of coal and contrasted it to the lower labor requirements and lack of democracy resulting from petroleum production.[15]

In contrast to petroleum, the companies that mine coal today have little name recognition among the public. For example, it's likely that only those involved in the coal industry itself would recognize the names Peabody Energy (United States), Rio Tinto (Australia/United Kingdom), Vostsibugol (Russia), or China National Coal Group Corp. (China), all members of the World Coal Association. Yet this global industry in 2011 produced 8,444 million short tons of coal that had a value of about $750 billion.[16]

PETROLEUM

Petroleum is perhaps the most important substance consumed in modern society.[17]

Petroleum provides heat and chemical materials for transport, heating, and manufacture of materials like plastics. *Petroleum* is an ancient word, from Latin (*petra*, "rock") and Old English (*oleum*, "oil"). Without plentiful supplies of "rock oil," the world would be a very different place.

Physical nature. Petroleum, according to most geologists, originated from photosynthesis by plants and the consumption of plants by animals during the past 600 million years. Plant and animal remains, buried in mud at depths of 7,500 to 15,000 feet and subjected to temperatures above 175°F (79.4°C), became liquid petroleum, a mixture of hydrocarbons: molecules predominantly of carbon and hydrogen with 5 to 20 carbon atoms. Below 15,000 feet, temperatures rose high enough to break the organisms down to smaller molecules. At one atom of carbon, the material became methane (CH_4), or natural gas.[18]

Contrasting theories, not widely accepted, claim that petroleum and gas originated from physical processes alone and that petroleum and gas are not "fossil fuels."[19] Whether petroleum comes from decayed biological organisms or not, its rate of formation is vastly slower than the rate at which we now burn it, so it is "nonrenewable."

Physical distribution. Petroleum, in contrast to the three fundamental resources for human life—air, water, and agricultural soils—lies in rock formations scattered unevenly across the earth. Some countries have huge amounts of it within their own boundaries, but others have little or none. Table 4.2 lists the known reserves of the top fifteen countries.

Two different methods dramatically demonstrate this uneven distribution: years of consumption based on known reserves and percent of yearly consumption obtained from the country's own supplies. Data from 2013 are the basis for the following portrait.[20]

TABLE 4.2. PETROLEUM PROVED RESERVES IN TOP 15 COUNTRIES, 2016

Area	Rank	Billion Barrels	Percent	Cumulative Percent
World		1697.6	100	
Venezuela	1	300.9	17.7	17.7
Saudi Arabia	2	266.6	15.7	33.4
Canada	3	172.2	10.1	43.6
Iran	4	157.8	9.3	52.9
Iraq	5	143.1	8.4	61.3
Russia	6	102.4	6.0	67.3
Kuwait	7	101.5	6.0	73.3
United Arab Emirates	8	97.8	5.8	79.1
United States	9	55.0	3.2	82.3
Libya	10	48.4	2.9	85.2
Nigeria	11	37.1	2.2	87.3
Kazakhstan	12	30.0	1.8	89.1
Qatar	13	25.7	1.5	90.6
China	14	14.5	0.9	91.5
Brazil	15	13.0	0.8	92.2

SOURCE: *BP Statistical Review of World Energy June 2016*, 65th ed. (London: BP, 2016), 6. Percent and cumulative percent calculated by author.

Consider first the years of consumption at current rates of use. The world as a whole has 50 years of consumption. Countries with more than 50 years of petroleum in reserve include Venezuela (1,040 years), Kuwait (610 years), Iraq (504 years), Saudi Arabia (247 years), and Iran (226 years). As a region, the Middle East has 270 years of reserves.

At the low end are countries like the United States, which has 5 years of consumption in reserve. The United States holds 33 billion barrels of reserves, but the country also consumes the largest amount of petroleum (6.9 billion barrels per year; 19 million barrels per day). Similarly, China (6 years) and India (4 years), the two countries with 37 percent of the world's population, have relatively few years of consumption of petroleum from their own territories.[21] Europe as a whole has 2 years of consumption, and 132 countries and territories have zero years of consumption. They have few or no reserves.

The number of years of consumption doesn't explain everything, because a country may have a very low rate of consumption. For example, Chad consumes only 1,870 barrels per day but has 1.5 billion barrels of reserves. Its years of consumption at current rates from its own reserves totals 2,198 years, highest in the world. In some countries, high reserves support substantial consumption. Venezuela, for example, holds the world's highest reserves (298 billion barrels) and a consumption rate of 286 million barrels per year. Population size matters: low populations increase years left of reserves, and large populations decrease the number of years, all other things being equal.

The percent of consumption derived from a country's own supply paints a similar picture of widely disparate rates of self-sufficiency. The world as a whole, by definition, produces 100 percent of its supply. A total of forty countries or territories produce more than they consume, leaving a surplus for export. For example, the Russian Federation produced 306 percent of its consumption, and Saudi Arabia produced 394 percent. In contrast, over 165 countries and territories do not produce enough from their own territories to supply their own consumption. For example, the United States produced 65 percent of its consumption, and China produced 43 percent. India produced 28 percent, and Europe as a whole produced 27 percent.

Petroleum industry. Known uses of petroleum from surface seeps or shallow wells date back thousands of years, but the material had only a few uses, such as waterproofing, lighting, weapons, and medicines. Production of small commercial quantities, such as near Baku, had gone on since at least the 1200s, as observed by Marco Polo.[22] Hand-dug pits increased production around Baku, then in the Russian Empire but now in Azerbaijan, in the 1820s. In 1859, Americans successfully drilled a well in the backwoods of Pennsylvania known for its oil seeps. Their objective was to find a cheap illuminant to compete with whale oil and coal oil.[23]

Despite the ancient origins of the word and uses, however, time travel back to 1860 would reveal that most people would know nothing or very little about petroleum. In a few cities lit by gas or oil, some might know the origins of the nice light that chased away the darkness. A few people would know that oil seeped to the surface of the earth in a few places. A subset would know about commerce in small amounts of oil. But the foundations of modern life? Well, that had nothing to do with oil in 1860.

Within a century, petroleum became "the most important substance," and nations proved quite willing to wage war with and for it. Moreover, extensive use of petroleum poses threats, sketched in chapters 6 and 7. Mobility and transport transformed a specialty product gathered in small amounts for thousands of years into the foundation of modern societies. Coal began the transformation of mobility through steam-powered railroads and ships, but petroleum eclipsed coal.

Here I briefly summarize the major twists and turns as a primitive, small industry turned into an industry valued in the trillions of dollars. Major production of oil began in Pennsylvania and Azerbaijan in the 1860s and 1870s, followed by Dutch companies in Indonesia in the 1890s. In the 1900s, further petroleum discoveries in many countries fueled the rapidly rising demand created by automobiles and other uses, such as making asphalt and plastics. Until the 1960s, however, the oil industry lay almost entirely within the domain of the "Seven Sisters," the eight international oil companies that dominated production, refining, and distribution of petroleum and its products. They all had bases in the United States and Europe.[24] Some of the eight were entirely in private hands, and all behaved like private companies, even if a government, like Britain or France, owned a part.

Rising nationalism and intense resentment of the meager shares of profits from the Seven Sisters led a number of countries to form their own national oil companies after World War II. With the establishment of the Organization of Petroleum Exporting Countries (OPEC) in 1959 came the recommendation that each member organize a national oil company and take control of the resources under its own soil. The original OPEC members were Iran, Iraq, Kuwait, Saudi Arabia, and Venezuela, later joined by Qatar, Indonesia, Libya, United Arab Emirates, Nigeria, Ecuador, Gabon, and Angola. (Indonesia left OPEC after becoming a net importer in 2008.)[25]

Today the national oil companies manage over 90 percent of the world's oil reserves. Sixteen of these government-owned, national companies are among the top 20 companies in the world. Saudi Aramco, the largest, owned by Saudi Arabia, has more than ten times the reserves held by ExxonMobil, the largest private company.[26] The worth of the national oil companies cannot be determined because they do not trade stock or publish financial details. Private companies, with publicly traded stock, however, are valued at enormous sums. In 2014, for example, the ten largest private oil companies in the world had a market value of $1.78 trillion.[27] After the collapse of the USSR, the oil deposits of the former Soviet Union's national oil company moved into private hands and opened to investment from foreign companies.[28]

The first markets for oil centered on lighting, lubrication, and paving. Kerosene burned brightly and cost less than whale oil. It competed well with candles and camphene, a by-product of turpentine from pine trees. Growing use of machines such as railroads required lubrication. Elimination of dust and mud by paving roads with asphalt proved popular. In the United States, oil moved to refineries and then to market first by railroad and then mostly via pipelines. Oil contained more heat energy per unit weight than coal, and its transportability through pipelines provided yet another advantage over the "first" fossil fuel.

Petroleum's breakthrough into becoming a foundation of modern life, however, did not stem from lighting, lubrication, paving, and pipeline transport. Instead, development of internal combustion engines (gasoline and diesel) made petroleum indispensable for both civilian and military mobility. Steam engines burned the coal *outside* the cylinder in which the piston went back and forth. Coal, a rock, could not be easily injected and ignited inside the cylinder. Liquid fuels from oil could, although experiments throughout the 1900s indicated the theoretical possibility of internal combustion engines burning powdered coal or biomass.[29] Heat of the burning fuel expanded the gases inside the cylinder and made the piston move. The new engines used air and fuel vapors as the expanding gas to do work rather than steam, in a way fulfilling Sadi Carnot's belief in air as useful for heat engines (see chapter 3).

A few simple numbers from the United States show the dramatic changes. In 1900, a total of 8,000 automobiles carried a few people, but most transport was by horses and steam-engine trains. No records exist for trucks in 1900, but 700 plied the roads in 1904. By 1920, the United States had 8.1 million automobiles (a thousand-fold increase in twenty years) and 1.1 million trucks. In 1950, the country operated 40.3 million autos and 8.6 million trucks. Horses on farms peaked at 21.4 million in 1915 and declined steadily thereafter to 3.6 million in 1957. Steam engines dominated rail transport until 1951. A total of 30,000 operated in 1890, and the peak number of steam engines reached 69,000 in 1924. The first diesel-electric locomotive appeared in 1925, and by 1952 diesel electrics outnumbered steam engines. In 1957 only 2,600 steam engines remained, but 29,000 thousand diesel-electric engines served the railroads.[30]

In 2013, the world's petroleum industry produced 91.33 million barrels per day. At the average spot price per barrel of approximately $106, the daily production of oil equaled $9.68 billion per day, or about $3.5 trillion per year, or about 5 percent of the world's gross domestic product ($75.6 trillion).[31] Petroleum derives most of its importance from powering the world's mobility, and this is reflected in its monetary value.

NATURAL GAS

Natural gas, the last fossil fuel to rise to prominence, poses the most challenges to produce, capture, store, and transport. Natural gas further differs from coal and oil in terms of its first major uses. Coal first transformed the production of process heat by replacing firewood and later gained fame by producing motion and mobility. Oil considerably expanded mobility. Gas, in contrast, derived its importance first from its use in lighting. In the early 1800s, gas manufactured from coal made it possible for the first time to light the interior of large buildings and streets at a reasonable cost. Never before had people been able to so overcome the darkness of night. Ultimately electric lights replaced gas lights, but Thomas Edison knew he had to beat gas to succeed (see chapter 2).

Despite electricity's prominence in lighting after the late 1800s, natural gas moved in the 1900s into many additional uses: space and water heating, heat for industrial processes, raw material for manufacturing other chemicals, generating electricity, and, in limited amounts, powering transport. In the early 2000s, natural gas acquired a new but contested function as the fossil fuel to "bridge" the movement from coal and oil to a renewable energy economy (see chapters 8–11). In the United States as of 2013, gas is second only to oil in the amount of heat energy provided,[32] and gas has major uses in the industrial, commercial, and residential sectors as well as in generating electricity.

Physical nature. Supplies of gas, primarily methane but sometimes including other substances with more carbon atoms, came first from manufacturing processes. Heating coal and later also oil products drove off the easily volatilized components—gas—that burned cleanly and brightly. First developed in the late 1600s and used well into the 1800s, virtually all gas was manufactured. Today processors make very little manufactured gas. After 1859, natural gas often associated with oil increasingly replaced manufactured gas.[33] The section "Gas Industry" below explains more, but I want to focus here on natural gas.

The dominant theory of the origins of natural gas pictures plant and animal remains buried under sediments and transformed into coal, petroleum, or gas, depending on temperatures, pressures, and times, which range from tens to hundreds of million years. Deposits at 3,000 feet or less most likely contain heavy oil with no gas. Deeper deposits have gas and oil, with the gas either on top of the oil or dissolved in it.

TABLE 4.3. TOP 15 COUNTRIES IN NATURAL GAS RESERVES, 2016

Area	Rank	Reserves (trillion cubic feet)	Percent	Cumulative Percent
World		6599.4	100	
Iran	1	1201.4	18.2	18.2
Russia	2	1139.6	17.3	35.5
Qatar	3	866.2	13.1	48.6
Turkmenistan	4	617.3	9.4	58.0
United States	5	368.7	5.6	63.5
Saudi Arabia	6	294.0	4.5	68.0
United Arab Emirates	7	215.1	3.3	71.3
Venezuela	8	198.4	3.0	74.3
Nigeria	9	180.5	2.7	77.0
Algeria	10	159.1	2.4	79.4
China	11	135.7	2.1	81.5
Iraq	12	130.5	2.0	83.4
Australia	13	122.6	1.9	85.3
Indonesia	14	100.3	1.5	86.8
Norway	15	65.6	1.0	87.8

SOURCE: *BP Statistical Review of World Energy June 2016*, 65th ed. (London: BP, 2016), 20. Percent and cumulative percent calculated by author.

At between 10,000 and 12,000 feet, gas in the oil generally produces 500 to 1,000 cubic feet of gas per barrel of oil. Below 17,000 feet, and certainly below 20,000 feet, deposits consist almost entirely of gas. Chemically "gas" often contains a variety of different substances, and production processes separate them.[34]

Physical distribution. Like coal and petroleum, natural gas deposits lie unevenly distributed across the world. The four countries with the highest reserves in 2016—Russia, Iran, Qatar, and Turkmenistan—together have 58 percent of the known reserves. The top fifteen countries have 88 percent (table 4.3).

Six of the top ten holders of gas resources also rank in the top ten producers (United States, Russia, Iran, Algeria, Qatar, Saudi Arabia). Canada (twentieth in reserves), Norway (seventeenth), China (thirteenth), and Indonesia (eleventh) are also among the top ten producers, despite not possessing reserves in the top ten. Consumption of gas per year shows even more disparity. The United States, Russia, Iran, and Saudi Arabia are among the top ten consumers. Six countries, however, consume great quantities but rank rather low in reserves: China (13), Japan (73), Canada (20), Germany (45), the United Kingdom (42), and Italy (57).

The world as a whole has 57 years of reserves at current consumption rates. Less than 57 years of consumption from reserves indicates reserves below the global average, and among the top ten consumers, only three countries have greater than 57 years of reserves: Saudi Arabia (81 years), Russia (107 years), and Iran (212 years). Seven of the top ten consumers have less, some considerably less, than the global average: Japan (0.2 years), Italy (0.9 years), Germany (2.1 years), United Kingdom (3.2 years), United States (13 years), Canada (20 years), and China (21 years).

All countries in the top ten produce more gas than they consume, but some have relatively small amounts in surplus. The United States consumes 86 percent of its production and Saudi Arabia 88 percent (2012 data). In contrast, some have large amounts in surplus for export. For example, Nigeria consumes only 10 percent of its production, Algeria 21 percent, and Russia 68 percent. Other countries have consumption vastly larger than domestic production and must import. Japan depends on imports, with consumption at 2,653 percent of production; Italy (871 percent), Germany (691 percent), and the United Kingdom (181 percent) are also among the most import dependent.

Gas industry. Heating coal, wood, peat, and other materials produced *manufactured gas*, the first use of gas at a commercial scale. Britain clearly led the way in manufactured gas, most of it for lighting. Several inventors in England, France, the Netherlands, and the United States began to make gas at a small scale in the late 1700s to light small areas. One of the best known, William Murdock (sometimes spelled Murdoch) (1754–1839), an employee of Boulton and Watt steam engine makers, lit his employer's factory, his own home, and a cotton mill in Salford in Lancashire.

Manufactured gas reached a new level, however, when the London and Westminster Gas Light and Coke Company began lighting streets in London in 1812, followed quickly by comparable efforts in Baltimore (1817), Essen (1818), Boston (1822), New York (1825), Hanover (1825), and elsewhere. By 1870, 340 plants in Germany manufactured gas, and this industry persisted into the mid-1900s. In Britain, for example, 1,050 plants manufactured gas in 1948 and served ll.3 million customers. Exploitation of offshore gas fields after 1965, however, eliminated manufactured gas in favor of natural gas. In a major way, the manufactured gas industry paved the way for natural gas.[35]

Industrial-scale gas manufacturing began in Britain, but the United States led the way in the use of natural gas. In 1821, a shallow natural gas well near Fredonia, New York, sent gas through a short pipeline made of hollowed logs to provide light to the small city. Petroleum production changed the situation, because gas frequently accompanied the oil. Lack of a way to transport the gas to customers, however, meant that until the mid-1900s drillers flared the gas as a waste product.

Pipelines made all the difference. Relatively short pipelines carried natural gas to Pittsburgh in 1883 for use in making steel and glass. In 1925, the Magnolia Gas Company of Dallas built the first long-distance pipeline, carrying gas 217 miles from Louisiana to Texas. The Natural Gas Pipeline Company transported gas over 1,000 miles from Texas to Chicago in 1931. After World War II, the U.S. natural gas industry entered a new era begun by conversion of an oil pipeline to carry gas from East Texas to Pennsylvania, thus opening the heavily populated and industrialized northeastern states to natural gas. By 1955, gas pipelines stretched 145,000 miles, and by 1966, natural gas reached the forty-eight continental states south of Canada.[36] Today gas has supplanted coal as the number two energy source for the United States, after petroleum.

Similar to the petroleum industry, the gas industry consists of three distinct segments. *Production* extracts the gas from the ground and prepares it for shipment as a commodity. *Transport* moves the gas, often for long distances in a pipeline. Sometimes transport occurs as liquefied natural gas (LNG) across oceans. Finally, *distribution* companies move the gas from reception depot to individual customers. In the United States, distribution companies often market both gas and electricity to customers. Customers of gas use it for an enormous variety of purposes, including space heating; hot water heating; cooking; and making plastics, fertilizers, and many other chemicals.

Each of these three segments of the industry may involve companies that are privately owned or owned by a government. In the United Kingdom, for example, the gas business was nationalized in 1948 but privatized again in the 1980s.[37]

One final point about the gas production industry that also affected the oil production industry: since Edwin Drake successfully drilled for oil in Pennsylvania in 1859, drilling technology has steadily improved. One development, dating to 1891, enabled drills to "turn corners" so that what started as a vertical hole turns horizontal. Although initiated in dental drills, horizontal drilling for oil dates at least to 1929. The use of horizontal drilling for both oil and gas increased strongly after 1980.[38]

A second development, this one dating from the 1940s, forced water mixed with various materials at high pressures into wells and fractured the rock formations in which oil and gas deposits lay. This technique, *hydraulic fracturing*, or fracking, made it cost-effective to retrieve energy sources that would otherwise be too costly to mine. George P. Mitchell, a petroleum engineer–geologist–independent oil producer, successfully developed fracking for gas in a formation known as the Barnett Shale near Dallas–Fort Worth, Texas, in the late 1990s. Later combinations of fracking and horizontal drilling reversed the declining production trends in both gas and oil in the United States.[39]

Great jubilation greeted the new technologies for mining gas and oil, but the rush of new energy supplies also generated controversy, particularly for unconventional gas produced by fracking. Feature-length films—*Promised Land* (2012) and documentaries *Gaslands I* (2010) and *II* (2013)—brought the issues before the public in dramatic ways. Fracking involves injection of high-pressure water, sand, and other chemicals into shale formations to release embedded natural gas.

URANIUM

Unlike coal, oil, and gas, no deposits of uranium lay conveniently on or near the surface, ready to burn. Also unlike the fossil fuels, all of which have complex mixtures of distinct chemical substances, uranium is a single chemical element, although it occurs in ores mixed with other elements, such as oxygen. Yet a further difference, uranium produces heat by fission, a completely different pathway from combustion (see appendix 2). At first no one recognized uranium as a fuel, but, after discovery of fission, it completed the Third Energy Transition.

In 1789, the German pharmacist-chemist Martin Heinrich Klaproth (1743–1817) studied pitchblende, a mineral found on the German–Czech Republic border and believed he had isolated a new element he named uranium.[40] Klaproth's findings probably encouraged use of uranium in glass making to give a greenish-yellow tint, but Roman glassmakers around 79 CE had already used these minerals for this purpose.[41]

Physical nature. For more than a century after Klaproth's work, uranium and pitchblende remained useful for glassmakers but a curiosity for science: a metal with high density, such as lead and gold. While investigating phosphorescence, Henri Becquerel (1852–1908) in France found between 1896 and 1903 that uranium emitted a new type of invisible radiation.[42] Ultimately, this discovery considerably broadened the concept of matter by supporting the discoveries of subatomic particles: protons, electrons, and neutrons.

Uranium mining produces an ore that, in contrast to coal, doesn't burn or even look like it could produce heat. Preparation of uranium as a primary energy source involves a series of steps, the *nuclear fuel cycle*. Smelting isolates the uranium, which undergoes enrichment in its content of ^{235}U. Once enriched, the material is made into fuel pellets, cylinders somewhat smaller than the tip of an adult's finger. Pellets fill long tubes to make fuel rods, generally about 3 to 4 meters long. Multiple fuel rods comprise a fuel assembly, and multiple assemblies in a reactor fission to generate heat to produce steam for turbines to make electricity.

Once fission has proceeded for a time, usually one to two years, cranes remove the assemblies to a cooling pool in which the heat from residual radioactivity declines over a period of years. At this point, reprocessing can remove the still usable uranium, plus plutonium made when a neutron hits a nucleus of ^{238}U to make ^{239}Pu; these materials can again be made into new fuel pellets. Reprocessing *closes* the nuclear fuel cycle. Alternatively, deposition of spent fuel rods in a depository isolated from all living creatures for over 100,000 years leaves the nuclear fuel cycle *open*. Chapter 7 assesses the geopolitical tensions and health issues arising at different points in the nuclear fuel cycle.

Physical distribution. The earth has abundant supplies of uranium, scattered widely over the inhabited continents and in the ocean. As with most minerals, however, commercial production comes from a limited number of deposits with high concentrations of uranium. The World Nuclear Association, a trade group, estimates known global reserves at 5.9 million metric tons, about 97 percent of which occurs in sixteen countries (table 4.4). Three countries alone, Australia, Kazakhstan, and Russia, hold half of the reserves.

Production is also highly concentrated. Two countries, Kazakhstan and Canada, produce over half of the world's uranium, and eight countries produce over 90 percent (2013 figures). Ten mines in just six countries produce over half the total output. One of the Canadian mines has ore with up to 20 percent uranium, far above the more usual concentration of 0.10 percent needed for commercial production.[43]

TABLE 4.4.   WORLD URANIUM RESERVES, 2013, BY COUNTRY

Area	Rank	Uranium (metric tonnes)	Percent	Cumulative Percent
World		5,902,900	100.0	
Australia	1	1,706,100	28.9	28.9
Kazakhstan	2	679,300	11.5	40.4
Russia	3	505,900	8.6	49.0
Canada	4	493,900	8.4	57.3
Niger	5	404,900	6.9	64.2
Namibia	6	382,800	6.5	70.7
South Africa	7	338,100	5.7	76.4
Brazil	8	276,100	4.7	81.1
United States	9	207,400	3.5	84.6
China	10	199,100	3.4	88.0
Mongolia	11	141,500	2.4	90.4
Ukraine	12	117,700	2.0	92.4
Uzbekistan	13	91,300	1.5	93.9
Botswana	14	68,800	1.2	95.1
Tanzania	15	58,500	1.0	96.1
Jordan	16	40,000	0.7	96.8
Other		191,500	3.2	100.0

SOURCE: World Nuclear Association, world uranium mining production, found at www.world-nuclear .org/info/Nuclear-Fuel-Cycle/Mining-of-Uranium/World-Uranium-Mining-Production/, May 4, 2015.

Uranium industry. The uranium industry differs from other energy industries, because uranium enriched in ^{235}U at low concentrations (3–5 percent) generates heat for electricity and at high concentrations (about 90 percent) fissions rapidly enough to produce a bomb. Moreover, fission of ^{235}U in a power reactor produces ^{239}Pu as a "waste" product when a neutron hits ^{238}U. As a result, nuclear electric power can't be completely isolated from nuclear weapons.[44] Consequently, development of the uranium fuel industry followed a distinctly different pathway from all the other energy industries.

Fossil fuels, for example, developed at first outside government, in the private sector. Landowners generally owned and mined coal, petroleum, and gas. Industrial firms procured the mineral fuel, refined it, and sold it to customers. Customers bought machines, like automobiles, to use the refined fuels. Governments may have promoted and regulated each step in the process, but private individuals and firms organized and administered manufacturing, processing, transport, sales, and uses.

The modern uranium industry started first in the United States, and from the beginning the federal government owned and controlled all aspects of it. Even after enlarging the scope for private industry in 1954, the uranium fuel industry remained largely a "ward of the state": both subsidies and regulation remained vital to its existence, and many reasons continue to underlie the imperative for state control and sponsorship. Consequently, the uses of uranium as a fuel, now and in the future, will continue, as in the past, to reflect the strategic politics, policies, and programs of nation-states. Even more than other sources of energy, little room exists for the entrepreneurial spirit of private companies. A brief overview of events in the United States highlights these close relationships between industry and government.

Before the recognition of fission in 1939, uranium—not then a fuel—had a minuscule market, primarily for glassmakers and others who wanted very dense material for purposes such as ship ballast. A few scientists bought minuscule quantities for research purposes. This situation changed drastically when President Franklin D. Roosevelt authorized the Manhattan Project in 1942, run by the U.S. Army Corps of Engineers. The far-flung efforts of the Manhattan Project successfully constructed atomic bombs by 1945, and the use of two of them against Japan quickly ended World War II.

German scientists had been the first to recognize uranium fission, so every reason existed to think that Germany's Nazi government might also attempt to make an atomic bomb. The American development of nuclear engineering thus began as a secret, strategic race to be first; use of the weapons in 1945 evaporated the secrecy. Nevertheless, the strategic overtones of nuclear science and engineering have never dissipated. In 1946, Congress created the U.S. Atomic Energy Commission (AEC) to own and manage the industrial and research facilities built by the Manhattan Project.

These facilities comprised one of the largest industrial manufacturing enterprises on the planet. Moreover, the war had destroyed much of the industry in Europe, the USSR, and Japan, so AEC's industrial capacity truly stood as a colossus.[45] In 1952, AEC employed 6,700 people directly, and its contractors employed 135,000.[46] Production of weapons clearly ranked as AEC's first priority, but Congress also wanted the commission to pursue the development of fission for electricity. In 1954, Congress opened development of civilian nuclear power to private industries, and the first nuclear electricity fed to the grid in the United States began in 1957.[47] The USSR and the United Kingdom had activated nuclear reactors for electricity for the grid by 1954 and 1956, respectively, ahead of the United States.[48]

Ultimately the United States licensed 127 reactors for power generation, more than any other country. In early 2015, 99 reactors still had valid operating licenses, and one additional facility had permission to continue operating pending consideration of license renewal.[49] In 2014, nuclear reactors generated 19.5 percent of U.S. electricity.[50] Other countries, such as France, Belgium, and Ukraine, built fewer plants, but nuclear electricity comprised a larger proportion of total electricity in these countries than in the United States. Despite different national policies regarding nuclear power, however, nuclear electricity always relied completely on favorable government policies for its existence.[51] Despite the best efforts of the United States to make nuclear energy a viable, free enterprise operation, it never achieved this status.[52]

ENERGY EFFICIENCY: THE INVISIBLE FOUNDATION OF THE THIRD ENERGY TRANSITION

The big-four nonrenewable fuels (coal, oil, gas, uranium), plus hydropower (renewable; see chapter 8), made up one of two energy building blocks of the Third Energy Transition. Modern states achieved global status and security through energy services secured by access to plentiful supplies of these primary sources of energy (see chapter 3). All too often, however, security of supplies occupied center stage in the development of energy and obscured the second building block, *energy efficiency*.

The origins of energy efficiency lay hidden in the history of the First Energy Transition (see chapter 1). Firewood was easily available in many places on the planet, but using it meant someone had to bring it into the shelter or to the workplace. Whatever the benefits of fire, it took work to procure and use it, so the ancestors of modern humans probably began to find ways to use as little as possible.

Learning to economize undoubtedly became crucial starting around 10,000 years ago, as people expanded energy services to make durable materials such as ceramics, copper, glass, bronze, and finally iron and steel. The manufacture of all these new materials required tremendous amounts of wood and of the first manufactured fuel, charcoal from wood heated without enough oxygen to burn. Hauling wood for heat and charcoal making involved a great deal of labor, which increased over time as forests diminished and production of new materials increased. Clever makers of these new materials found new ways of making their products with less heat and less fuel.

The driving force of energy efficiency was often either competition between companies selling the same product or declining supplies of fuels. Companies paid cash for energy, and they wanted to pay as little as possible. Energy was a cost of production, and companies that prospered did so by minimizing production costs. As firewood became more scarce in the 1500s in England, customers turned to coal (see chapter 1).

These simple facts in the 1700s established energy efficiency as an important problem for engineers. For example, Newcomen's engine used cheap coal at the mine, so the amount of coal consumed posed few problems. Later, engineers, most famously James Watt, tinkered with Newcomen's design and stepped up the efficiency of the engines. Watt and Boulton priced their machines based on their customers paying them one-third of the savings in fuel costs of the new engines compared to the Newcomen engines (chapter 1).

English production of iron provides another example. From 1540 to 1760, charcoal powered iron production as the woodlands of Britain declined. Ironmasters reduced the charcoal needed to make a ton of pig iron from 5.5 to 2 "loads" (one load probably was equivalent to between 675 and 875 kg). For production of bar steel, the need for charcoal declined from 16 to 4 loads of charcoal per ton.[53]

After 1718, coke (coal heated without sufficient oxygen to burn) replaced charcoal for iron and steel making, and efficiency in the use of coal increased dramatically. In about 1760 English ironmasters needed about 300 gigajoules (GJ) of heat to produce one ton of pig iron.[54] In 2013, about 24 exajoules (EJ)[55] produced about 1.167 billion tons of pig iron,[56] or about 21 GJ per ton. This was about 93 percent less than the amount of heat needed in 1760.

Put another way, had the world iron industry produced 1.167 billion tons of pig iron in 2013 with eighteenth-century efficiency, the heat energy needed would have equaled about 350 EJ, or about 70 percent of the total world energy actually used (about 500 EJ). Instead, making pig iron required only about 5 percent of the world's energy. This is still a significant amount, but 1700s methods could not have made as much iron as the world now does.

Energy efficiency measures developed in the making of pig iron between 1760 and 2013 produced the equivalent of 326 EJ per year, or 65 percent of the world's energy consumption. This amount was larger than any one fuel provided and approximately equaled the heat available from coal and oil combined in 2012 (about 60 percent of total world energy consumption).[57]

Efficiencies in making electricity, mobility, lighting, and other services provided comparable magnitudes of energy savings. Had the United States relied on the energy technologies and market structures of 1970 in 2008, the country would have consumed about 211 quads (one quad equals 10^{15} Btu) of energy instead of about 104.[58] Developments in efficiency helped make the equivalent of about 107 quads of energy within about forty years.

Without exaggeration, the Third Energy Transition rested as much on energy efficiency developments as it did on the development of energy supplies. Improvements in efficiency will continue their key contributions in the Fourth Energy Transition. It's harder to see than the mammoth installations to generate supply, but lack of efficiencies stymies the quest for sustainable energy economies.

CONCLUSION

Four nonrenewable fuels (coal, oil, gas, and uranium), supplemented by significant amounts of renewable hydropower, powered the energy services of the Third Energy Transition, and energy efficiency made these services possible with less fuel. Big industries produced and marketed these fuels, and modern states based their economic and political power on them.

From time to time, between the mid-1800s and the present, angst arose about the sufficiency of these fuels to continue the prosperity they enabled. Even in times of trouble or controversy, however, problem solving has tended to focus on one fuel at a time, not the energy system as a whole. Neither leaders nor citizens nor customers have seen energy collectively as a problem, but new kinds of problems in the 1970s required a new focus on energy as a whole. These serious problems threaten to undermine the benefits of modern energy services, a topic discussed in the next three chapters.

Connections with Everyday Life

For most readers, this chapter describes an entirely new way to see and understand energy.

Navigating everyday life, most consumers encounter only a few primary energy sources, one fuel at a time, mostly gasoline and natural gas. In some areas, they may deal with firewood on a regular basis, and if they have an outdoor grill, maybe charcoal. Daily life also includes electricity, but most consumers have little idea of the fuel needed and used to produce their electric power.

As important as these individual encounters with primary energy sources may be, they don't shed light on the energy economy or energy profile of regions and nations. They also don't invite consumers to unify their various purchases under the umbrella of "energy." Here we step away from "everyday life" to look at energy profiles that provide energy services collectively to large populations.

CHAPTER 5

Energy Systems

In the 1970s, the United States—the world's largest producer and user of energy at the time—faced new issues and dilemmas about energy and developed a new, visual framework for assessing them holistically rather than by each fuel individually. This new framework focused on all energy uses in a specified geographic area and displayed the heat content of each fuel, or primary energy source. Over time, holistic energy analysis spread beyond the initial methods and beyond the United States, and the more inclusive framework made it easier for both energy specialists and nonspecialists to understand energy dilemmas. This chapter presents the new methods for holistic analysis. Chapters 6 and 7 delve into the issues prompting them.

LEARNING TO SEE THE ENERGY FOREST, NOT JUST THE FUEL TREES

Before 1970, analyses of energy focused on one fuel at a time, and different assessments used many different units to indicate quantities. Metaphorically, these methods emphasized fuel trees, but they left the energy forests invisible. The Third Energy Transition utilized several mineral fuel trees added over a period of about 250 years, from 1700 to 1950. First, coal replaced firewood, and in the 1800s, oil, gas, and hydroelectricity joined coal. Uranium completed the roster of the major

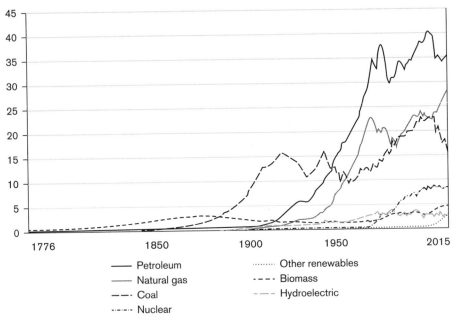

FIGURE 5.1. Historical pattern of energy use in the United States during the Third Energy Transition (energy in quads [quadrillion Btu]).

SOURCE: Energy Information Administration, "Today in Energy: Fossil Fuels Still Dominate U.S. Energy Consumption Despite Recent Market Share Decline," July 1, 2016, www.eia.gov/todayinenergy/detail.cfm?id=26912#, September 28, 2016.

fuels after 1945. Figure 5.1 shows the changes in fuel use during the Third Energy Transition in the United States.[1]

From time to time during the transition, supplies, prices, and pollution created concerns for governments, industries, and consumers, but resolution framed these concerns as fuel problems, not energy problems. In the early 1970s in the United States, however, new situations arose that involved multiple issues and multiple fuels, and resolution required framing matters as energy instead of a single fuel that provided it. Specifically, analysts simultaneously faced (a) air pollution from coal for generating electricity and making iron and steel, (b) oil supplies increasingly dependent on foreign supplies, and (c) an emergent nuclear power industry confronted by strong local opposition to siting of plants and concerns about safety. The Congressional Joint Committee on Atomic Energy addressed these energy dilemmas and, in collaboration with the U.S. Atomic Energy Commission, created new methods for studying the energy profile of the United States as a whole.[2] "Energy profile" is also referred to as energy budget, energy economy, or energy system; this book uses these terms interchangeably.

The major accomplishment of the new analytic methods was depictions of total energy use in specified geographic areas, with uniform units measuring heat content provided by each primary energy source. All primary energy sources, measured with common units, made it possible for the first time to see the energy forest, composed of different fuel trees.

Exactly what is an energy profile? This term seldom appears in newspapers, on TV programs, or even on websites devoted to energy issues. Most discussions about energy in the media and in political campaigns still focus on certain fuel trees, not the complete pattern of energy use in the energy economy. Sometimes talking about individual trees helps solve a specific problem, such as cost, but understanding the more serious problems posed by energy requires looking at the energy economy as a whole.

In 1898, Riall Sankey, an Irish engineer, developed a diagram, now called a Sankey diagram, to help him understand the flows of energy and materials in a steam engine. These diagrams give a simple, visual picture of a complicated situation that involves numbers, energy, and materials. Figure 5.2 shows a Sankey diagram or energy profile of the United States in 2015, the latest year available at the time of writing.[3] It traces energy flowing through the U.S. energy economy from primary energy sources through to end uses in the industrial, commercial, residential, and transport sectors. These diagrams change slowly over time, so one from a specific year approximately equals the diagrams for several years past and several into the future.

The figure invites you to think of the United States as a big machine that consumes fuel and produces energy services plus waste heat. Alternatively, the United States operates on an energy budget of about 97.5 quads per year to turn the heat equivalent of nine fuels on the left to services and waste heat on the right. Like a bank account that supports expenditures, the energy budget supports running all the motors and making heat, light, and electricity, or the energy services. Looking closely at the top of the figure, we see that the United States used 97.5 quads in 2015. A quad is one quadrillion Btu. Let's unpack this definition.

One Btu is easy to grasp: if you put a pound of water (one pint) with a thermometer on a gas stove, turn on the burner and watch the thermometer go up 1 degree Fahrenheit, then the heat from the burning gas is 1 Btu. Utility companies sell gas by "therms"; 1 therm is 100,000 Btus. A therm can heat 100,000 pints of water by 1 degree Fahrenheit

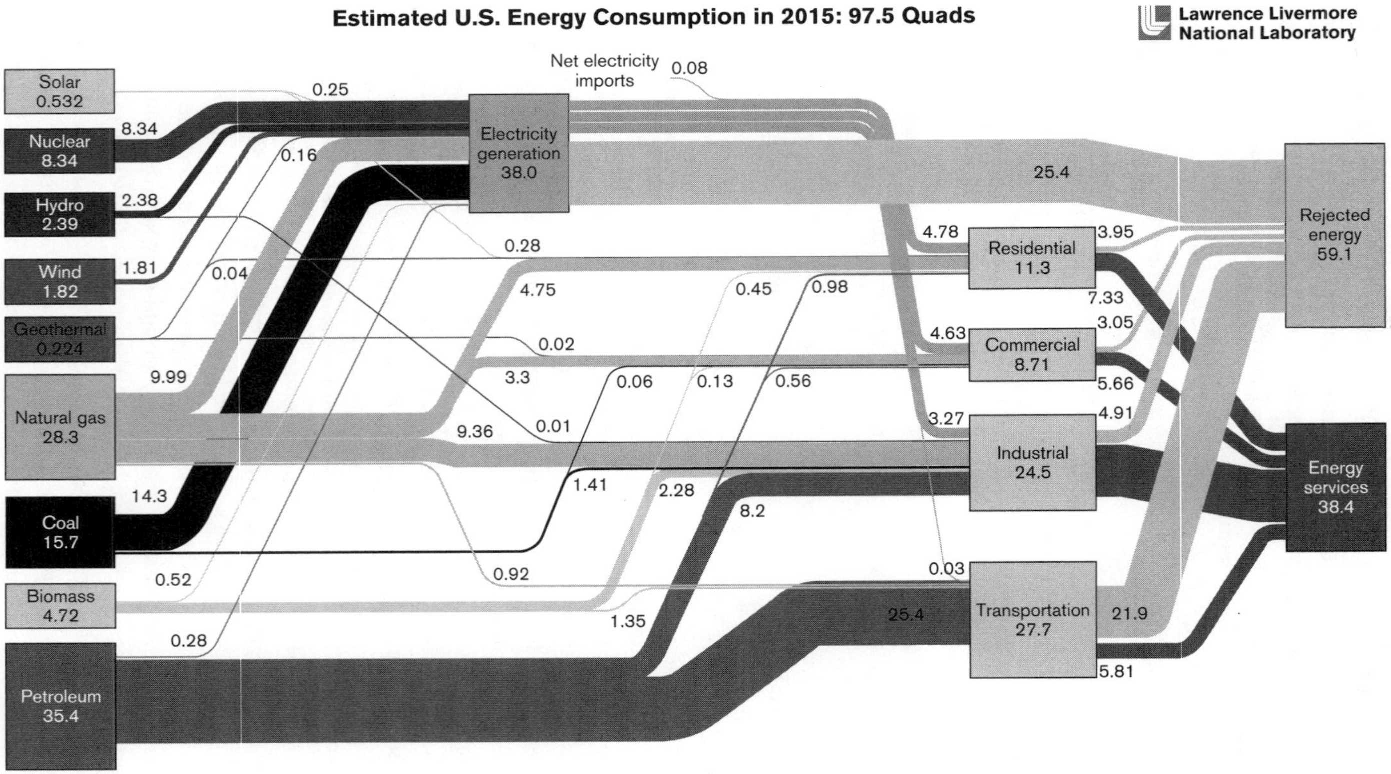

FIGURE 5.2. Energy flow, United States, 2015.

SOURCE: Lawrence Livermore National Laboratory, courtesy of Anne Stark.

or 50,000 pints by 2 degrees. One can think of a Btu, therefore, as an amount of heat carried by a certain amount of fuel. Now for "quadrillion." One quadrillion is 1 followed by 15 zeros, or 1×10^{15}. A quadrillion could also be called "one million billions" or "one billion millions." Therms measure energy used by households and quads measure energy used by large countries.

Btu is only one unit for measuring energy. Figure 5.2 lists the energy in all nine primary fuels in quads; the heat units could also be listed in the metric system in joules. For example, energy supplies for country-level energy are best expressed in petajoules (PJ) or exajoules (EJ) (peta = 10^{15}; exa = 10^{18}). It turns out that 1 quad approximately equals 1 EJ, so approximate conversions of quads to EJ are simple: the United States used approximately 97.5 EJ in 2015.

Appendix 1 explains the common units of measurement in both the American and metric systems. The important thing to know is that *every energy unit can be converted easily into each of the other units.* This conversion capability reflects the first law of thermodynamics (see chapter 2). Each unit has specialized uses, and energy statistics appear in different kinds of units, depending on the source and purposes of the respective units. Assessing energy economies requires the facility to convert units from one to another.

Figure 5.2 indicates that in 2015, U.S. purchasers bought or used about 35.4 quads of petroleum, 4.7 quads of biomass, 15.7 quads of coal, 28.3 quads of gas, and so forth. The big-four, nonrenewable fuels (petroleum, coal, gas, and nuclear) together provided about 87.7 quads, or about 90 percent, of the country's energy. Renewable fuels (solar, hydropower, wind, geothermal, and biomass) provided about 9.7 quads, or about 10 percent of the total, in 2015.

The number of quads from each fuel is currently shifting: coal use is dropping, and the use of natural gas, wind, and solar is increasing. This is like a mixed forest of evergreens and hardwood trees. Over time the proportion of evergreens may rise or fall, but the forest remains, at around 100 quads per year.

This book argues that renewable energy must make up a much larger proportion of the energy budget to have a sustainable energy economy. In other words, the energy forest we currently have must change. The Sankey diagrams for years after 2015 will provide a visual way to follow progress in this transition.

The next important feature in figure 5.2 is what happens to the quads in the primary energy sources shown on the left. About 38 quads manufacture electricity, or 39 percent of all U.S. fuel use. Production of electricity, however, operates at only about 33 percent efficiency, so the electricity output is only 12.6 quads to make light, provide heat, run motors, and power electronic communications. Another way of seeing this part of the forest is to say that 39 percent of U.S. quads are used to produce about 13 percent of its usable energy. The other 25.4 quads going into electricity production vanish as waste heat, the second law of thermodynamics in operation.

In fact, as shown on the right, about 59.1 quads of the original 97.5 end up as waste heat. This means that about 61 percent of the quads used do nothing useful for us. All the benefits we receive from the 97.5 quads come from about 38.4 quads (39 percent). Overall, over half of the quads in fuels used do nothing useful. This heat is the increase in entropy associated with every fuel use, but entropic heat can be reduced by energy efficiency.

Figure 5.2 provides useful information about what services come from the 38.4 quads doing useful work. Homes, primarily for heating, cooling, and lighting, consume about 11.3 quads (12 percent) of the total 97.5 quads. Commercial establishments consume about 8.7 quads (9 percent), and industry takes about 24.5 quads (25 percent). Finally, about 27.7 quads (28 percent) are used to run automobiles, buses, trains, ships, and airplanes. It's important to note that transportation, like generating electricity, is not very efficient. Of 27.7 quads burned in engines, only about 5.81 quads, or 21 percent, show up in the form of a vehicle in motion. The other 21.9 quads disappeared as heat. (People who keep outdoor cats may have noticed where cats head when you drive home on a cold day: straight for the car hood. Cats don't like to see heat wasted, and they know where to warm up)

Another important insight into the U.S. energy forest shown in figure 5.2 is the highly specialized uses of certain fuels. Petroleum, for example, goes almost entirely into transportation, and coal primarily goes to electricity plants. Some fuels, like nuclear and wind, serve only to generate electricity. With current practices, only gas is highly flexible; it goes to make electricity and to provide heat in residences, commercial establishments, and industry. Small amounts of gas go to transportation, and more could with modifications to vehicles.

Despite specialization among fuels, in theory any fuel can be used in any of the four service areas. Even nuclear and solar energy, for example, could be used to power electric vehicles. Today, however, the bedrock of transport consists of petroleum-based fuels and the internal combustion engine. Electric vehicles comprise a minute but increasing fraction. In theory, "any fuel for any use," but important constraints and barriers block practical implementation.

The final insight that can be gained from figure 5.2 requires distinguishing *efficiency* from *conservation*. In this book, I follow the practice of most energy analysts. Efficiency means doing the same amount of work with fewer quads. Conservation means doing less work and therefore needing fewer quads. Figure 5.2 suggests a powerful potential for efficiency. If 61 percent of the quads used in the United States is waste heat, then maybe it's possible to put some of those 59 quads to use for something besides making cats warmer on a chilly day. Indeed, multiple ways for increasing efficiency exist, and efficiency is the single most important ingredient of a new sustainable energy economy.

Conservation is also vitally important but perhaps more controversial. Conservation means not doing some things we currently do, which leads to the vexing question, Exactly who is supposed to give up what? The key to making conservation palatable is clever design of communities, buildings, and processes. For example, if urban design locates schools, commercial establishments, and mass transit convenient to houses, then people will drive fewer miles in automobiles. They can walk, bike, or take mass transit. In the end, people will have given up driving more miles, but the quality of their lives may have improved, not declined. Achieving conservation by improving quality of life takes the sting away from conservation.

DYNAMISM IN ENERGY SYSTEMS

Figure 5.2 shows a snapshot of the flow of energy through the United States in 2015, and over time these diagrams change slowly. Specialists in the energy industries, however, have always seen constant flux, and all changes require new investments of money.

First, new investments regularly had to replace mining equipment and equipment for refining and using fuels. Second, extraction of the big-four fuels depleted deposits and forced exploration for new ones. Competition among the energy industries drove yet other changes as customers substituted one fuel for another. In addition, the geographic spread of energy services across the world steadily increased energy consumption per person and in total.

And third, two contextual factors constantly stimulated change. Population growth, supported by fertilizers and irrigation in agriculture, created more demand for energy, and the projected population increase to over 9 billion by 2050 will continue this process. In addition, climate change has already altered the earth's environment (see chapter 6), and some of these changes will affect the energy economy. More intense, longer heatwaves will increase the demand for air-conditioning, decrease efficiency of electric generating plants, and force curtailments of operation of power plants to prevent thermal pollution. Rising sea levels may flood or threaten to flood power plants located at low elevations.[4]

Even if no flux existed, the infrastructure of the energy economy requires rebuilding every forty to sixty years, the maximum life span of most machines involved. Although a few devices such as hydroelectric dams operate for longer periods, new investments must constantly replace existing infrastructure. Constant rebuilding combined with the fluxes noted above create problems for managers but opportunities for change. Achieving an energy economy based on energy efficiency and renewable energy will follow if ongoing investments in new infrastructure flow to these technologies. Debates about energy futures, framed in this way, are really debates about ongoing, mandatory investment patterns.

A multitude of governments, business organizations, and individuals in many countries make investment decisions. Criteria for assessing strengths and weaknesses of different fuels and energy technologies thus naturally play a role in decision making about investments (see chapter 9).

INVESTMENTS AND POLITICAL ECOLOGY

The energy forest changes through time. What the United States did in 2015 with energy is different from the situation in 1950 or 1960, and the energy flow diagrams of 2050 or 2060 will also be different.

A better understanding of the energy forest, therefore, requires imagining the energy flowchart sliding along a horizontal rail, representing time. One can begin with an energy flowchart for 1950 and slide it to the right through to the present. As the chart slides from year to year its shape changes, and, as happened in the United States, the total number of quads used grew. In 1950, the country used about 33 quads; by 2010, about 100 quads. Only petroleum, coal, gas, and hydropower played a significant role in 1950. Over a period of sixty years, the energy forest grew larger and its trees changed in size and kind.

But viewing energy flow in this way doesn't capture the need for continual investment. No production or use of energy occurs without investments, and these investments must occur repeatedly. It's useful to divide investment steps into *upstream, midstream*, and *downstream* investments as multiple companies move energy sources from raw fuels to consumer products at the right place and the right times. Upstream investments and processes harvest the raw fuel, and midstream processes refine the raw fuel into a usable form. Downstream processes move the refined energy to delivery points for customers.

Users, too, must make investments to take advantage of energy services. Gasoline is merely a smelly, oily, dangerously flammable liquid without an internal combustion engine attached to a vehicle. Electric wires do no useful task without feeding the energy into a motor, light fixture, heater, or communication device.

All these investments, from upstream to use, must occur at regular intervals. None of the machines or devices employed to find, make, transform, or use energy lasts forever. At each replacement, the investor must make a choice: same again or something new? In other words, understanding the energy forest and how it changes over time requires not just understanding individual snapshots of energy flows but also the constant replenishment of the machines, processes, and devices making up the energy forest. Otherwise, the forest disappears as all the trees vanish.

A simple question arises from these considerations: How much money flows into investments to find, harvest, prepare, transport, and use energy? A recent study by the International Energy Agency provides at least a rough sense of the magnitudes of money involved, which are indeed huge. In 2013, $1,600 billion ($1.6 trillion) went to build, rebuild, and perpetuate the world's energy economy. The world's total economy that year was about $75,467 billion ($75.5 trillion).[5] In other words, investments in energy required allocating 2 percent of the wealth produced into perpetuation of energy services, mostly into replenishing the infrastructure of the Third Energy Transition.

This invested sum represents about 9 percent of the global capital formation in 2014,[6] which measures the outlays to add to fixed assets and increase inventories. Thus the amounts invested to keep energy services flowing requires a significant chunk of the world's investments to form new capital assets. Chapter 11 briefly returns to the question of investments, but here we turn to the *political ecology framework*, a way of envisioning the investment processes.

Political ecology visualizes the different components of investment decisions in technology. It begins with a simple premise: technology is knowledge that enables people to access natural resources in the environment to obtain the materials and energy they need and want. For example, farming technology enables people to use soil, plants, animals, sunlight, and water to grow crops and livestock. Similarly, iron making involves using knowledge to chemically alter iron ore into pure iron for tools and machines. Individuals must have the right knowledge to make these processes work correctly.

All technology requires an investment: people must learn, they must extract raw materials from the earth, and they must make the right machines in order to fulfill their material needs and wants. Figure 5.3 shows the relationships between human needs and wants, technology, natural resources, and investment.[7]

Investments can be large or small, and they come from governments, companies, or individuals. Government may want to promote investments in a technology indirectly by subsidies or other policies to encourage companies and individuals to make investments. All investments in technology aim to use some natural resource to make things to sell to satisfy needs and wants. The investor wants something in return: profit and/or a better life. The technology changes the environment by making pollution or other impacts. Investments always have risks: the technology might not work, it might be dangerous, and it might harm other animals or plants.

The political ecology framework indicates that risk and safety influence decisions. When risk becomes too high, the investor may back out, or government may regulate to modify or prohibit the technology. Risk and safety concerns can also affect the environment and sometimes destroy the very natural resource the investment seeks to use. For example, farming technology that causes soil erosion can destroy the soil resource and make future farming impossible. Similarly, careless pumping of oil can destroy the ability of the oil field to yield a major portion of its oil.

Understanding the energy forest, therefore, involves making a snapshot energy flowchart, imagining it sliding along a track and changing through time, and then overlaying the political ecology framework for understanding investment decisions. Investment patterns, too, change over time, reflecting the competitive costs of different technologies, government subsidies, judgments about risks, and government regulations. As investments grow and change, so too will the energy forest grow and change.

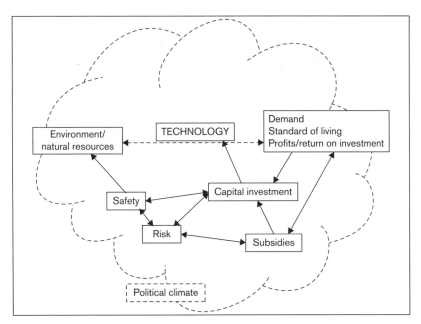

FIGURE 5.3. Political ecology: a framework for assessing relationships between technology, the environment, and the political economy.

Political ecology focuses on a technology enabling the use of resources in the environment to satisfy human economic demands, generate standards of living, and create profits and returns on investment. The framework underscores the relationships between economic factors (subsidies, capital investment, the need to generate profits or achieve a set return on investment), concerns about safety and risk (environmental, technological, investment related), natural resources, and environmental impacts. Political power affects all factors. Double-headed arrows suggest a potential two-way influence; a unidirectional arrow reflects the one-way nature of those influences.

Figure prepared by Kathleen Saul and used by permission. Explanation drawn from Saul as modified by author. Framework derived from John H. Perkins, *Geopolitics and the Green Revolution* (New York: Oxford University Press, 1977); Saul, Kathleen M. *The Renewed Interest in New Nuclear Construction in the United States: Lessons from History, the Media, and Interviews* (Olympia, WA: Evergreen State College, 2009); and Kathleen M. Saul, "Approaches to Energy Education: Political Ecology," paper presented at National Energy Education Summit, Washington, DC, January 26, 2015.

CONCLUSION

Until the 1970s, little effort was made to understand energy profiles holistically, but increasing problems with pollution, perceived shortages, and controversies about nuclear power all combined to stimulate new methods for understanding a multitude of dilemmas associated with energy. Sankey diagrams of energy flows opened new avenues to help everyone understand energy systems as a whole, not just one fuel at a time.

A more complete understanding of energy economies, however, required recognizing dynamism in them. Also required was a framework for understanding the links between the natural environment, energy technology, and the dynamics of investments in energy. Political ecology unifies an understanding of investments over time to create changes over time in energy profiles. Chapters 8 through 11 discuss the need for constant investment as the avenue toward changing energy profiles, but first chapters 6 and 7 explain the troublesome factors now demanding that current energy profiles change.

Connections with Everyday Life

Climate change, like energy profiles of regions and nations, seems to be distant from day-to-day activities. People are, however, aware of the issue. Since 1990, 50 percent or more of Americans in every Gallup Poll on the subject have stated they worry a "great deal" or "fair amount" about climate change, and a poll in March 2016 found 64 percent with these levels of concern. At the same time, no more than 41 percent of Americans have said that they expect climate change to pose serious threats in their lifetimes.[1]

Stalemated, partisan political battles about climate change rage, despite long-standing concerns about climate change among most of the U.S. population. President Barak Obama pushed the federal government toward action on climate change, against the unwillingness of Congress, and similar efforts have progressed in some states. For actively engaged political and environmental partisans, these debates have had greater impacts on their lives than have the actual effects of climate change so far.

This chapter steps back from both the scientific details of the risks posed by climate change and the unending political debates. Instead, it explores the logic of the science underlying climate change. Why did the vast majority of the scientific community converge on the conclusion that anthropogenic climate change was real, happening now, and posed grave risks for the future?

For the study of energy and energy services, it is important to understand this logic and to recognize the multifaceted nature of the conclusions in climate science. This logic undergirds the conclusion that the energy profiles now in place threaten the sustainability of the many energy services and benefits on which people depend.

CHAPTER 6

Climate Change

If only benefits flowed from the massive heat supplied by the big-four fuels (coal, oil, gas, uranium), the rest of this book would be unnecessary. Unfortunately, science has steadily uncovered dark effects of the Third Energy Transition. These consequences suggest the possibility of truly cataclysmic impacts globally, even in areas in which people did not enjoy the benefits of the Third Transition.

This chapter turns to climate change, the best known and most discussed challenge posed by the backbone of the Third Energy Transition, the fossil fuels (coal, oil, and gas). The risks of climate change alone require phasing out fossil fuels over time. Unfortunately, much needless misinformation and confused information surrounds the scientific evidence about climate change. This chapter summarizes the evidence and logic of climate science and suggests the likely impacts of climate change.

EMERGENCE OF A SCIENTIFIC CONSENSUS

Carbon dioxide and methane emissions associated with the energy economy are the biggest contributors to climate change. The association of carbon emissions with the energy economy shows clearly in a Sankey diagram of the United States in 2014 (figure 6.1). [2]

Burning fossil fuels (coal, oil, and gas) releases carbon dioxide (CO_2) into the atmosphere, as do deforestation and other land use changes. Once there, this gas traps heat and thus alters climate. In addition,

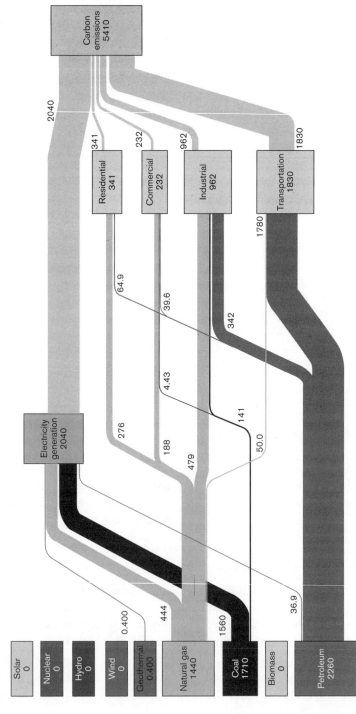

Estimated U.S. Carbon Emissions in 2014: ~5,410 Million Metric Tons

Lawrence Livermore
National Laboratory

FIGURE 6.1. Carbon emissions of the United States from the energy economy, 2014.
SOURCE: Lawrence Livermore National Laboratory, courtesy of Anne Stark.

methane (CH_4, natural gas) traps heat and alters climate if released to the atmosphere. Other gases—collectively called greenhouse gases—also trap heat, but I focus on the problems associated with the fossil fuels.

Questions about energy and climate change became increasingly important beginning in the 1970s and have grown more serious and urgent over the past four decades. Since the 1800s, scientific studies had recognized the potential for increasing concentrations of carbon dioxide from fossil fuels to warm the global climate, but during the 1900s, increasing industrialization and widespread agriculture, plus occasional volcanic eruptions, threw various pollutants into the air that might cause cooling. What was the balance between these two opposing effects, more warming or more cooling?

Conclusions shifted toward more warming in the late 1970s. Since then, continuing research has convinced close to 100 percent of the scientific community that earth's surface temperatures have warmed as a result of human influences, especially the use of fossil fuels. Most scientists also agree that without reductions in emissions of CO_2 and CH_4 from fossil fuels warming will continue and that projected climate changes could drastically, perhaps catastrophically, alter climate. The exact timing and amount of warming and other attendant changes remain under research and debate, but the overwhelming scientific consensus predicts dangers ahead.

The Intergovernmental Panel on Climate Change (IPCC), organized under the auspices of the United Nations, stated the following in 2013:

> Warming of the climate system is unequivocal, and since the 1950s, many of the observed changes are unprecedented over decades to millennia. The atmosphere and ocean have warmed, the amounts of snow and ice have diminished, sea level has risen, and the concentrations of greenhouse gases have increased. . . .
>
> The atmospheric concentrations of carbon dioxide (CO_2), methane, and nitrous oxide have increased to levels unprecedented in at least the last 800,000 years. CO_2 concentrations have increased by 40% since pre-industrial times, primarily from fossil fuel emissions and secondarily from net land use change emissions. . . .
>
> The total radiative forcing is positive, and has led to an uptake of energy by the climate system. The largest contribution to total radiative forcing is caused by the increase in the atmospheric concentration of CO_2 since 1750. . . .
>
> It is extremely likely that human influence has been the dominant cause of the observed warming since the mid-20th century.[3]

IPCC's statement outlines the theory of human-caused climate change, or *anthropogenic climate change* (ACC), but its implications for public policy require addressing two further points: What is the evidence for ACC, and what likely consequences will follow it?

EVIDENCE FOR ACC

The idea that humans can alter the climate of earth rests on scientific conclusions reached between the 1820s and today. They compel attention and argue for changes in the energy economy.

ACC, however, resists an easy, intuitive understanding. It asserts that slight changes in the amount of carbon dioxide and methane in the atmosphere can unleash all sorts of unanticipated and unpleasant consequences. How could that be? Can a person who is not a scientist grasp why the vast majority of scientists believe the theory is true?

Consider the following fictional statement, which some intelligent, skeptical, and slightly irascible citizen might make.

> OK, so I burn gasoline in my automobile and I use electricity made from burning coal. I heat my house with gas, and I buy lots of stuff, which I know took energy to make. I also know a lot of this stuff was made in China, and some ship burning diesel fuel brought it to me. I also know that my great-great-grandparents homesteaded in Ohio and cleared the forest to plant corn. And OK, I'll accept that because of all these things, done by people over the whole world, the amount of carbon dioxide in the air changed.
>
> But is this really a big deal? Since 1700, the change has been almost nothing. It used to be about three molecules of carbon dioxide for every 10,000 molecules of the other things in air, and now it's about four, and OK maybe it's headed to five or six. Out of 10,000! What's the big deal? You want me to believe that small changes in amounts of carbon dioxide have warmed the air and oceans, diminished snow and ice, and raised sea levels? And even if it has, is this bad? Come on, how can such a little bit of carbon dioxide do all these things?
>
> And isn't carbon dioxide something that occurs naturally? Don't we exhale it? Don't green plants use it to grow? Don't we put it in our soda pop to make it fizzy? How can a teeny-tiny increase of this stuff possibly cause real trouble?
>
> Why should I believe such a story that sounds so cockamamie?[4]

The doubts raised by our fictional, skeptical citizen resonate with the fact that the theory of human-induced climate change rests on the importance of a seemingly small change in the composition of the atmosphere. Scientists participating in the IPCC understood the wonderment that such small changes could be important, and the theory became convincing only because of multiple conclusions about the earth's climate system, developed over the course of two hundred years and summarized below.

1. *Radiation and heat balance of the earth.* The sun warms the earth, and in the ensuing night the area in darkness cools. During the year, the sun's warmth waxes and wanes through the seasons. Early in the early nineteenth century, Joseph Fourier of France calculated that even though the sun was the prime source of the warmth, the atmosphere affected the earth's average temperature. Something in the atmosphere kept earth warmer than it otherwise would be.[5]

Fourier based his ideas on the notion that all bodies, including the earth, emit heat. In other words, heat from the sun during the day dissipates during the night. This emitted radiation is now called infrared radiation, which most people know from things such as infrared heat lamps. It is invisible to the eye, but some molecules absorb infrared radiation. When they do, they move faster and we perceive heat or warmth (see chapter 2).

Stable climate depends on the balance between the incoming and outgoing heat. Over each day and each year, the earth has a steady average temperature, or equilibrium temperature: warmer in the day and in the summer, cooler at night and in the winter. Only scientists tend to think of earth as an "infrared radiator," but the validity of this concept—that all things with a temperature above absolute 0° (0 K) emit heat as infrared radiation—meshes with all of modern physics.

Climate change theory begins with the notion that some molecules in the atmosphere reduce the amount of infrared radiation lost from the earth, which in turn changes the heat balance and increases the earth's temperature. Much of climate science, therefore, focused on the question, What happens as the atmosphere changes? That was an easy question to ask, but finding good answers took a great deal of hard work and considerable debate.

2. *Carbon dioxide and several other gases absorb infrared radiation.* Why did the earth not shed all of its warmth and cool to the frigid temperatures as Fourier had calculated? In 1859, John Tyndall in England found an answer. He knew that most of the atmosphere was nitrogen and oxygen, and he found that visible light from the sun passed right through them; they were colorless and invisible. Water vapor and carbon dioxide also allowed visible light from the sun to reach the surface of the earth, and they, too, were colorless and invisible. When light

reaches earth, much of it is absorbed and the energy of the absorbed light warms the earth. The warmer earth emits more infrared radiation.

Water vapor and carbon dioxide, however, absorb infrared radiation. These gases present at low levels in the air block the passage of infrared and in so doing become agitated, vibrate, and move faster. The air "warms up," as detected by thermometers and our bodies. Tyndall thus answered Fourier's question: the warmth kept in the atmosphere by water vapor and carbon dioxide makes the earth warmer than it otherwise would have been.[6] As long as the amounts of water vapor and carbon dioxide don't change, heat coming in and heat leaving will equal each other, temperature will go up and down each day and over the seasons, but the equilibrium temperature will remain steady, and climate will be stable. Fourier's and Tyndall's work underpins all of modern climate science.

3. *Earth's climate varies immensely over long periods.* Most people, from everyday experiences, think of climate as stable. Some years have more cold, snow, and ice, and others have more heat, but overall not much changes. New England is cold compared to Florida, and Sweden is cold compared to Kenya; these observations are stable.

In 1837, Louis Agassiz, a Swiss naturalist, boldly upset this common perception by proposing that glaciers from the north had once spread over much of Europe, Asia, and North America. Despite skepticism from many, Agassiz and others pieced together the evidence that ultimately made the theory of repeating ice ages common knowledge.[7]

Agassiz opened entirely new questions: What caused normal weather patterns to go so off kilter that glaciers covered much of the earth? When? Why? Would a new ice age come? Many theories emerged as geologists, oceanographers, biologists, and astronomers each contributed ideas about what might govern earth's climate. Changes in water vapor in the atmosphere? New mountains or volcanic eruptions? Changes in ocean currents like the Gulf Stream? Variations in the sun's output? Variations in the earth's orbit?[8]

The Swedish physicist Svante Arrhenius advanced a new theory in 1896. He concluded that changes in carbon dioxide concentrations in the atmosphere can stop and start ice ages. Arrhenius juxtaposed his conclusions to findings from geology that carbon dioxide levels in the atmosphere can change enough to affect climate.[9]

Arrhenius focused on triggers for ice ages, and he thought countries like Sweden might benefit if carbon dioxide from fossil fuels warmed things up. His work did not gain rapid acceptance, because many considered his calculations wrong. Others thought increases in carbon dioxide would be locked up in rocks and the ocean, and new additions from fossil fuels would leave the atmosphere and not trap heat on earth.

Not until the 1930s did new work resurrect Arrhenius's findings and direct questions away from ice ages toward the idea that burning fossil fuels might warm the climate. Gilbert Plass in the 1950s in the United States noted that studies since Arrhenius clarified the infrared absorption patterns of carbon dioxide and concluded that increasing levels of carbon dioxide could absorb increasing amounts of infrared radiation and thus increase the temperature of the earth.[10]

4. Human actions can alter the earth in major ways. Agassiz, Arrhenius, and Plass had opened the door for ideas about long-term changes in climate, but another barrier still hindered ready acceptance of ACC. For most of human history, human beings considered themselves subject to major natural events over which they had no control. Cold, heat, droughts, floods, plagues, earthquakes, and volcanoes happened, often producing great destruction and hardship. Natural disasters came from angry spirits or an angry god, possibly in retribution for human evil. People identified, named, and feared natural calamities, but people could not change climate, especially over short periods of time.

George Perkins Marsh's book, *Man and Nature* (1864), had argued that humans could significantly affect the earth, but such ideas gained little traction until after 1945, when two books reflected the change in perspective. Fairfield Osborn's *Our Plundered Planet* (1948) and William Vogt's *Road to Survival* (1948) both saw expanding human populations with new technologies that could bring both benefits and massive, unwanted, destructive change.[11] For example, testing nuclear weapons led to the far-flung distribution of strontium-90, and DDT insecticide showed up everywhere, even in penguins in Antarctica, a place where no one used the chemical. Rachel Carson's widely acclaimed book, *Silent Spring* (1962), saw the parallels between weapons testing and pesticides and left the indelible conclusion that human activities really could rapidly alter the entire earth in ways never before imagined.[12]

5. Climate change may not be slow. Theories about ice ages envisioned temperature and climate changes in cycles of thousands of years. Findings of human impacts on a global scale could clearly occur within a time of a few years, but what about climate? Could climate changes also occur within a small time period? Did it matter if the changes were natural or anthropogenic?

By the 1960s and early 1970s, scientists such as Reid Bryson of the United States raised questions about impacts on climate of dust and aerosols as well as carbon dioxide. Bryson, a climatologist, also worked with anthropologists, archaeologists, and botanists, and they found that Native American groups had suffered from rapid shifts in climate before European contact. Not only was climate changeable in either direction; its changes could be rapid, perhaps within one human life span.[13]

These new perspectives on long-term changes, short-term human-caused global changes, and short-term shifts in climate all shaped research in the late 1960s and after. A power plant or factory burning coal and releasing carbon dioxide, and dusts, previously seen as a benefit of modern engineering and inconsequential on a global scale, became an agent of potentially earth-altering importance.

But was carbon dioxide the driving force? And what about the dust? Where did the balance lie between factors favoring cooling compared to heating? Scientists with different opinions about these matters raised concerns about warming and cooling, but scientific uncertainty ruled. In 1975, the U.S. National Academy of Sciences could not resolve competing claims and urged more research.[14]

6. Carbon dioxide and temperatures increased in the past two centuries. In the mid-1950s, no scientific research could say what was happening to levels of carbon dioxide and temperature. The fact that fossil fuels released carbon dioxide meant only that levels of carbon dioxide in the atmosphere *might* be increasing. But were they really increasing, or was the gas going into vegetation, the oceans, or mineral deposits? If so, then it wouldn't block infrared radiation from leaving the earth. And if the levels of the gas were changing, was it really due to burning fossil fuels? And what about temperature? Did temperature track levels of carbon dioxide? Between the 1950s and now, new studies answered these questions: carbon dioxide and temperatures have increased, and increased temperature can be traced to increased carbon dioxide.

David Keeling of the United States found direct evidence of changes in carbon dioxide levels in the atmosphere starting in 1958 with new instruments on top of Mauna Loa in Hawaii, a spot in the Pacific Ocean very far from the industrial centers of the world that burned most of the fossil fuels. Keeling believed that maybe here he could measure carbon dioxide after the gas had been uniformly mixed and thus would repre-

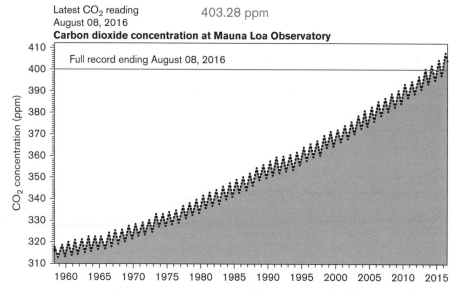

Latest CO$_2$ reading
August 08, 2016

403.28 ppm

FIGURE 6.2. Keeling curve showing increase of carbon dioxide in the atmosphere since the late 1950s.

SOURCE: Scripps Institute of Oceanography, https://scripps.ucsd.edu/programs/keelingcurve/wp-content/plugins/sio-bluemoon/graphs/mlo_full_record.pdf, April 27, 2016.

sent "average" levels.[15] Since the late 1950s, the concentration of carbon dioxide in the atmosphere rose steadily, a pattern not yet changed (figure 6.2).[16]

During each year, carbon dioxide levels rose in the northern hemisphere's winter and dropped the following summer. Despite the annual drop, the average levels climbed steadily through the years, and now the minimum is about 400 parts per million (ppm). Keeling explained the summer drop as the absorption of carbon dioxide by green plants in the northern hemisphere.

In the 1980s, Swiss scientists analyzed the carbon dioxide found in air bubbles in ice cores taken in Antarctica. They found that the concentration of carbon dioxide in 1750 was about 280 parts per million and that the concentration increased steadily after that time. Moreover, the composition of the carbon dioxide found in the ice indicated the increase came from burning fossil fuel and deforestation.[17]

What was happening to temperatures while carbon dioxide was increasing? Significant numbers of direct readings of temperature at the earth's surface, that is, with a thermometer, exist only since about the mid-nineteenth century. Temperatures rose from this time until about 1940, then fell slightly until the mid-1960s, and then began to rise steadily again after that.[18]

Carbon dioxide levels rose steadily during this time period. But was the concomitant rise of both just an association or was it a cause-and-effect relationship? If the latter, did the increase in carbon dioxide cause temperature to rise, or was it the other way around? Then there was the dip between about 1940 and the mid-1960s: if carbon dioxide was rising steadily and trapped heat, why was there a downturn in temperatures?

This tangle of questions stimulated intensive research that pointed to water vapor, carbon dioxide, methane, nitrous oxide, ozone, chlorofluorocarbons, and other gases as heat trappers (greenhouse gases) that would increase temperatures. Clouds, soot particles, volcanoes, and sulfate aerosols could reflect solar radiation away from earth and cause cooling. The dip in temperatures from 1940 to 1965 may have represented a period in which the cooling constituents of the atmosphere dominated heat trapping. After 1965, increasing uses of fossil fuels and fertilizers, deforestation, and increasingly intensive livestock and rice farming, plus other factors, steadily pushed the heat trapping into first place.[19]

As temperatures rose, positive feedback loops exerted a magnifying effect. For example, warmer temperatures put more water vapor into the air, and water vapor is an important heat-trapping gas. Rising temperatures melted ice and snow, which caused less reflection of incoming radiation, and thus the earth warmed more.

7. *Reconstructions of paleoclimates added perspective and confidence.* Paleoclimates are climates that existed before there were direct human records of climatic patterns, that is, before a few centuries ago. Scientists reconstruct paleoclimates with proxy measurements of temperature, carbon dioxide levels in the air, ice volumes, and other factors that affect climate or reflect climatic conditions.

Proxy measurements rely on numerous physical markers known to depend on temperature and/or carbon dioxide levels in the atmosphere, such as the width of tree rings, growth rings in coral reefs, the composition of shells of microscopic marine organisms in ocean floor sediments, the composition of water in ice cores from polar regions, and the concentration of carbon dioxide in tiny bubbles trapped in different layers of these ice cores.[20] Proxy measurements based on multiple methods done by multiple groups in many countries showed substantial agreement and

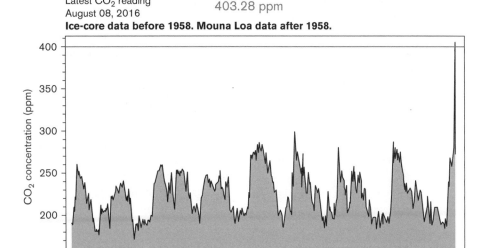

Latest CO$_2$ reading
August 08, 2016
403.28 ppm
Ice-core data before 1958. Mouna Loa data after 1958.

FIGURE 6.3. Carbon dioxide in the atmosphere for the last 800,000 years, showing current levels unprecedented in that time.

SOURCE: Scripps Institute of Oceanography, https://scripps.ucsd.edu/programs /keelingcurve/wp-content/plugins/sio-bluemoon/graphs/co2_800k.pdf, March 25, 2016.

allowed reconstruction of climate and carbon dioxide concentrations for periods extending to 800,000 years before the present (figure 6.3).[21]

An important point is clearly shown in figure 6.3: the level of carbon dioxide in the atmosphere now stands at about 400 parts per million, a level unprecedented for 800,000 years. *Homo sapiens* had not even evolved 800,000 years ago. In other words, modern humans have increased carbon dioxide levels to amounts never seen since before they appeared on earth.

Ice cores suggested a resolution of the nineteenth century debate about the causes of waves of ice ages. The physical composition of water in the ice cores indicated a regular pattern of warmer and cooler temperatures, and the low temperatures caused the ice to expand from the polar regions. These patterns matched the changing distances between earth and the sun predicted by slight irregularities in earth's orbit around the sun. These changing distances occur regularly in cycles—the Milankovitch cycles—calculated by the Serbian astronomer Milutin Milankovitch in the early 1900s. Ice ages began and ended based on the very slight cooling and warming caused by earth moving farther from or closer to the sun.[22]

Ice cores also yielded samples of ancient air in bubbles trapped in the ice. From these bubbles, scientists learned the concentration of carbon dioxide in the air at the time the air bubbles froze in solid ice. Thus ice cores yielded both temperature and carbon dioxide data, which revealed that temperatures and carbon dioxide rose and sank approximately in synchrony.

Did an increase in temperatures before carbon dioxide was a factor mean that something else caused the warming? Was increased carbon dioxide simply a result of warmer temperatures, not a cause? If ends of ice ages really didn't depend on increased carbon dioxide, did that mean that evidence suggesting carbon dioxide caused temperature increases in the past two centuries might be wrong? The fact that carbon dioxide absorbs the energy in infrared radiation meant that the absorbed energy no longer radiated from the earth and thus warmed the atmosphere. This trapping of energy gave a powerful reason to suspect that increased temperatures stemmed at least in part from increased carbon dioxide.

Research on the triggers that started and stopped ice ages is still under way. For example, an important reconstruction of the end of the last ice age, about 11,500 years ago, indicated that first the northern hemisphere warmed and melted ice, which caused sea levels to rise. Freshwater from ice melt interrupted ocean currents that carried cool water to the southern hemisphere, which caused warming in Antarctica. Various processes, still not entirely understood, led to increased carbon dioxide levels. One possible mechanism is loss of carbon dioxide from the oceans by "degassing," like a carbonated soft drink will lose its fizziness as it warms. Researchers argued that increased carbon dioxide was a key mechanism in warming the earth and ending the last glacial period.[23]

Reconstruction of paleoclimates based on proxy measurements added the perspective of long periods and cycles through many climatic changes. Knowledge of past climates also showed that the warming observed in the twentieth century had no counterparts in the historical past. For example, claims of a "medieval warm period" (1000–1400) in England did not represent global conditions, nor were these claims derived from replicable proxy data. Most important, data suggesting a warm period in the Middle Ages, followed by cooling and then warming again in the late nineteenth century, did not capture the highly significant increases in average global temperatures after 1950.[24] Early

Holocene (10,000–5,000 years before present) global average temperatures, however, exceeded today's. Nevertheless, temperatures predicted by climate models will soon exceed the early Holocene temperatures substantially.[25]

8. *Climate models can reproduce past events with increasing accuracy.* First, a word about distinctions between *weather* and *climate.* A scientist who can accurately tell a person whether or not to take an umbrella tomorrow is a weather forecaster using weather models. Weather focuses on exactly what temperature, rainfall, and wind patterns will prevail at a specific place, generally the size of a city, at a specific time, generally for one to ten days in the future.

Climate forecasts, in contrast, focus on average and extreme temperatures, average and extreme rainfall and droughts, and average and extreme winds over large areas (from region or state to world) and over longer periods (years to centuries). A climate forecaster tells farmers, for example, that reliable rains received almost every year will start to fail half of the time starting about fifty years from now.

Scientists began more intensive efforts to model weather after World War II, and an offshoot of these efforts turned to building climate models. The two types of models share some characteristics, the most important being that both types use equations to predict movements of air and moisture at specific places and times caused by heat inputs from the sun. Climate models, especially general circulation models, require an enormous number of calculations, and parallel development of rapid computers provided the only possible way to use the more complex climate models.[26]

Weather forecasting and climate predictions, however, differ in important ways. Weather forecasters receive rapid feedbacks about the accuracy of their predictions, often from wet citizens who received a forecast indicating no need for an umbrella. Long-term climate forecasting receives no immediate feedback on accuracy, and the definitive test of accuracy may not occur until the climate scientist is dead.

As a result, climate scientists check the likely accuracy of their models by reconstructing past, known conditions. If, for example, in 2010 they took the conditions of the atmosphere in 1900 as "given," could they accurately re-create the climatic patterns that actually occurred between 1900 and 2010? If yes, then they gained confidence that they could take the conditions of 2010 as given, make assumptions about the future inputs of carbon dioxide and other materials into the atmosphere, and

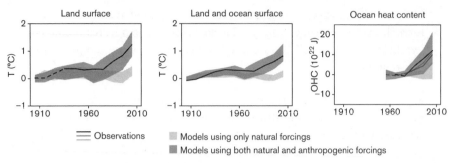

FIGURE 6.4. Models of past changes reproduce observed changes only if model includes CO_2 in atmosphere.

SOURCE: Figure SPM.6 (bottom panels) from IPCC, 2013: Summary for Policymakers. In *Climate Change 2013: The Physical Science Basis. Working Group I Contribution to the Fifth Assessment Report of the Intergovernmental Panel on Climate Change* [Stocker, T. F., D. Qin, G.-K. Plattner, M. Tignor, S. K. Allen, J. Boschung, A Nauels, Y. Xia, V. Bex, and P. M. Midgley (eds.)]. Cambridge University Press, Cambridge, UK, and New York, USA. Used by permission.

with their model predict what the climate conditions would be in 2050 or 2100.

Not surprisingly, the first climate models could match past conditions only approximately, and climate modelers believed therefore that they could make only approximate predictions for the future. Nevertheless, climate models of the late 1970s were sufficiently good to cause a committee of the National Academy of Sciences to conclude that if concentrations of carbon dioxide doubled from preindustrial times, the average global surface temperature would increase from 2° to 3.5°C. Higher latitude regions near the poles would warm even more. Significant socioeconomic results would follow, but the committee could not predict them.[27]

This brief report in 1979 began to settle the uncertainty among scientists about whether the cooling effects of aerosol pollutants would outweigh the warming effects of carbon dioxide. In the following years, 1980 to the present, the consensus in the scientific community on warming steadily increased to near-unanimity.[28]

By 2013, models replicated the previous century's changes in surface temperatures over land and ocean, ocean heat content, and sea ice at the poles if and only if the simulations included human factors, especially increases in CO_2. Simulations with natural factors alone, such as volcanos and changes in solar radiation, did not replicate the observed changes (figure 6.4).[29] Reproduction of already known changes increased confidence in the accuracy of future predictions.[30]

LIKELY IMPACTS OF CLIMATE CHANGE

Points 1 through 8 above identified and explained climate events that have already happened: (a) the atmosphere has always kept earth at a warmer temperature than it would be without it, the "greenhouse effect"; (b) temperatures on earth have risen since the mid-1800s, sea levels have risen since 1900, the heat contained in the upper 700 meters of ocean has risen since 1950, and Arctic summer sea ice has decreased since 1900; (c) CO_2 and CH_4 increased in the atmosphere after the eighteenth century, mostly from uses of fossil fuels; (d) the levels of CO_2 in the atmosphere have reached levels unprecedented since before modern *Homo sapiens* evolved; and (e) CO_2 and CH_4 have contributed most of the additional warming of the earth. In addition, CO_2 dissolved in the ocean has made seawater more acidic; that is, it has lowered its pH.[31]

Delving deeply into the methods and conclusions of the IPCC goes far afield from the study of energy.[32] Suffice it to say that projections of the new global average temperatures likely to be reached by 2100 will be higher by 1.5°C, compared to temperatures in the period 1850–1900. If humanity does not curtail emissions of greenhouse gases, temperature increases will likely be greater than 2°C, with a fifty-fifty chance of exceeding 4°C (7.2°F).[33]

These changes may seem small, especially because from day to night and from summer to winter people and ecosystems already experience changes each day and each year far larger than these predictions. It's easy to imagine our intelligent, skeptical, and somewhat irascible citizen asking, "OK, maybe it's going to be a bit warmer, but do you really want me to get excited by a temperature rise of just a few degrees? That doesn't look like anything to worry about based on normal changes already taking place all the time."

Unfortunately, the scientific projections of global *average* temperatures hide a great deal. First, the projections generally mentioned in the media generally go no further than the year 2100, at most. Suppose, however, people continue using fossil fuels in the future at the same rates they do today. With this assumption, the IPCC projected that after 2100, global average temperatures would rise more than 4°C (7.2°F) and perhaps up to 7.8°C (14°F), with a small chance of rising 12.6°C (22.7°F).[34] Thus the rises projected today are *minimums*, not what temperatures might eventually reach if societies make no changes in energy systems.

Second, even the seemingly small changes in temperature projected by 2100 will adversely affect people in three major ways: more heatwaves will occur, precipitation patterns will alter, and sea levels will continue rising. More heatwaves will bring discomfort and probably lower agricultural production. Altered precipitation patterns will lower agricultural yields by either too much or too little rain. Rising sea levels will inundate coastal areas and many of the world's biggest cities. Some of the increased CO_2 in the atmosphere will dissolve in the oceans, make it more acidic, and adversely affect many ocean species.[35]

Changes that have already occurred, consistent with projections, raise anxiety rather than reassure. For example, in 2012, James Hansen, an American climate scientist, and his colleagues compared seasonal (three-month) temperatures during 1980–2010 with those during 1951–1980. The global average temperature went up about 0.5° to 0.6°C (about 1°F). Not very much, at least on the surface! Unfortunately, even these small global averages had apparently increased the areas with hot summers. In 1951–1980, hot summers covered about 33 percent of the area. In 1980–2010, these hot summers covered about 75 percent. At the same time, areas having a "cold" summer decreased from 33 to 10 percent.

More important, "extremely hot" summers emerged as a new category. They covered at most a few tenths of 1 percent of the land in the early period, but now they occupy about 10 percent of the land area each summer. The extent of land area now covered by very warm summers also surpasses the very hot years of the 1930s, a period of severe heat and drought in the United States. Some of those years had temperature extremes as high as those observed recently, but the maximum area covered by an "extremely hot" summer never exceeded 2.7 percent, compared to the recent average of about 10 percent and some years as much as 20 percent.[36]

These extremely hot summers already show evidence of clobbering agricultural yields over large areas, enough to seriously impair food reserves of entire nations, even if to date they have not induced actual famines. Heat waves in Western Europe in 2003 and in central Russia in 2010 were extraordinary by any standard, and Hansen's group argues that such events almost surely derived from human-induced climate change.

In 2003, Western Europe experienced a pronounced heat wave. Temperatures soared far higher than normal in June, July, and August. Highest temperatures arrived in the first two weeks of August and ranged from 35°C to over 40°C (95°–104° F). Drought also affected the area as 300 mm (about 12 inches) of "normal" rain failed to fall.[37]

Scientific analysis of this heat wave demonstrated that the average temperatures observed for the three months were exceptionally unusual. The summer of 2003 was probably the hottest in Europe since, at latest, 1500.[38] The unusualness of this summer was so extreme that it was virtually impossible in the climate of pre-1980.[39]

High heat caused excess mortality of perhaps 35,000 people. Extensive wildfires roared through the dry, hot landscape, particularly in southern Europe. In Portugal over 5 percent of the forested land burned. The Danube, Po, Rhine, and Loire Rivers ran low, which interfered with navigation, irrigation, and the operation of electric generating plants.[40]

Mortality, fires, and impaired river services stretched the abilities of governments to respond, but in a subtler way the heatwave's impact on agriculture posed the largest long-term risks. Wheat and maize, two major European crops, both suffered significant yield losses. Overall national yields of wheat in France and maize in Italy both dropped by about 21 percent compared to 2002.[41]

Wheat suffered somewhat less, because at the time the heatwave began winter wheat had mostly finished it growth and flowering. Maize was in the middle of growth and flowering, so its productivity suffered more from the excessive temperatures. In the areas most affected by the heat, productivity of maize dropped 36 percent, for example, in the Po Valley of northern Italy. Yields in Western Europe declined mostly due to the excessive heat, while in Eastern Europe (Ukraine and Romania) they declined mostly due to drought.[42]

The summer of 2010 also proved to be exceptionally warm in Europe, but in that year the center of the excessive heat lay over central Russia. Temperatures in Moscow in July broke the previous record by 2.5°C.[43] They reached 38.2°C (100.8°F), and nighttime lows in the region were also high (25°C, or 77°F, in Kiev, Ukraine). A close examination of exceptionally high daily maximum temperatures over all of Europe during the entire summer indicated that for one-week and two-week periods, the average maximum temperatures in central Russia exceeded the average for 1970–99 by 13.3°C (23.9°F) and 12.3°C (22.1°F), respectively. For the statistically minded, these average maxima were 4.4 and 5.2 standard deviations, respectively, above the long-term average.[44] In simple English, this means that these temperatures were extremely unlikely to occur by chance alone, unless the climate had changed.

Climate scientists believed that the 2003 and 2010 events in Europe presaged the weather patterns that would become more common in the years ahead, based on the warming trends already visible from increasing levels of greenhouse gases.[45] A group of scientists at the National Oceanographic and Atmospheric Administration argued that the 2010 heatwave in Russia was not entirely unprecedented and that this episode was not necessarily by itself a signal of climate change. Nevertheless, their model projections indicated that 2010 was the type of event that would move from very rare to quite common in the years ahead.[46]

Physically, therefore, the summer of 2010 displayed highly unusual features. Did it affect Russia? Without a doubt, it did. The heat caused an estimated 55,000 excess deaths, caused crop failures of 25 percent, aggravated fires over one million hectares, and led to economic losses of $15 billion, one percent of Russia's GDP.[47]

Both yield and total production of grains took serious hits. Yields of wheat, the most important grain, for example, dropped to 78 percent of yields in 2008. Barley, the second most important grain, yielded only 69 percent in 2010 compared to 2008. Rye, a minor grain, yielded only 57 percent of what it had in 2008. Total grain production dropped from 99.8 million tons in 2008 to 56 million tons in 2010. The land harvested was only 73 percent of the harvested area in 2008.[48]

Loss of yield, based on the raw numbers, looks serious. Compared to 2008, the grain harvest in 2010 lost about 44 million tons. To put this large number in perspective, Russia consumes about 77 million tons per year of grain for people and livestock, so the losses from the heatwave represented three-fifths of the country's normal consumption. Russia did not suffer a famine due to the heatwave, but the losses surprised and significantly disrupted the government's economic planning. Since the dissolution of the USSR in 1991, Russia has revamped its agriculture to increase production, decrease imports, and increase exports of grain, plus increase livestock production. At the time of the heatwave, the country had good reserves of grain. Thus the losses, viewed at the national level, appeared largely as a 30 percent drop in exports and a 65 percent drop in reserves from 2008–9 to 2010–11.[49]

These examples, however worrisome they may be, may or may not indicate the ultimate global impacts of climate change. Moreover, physical events, the easiest to predict, will in turn have economic and cultural consequences, which are more difficult to predict and subject to mitigation by adaptation. The exceptionally thorough studies of the International Panel on Climate Change and these detailed case studies of Europe, however, suggest potentially serious, adverse effects on human societies and economies.

In other words, climate science predicts *risk*, the probability—but not 100 percent certainty—of harm. Humanity thus faces the task—now—of deciding whether the predictions and their consequences look acceptable, tolerable but not ideal, or unacceptably damaging. The inevitable follow-up task is deciding what is to be done. Something? A little? A great deal? Nothing? Does the best course of action lie in accepting the most thorough scientific studies? Or should we bet that things won't really turn out too bad, a bet that has no significant support from the scientific community?

The urgency of decision making about climate change leaps from some simple numbers in the latest report of the IPCC. By 2011, humanity had already released about 515 GtC (gigatons of carbon; giga = billion, or 10^9) and was adding about 10.4 GtC per year in 2011,[50] over 90 percent of which (9.5 GtC per year) from burning fossil fuels. In order to have a good chance (greater than 2 out of 3), of keeping global average temperature rise less than 2°C, the total amount of carbon released into the atmosphere cannot exceed 1,000 GtC.[51]

Put simply, humanity has already emitted over 50 percent of the "carbon budget" required to have a good chance of minimizing rising temperatures. After 2011, humanity could emit an additional 485 GtC (1,000 − 515 = 485), and human activities were doing so at about 10.4 GtC per year, a rate that, if held steady, would total 485 GtC in forty-seven years. It is projected that 2058 will be the year that the amount of carbon released into the atmosphere will exceed the budget intended to have a good chance of keeping temperature increases below 2°C.

It's important to note that keeping the global average temperature increase below 2°C is not a guarantee of safety. This target figure was merely a rough estimate that increases beyond 2°C were likely to create climate effects increasingly detrimental to human welfare. The increase of average global temperatures had reached 0.85°C by 2012,[52] but the discussion above indicated that societies have already been affected adversely by this rise, which is less than 2°C.

Despite strong scientific consensus about the seriousness of climate change and the urgency for reducing carbon emissions, the United States has since the 1980s experienced an ongoing argument between (a) the majority of scientists and their professional associations and (b) a small minority of scientists plus a host of political, business, and media leaders. The latter group has questioned the existence of climate change; the role of natural climate variability compared to human-caused variations; the rate, magnitude, and seriousness of change; and the feasibility of successful mitigation actions.

This skepticism in its various forms has little relationship to an understanding of climate science; instead it rests on a political ideology that is opposed to regulations limiting the uses of fossil fuels. This climate change countermovement has resisted and blocked essentially all proposals to regulate these energy sources or reduce their uses in the United States.[53] The disputes have acquired a distinctly partisan flavor, with Republicans generally opposed to the conclusions of scientists and Democrats generally supportive and a few from each party breaking ranks with their respective majorities. As a result, an alleged controversy about the science, a controversy that has not existed in the scientific community for nearly a generation, has crowded public debate with political disagreements masquerading as scientific ones.[54] Resolutions of these debates could be assisted by better and more comprehensive risk analyses, better systematic decision making, and improved models for educating the public about the risks of climate change.[55]

CONCLUSION

Scientific evidence, based on multiple methods and lines of reasoning, indicates that human activities have already increased the global average temperature above what it would have been absent those activities. Models that correctly identify the human role in past temperature increases predict serious consequences from continued climate change if human activities continue to inject carbon dioxide, methane, and other greenhouse gases into the atmosphere. Consequences of climate change have already occurred, and the "new climate" shows many signs of being detrimental and virtually no sign of being benign.

Addressing climate change requires embracing a strategy of rebuilding the energy systems powering the world, because carbon emissions have caused the most change and most carbon emissions come from fossil fuels. As explored in chapter 7, climate change is only one of four factors supporting development of a clear strategy to change energy. Geopolitical tension, health effects, and depletion of easily procured resources also indicate the need to change. Together chapters 6 and 7 support this book's strategic challenge: to phase out the Third Energy Transition in favor of the Fourth, to which we turn in chapter 8.

Connections with Everyday Life

Climate change is the most prominent factor driving the need to change primary energy sources, but it is not the only one. Dependence on the big-four fuels also contributes to geopolitical tensions and health and environmental effects. Depletion of these fuels threatens to undermine economies and societies dependent on them.

Wars and concerns about possible wars affect everybody, especially those individuals and families participating in the armed services. Only a few national leaders (those of the worst sort) would even suggest that a war's objectives might include controlling the petroleum reserves of another country, but it's difficult to explain some wars without recognizing this reality. The use of nuclear weapons has so far been confined to just one war, but the threat of widespread nuclear confrontation throws a constant pall over the world. Unfortunately, it is impossible to separate nuclear power from nuclear weapons.

Stories about health and environmental effects of the big-four fuels, like air and water pollution or mining impacts, are reported in newspapers and on television daily. Many people reading this book will have experienced these impacts firsthand.

Depletion of fuel resources reaches everyday life in the muted form of higher prices or the health and environmental impacts resulting from mining activities in difficult places. Those who live near ocean shores, for example, may know of coastal oil spills from offshore oil rigs. What is not often recognized is that oil companies would not choose to drill offshore if good resources on land existed. Depletion thus lies behind some pollution events.

This chapter highlights the seriousness of these three factors, which add to the risks posed by climate change.

CHAPTER 7

Geopolitical Tensions, Health and Environmental Effects, and Depletion

Three risks in addition to climate change plague all big-four fuels: geopolitical tensions, health and environmental effects, and depletion. Oil and uranium figure prominently in geopolitical tensions. All four pose health and environmental hazards. Finally, many of the most easily obtained sources of oil and gas have already been depleted; in some areas, depletion also has affected coal and uranium. Abundant physical supplies remain buried in the earth, but extracting some of them has become more expensive and increasingly risky.

The challenges of dealing with these four hazards entail trade-offs between benefits and risks. For climate change, risks threaten after 2050 to increasingly overwhelm benefits globally. For the other three risks, damaging changes occur primarily at the local and regional levels. The regional and local risks sometimes reinforce each other and sometimes interact with climate change.

GEOPOLITICAL TENSIONS

Two of the big-four fuels, oil and uranium, have provoked the most geopolitical tensions, but their respective pathways to unrest and—in some cases—war differ substantially. Disparity between the geographic locations of petroleum deposits and the locations where they are consumed has provoked rivalry and war on many occasions. Some nations with higher consumption than production have sought secure access by means of intimidation, imperialism, aggression, and war.

For uranium, tensions arose because nuclear fission can make both horrendous bombs and electricity. Mastery of fission technology supported both weapons and electricity, so uranium fuel inevitably spawned geopolitical tensions.

Oil and Geopolitical Tensions

Coal, not oil, prompted the first concerns about an energy resource. William Stanley Jevons (1835–82) argued in *The Coal Question* (1865) that the exponential rise in English coal consumption after 1700 could not last forever. As British mines delved deeper into the earth, production costs would rise, and the advantage then held by Britain would dissipate. For Jevons, perhaps the first strategic thinker about energy, the consequences of inevitable decline had to guide national political discussions.[1]

Nevertheless, between 1910 and 1912 oil replaced coal in the strategic arena as the United States and the United Kingdom both switched from coal to oil to power naval warships. Oil had about 50 percent more potential heat energy than coal on a kilogram-for-kilogram basis, and it outperformed coal technically.

By this time the United States already had the world's largest oil production and refining industry, but the change posed special challenges to Britain. At the time the United Kingdom had no domestic sources of oil, but nevertheless Winston Churchill, then Britain's naval leader, pushed the Royal Navy to make the switch.[2] Secure access to oil thus became a critical necessity. As Sadi Carnot had noted nearly a century earlier, coal and the Royal Navy were the bulwarks of British power. Now the Royal Navy's oil supply had to be secure.

British negotiations with the Persian Empire (now Iran) led to founding of the Anglo-Persian Oil Company in 1909, but Churchill had to persuade Parliament to inject £2.2 million into the company in return for 51 percent of the company's stock.[3] Britain was already the world's leading imperial power, and now through Anglo-Persian the British government had its own oil company to ensure naval supremacy.

World War I (1914–18) came too soon for the switch to oil by the Royal Navy to make a strategic difference in the war's outcome, but in 1918, military and political leaders in Europe, America, and everywhere else knew that future wars would hinge on oil.[4] The "British pattern" of securing access to another country's oil became the model after 1918.

Britain and France dismembered the Ottoman Empire, which had allied with Germany, and reduced it to modern Turkey. The spoils of war included the Turkish Petroleum Company, formed in 1912 by British, German, and Dutch interests.[5] The settlement created the Iraq Petroleum Company from the Turkish Petroleum Company, and France received the German holdings. Exploration confirmed oil in Iraq in 1927, and the carving up by imperial powers of Middle Eastern oil had begun.[6]

Yet more resentments remained in Germany. Not only had the Germans forfeited the Turkish Petroleum Company, but the country had no access to oil, domestic or foreign. Germany had substantial coal, but the war had taught it a very hard lesson. Without access to oil, Germany saw itself trapped by the oil industries of the United States, the United Kingdom, and the USSR.

The Soviets had developed their oil facilities at Baku, the Americans dominated world trade from oil produced on their own territories, and the British and French had locked up the promising oil fields of the Middle East. After Hitler's rise to power, the Nazi government organized the Continental Oil Company, "Konti," in 1941. Konti produced some oil in Romania, but Germany sought big payoffs, the oil areas of Baku and the Middle East.[7]

The outcome, of course, we know. After failing to capture Moscow in late 1941, the German army steadily lost ground in the east and after 1944 in the west. By 1945, Konti and the German Third Reich lay in ruins, reflecting Germany's lack of its own oil and of its failure to gain secure access to another source. The Soviet and Anglo-American companies had dominated oil production, and those three countries destroyed the Nazis.

A similar situation prevailed in the Pacific. Japan lacked its own oil and sought guaranteed access elsewhere. After attacking Pearl Harbor in 1941, Japan seized the Dutch oil fields of what is now Indonesia. Japan endured the same result as Germany. After recovering its naval strength, the United States relentlessly destroyed Japan's effort to replace Dutch imperialism with a Japanese brand. The nuclear weapons that devastated Hiroshima and Nagasaki in 1945 attracted the most attention, but in fact the Japanese had little chance of prevailing once their access to oil ended.

American and British operations in World War II always focused sharply on maintaining oil supplies and denying enemies access to them. After 1943, however, a new chapter on the strategic quest for oil opened. The United States since 1860 had dominated world oil production from domestic deposits. By the early 1940s, however, analysts warned that oil production in the United States might soon be unable to keep pace with the growing use by automobiles and trucks, let alone have the ability to continue supplying foreign markets. At the time, American oil fueled both the American and British war efforts.

In 1943, President Franklin D. Roosevelt announced that the defense of Saudi Arabia was vital to the United States,[8] and to this day no American president has seen the situation in any other light. Why? The answer was oil. Standard Oil of California and the Texas Oil Company (now Chevron and Texaco, respectively, merged in 2002) had found oil in 1938 and formed Aramco.[9] Roosevelt met with the Saudi king in 1945 and assured America's support in return for guaranteed access to Saudi oil.

The strategic importance of oil to the United States shifted in the years after 1945. Until 1943, the country produced more oil each year than it consumed, but from 1944 on, consumption outpaced production. Net exports continued through 1948, but by 1949 imports exceeded exports. The proportion of oil imported compared to consumption steadily increased, and between 1995 and 2011 over half of American oil came from imports.[10] After 2012, U.S. production rose based on hydraulic fracturing, but dependency on significant imports remained.[11]

Geopolitical tensions surrounding oil after 1945, however, stemmed from more than U.S. dependency on foreign supplies. As postwar events unfolded, the United States organized a new coalition to contain the USSR, with Western Europe and East Asia as the front lines but without domestic oil supplies. U.S. domestic production could no longer support exports, so American strategic planners guaranteed access to oil in the Middle East as an integral part of the Cold War strategy. That fateful choice remains in place today, and two major factors stand out.

First, in the United States every president since Roosevelt—Republican and Democrat—has embraced and expanded Roosevelt's stance. Middle Eastern oil remains a vital strategic interest of the United States. Second, the attempt to keep friendly Middle Eastern governments in power rested on the readiness of the United States to use covert intrigue and force and/or military action to achieve its ends.

The overthrow of Mohammed Mosaddegh, prime minister of Iran, in 1953, a U.S.-British maneuver, marked a turning point in the tortured history that had begun with formation of the Anglo-Persian Oil Company in 1909. Iranians resented both continual British intrusion and the meager split of oil revenues. American intervention removed Mosaddegh from power and installed Mohammad Reza Shah Pahlavi (1919–80) as monarch of Iran. American and British officials argued that Mosaddegh had edged closer to the Soviet Union, and the United States would not tolerate Iranian oil fields moving under Soviet influence or control.[12]

From 1953 to 1979, U.S.-Iranian relations remained cordial, but resentment inside Iran remained. Pahlavi fell in the Iranian Revolution of 1979 as Islamists seized control and pushed the United States out. Relations between the two countries have verged close to violence ever since, although a deal to prevent Iran from acquiring nuclear weapons in 2015 may have reduced the possibility of war. Would the United States be so concerned without the oil? This is a speculative question, but there is no better explanation.

The United States launched full-scale military invasions of Middle Eastern countries if events provoked the president sufficiently. In 1990–91, the United States invaded Kuwait to release the country from control by the Iraqi army. Iraqi nationalists had in 1958 overthrown the monarchs installed by the British at the end of World War I. Relations between Iraq and the United States remained tense after 1958, but open warfare did not erupt until 1990 when Saddam Hussein, the Iraqi president, seized Kuwait and its oil in August. An American-led coalition forced Iraq from Kuwait.

In 2003, the United States again went to war against Saddam Hussein, this time over allegations that Iraq was developing nuclear and chemical weapons and that it had supported the attacks on the United States in September 2001. American-led forces conquered Iraq, a war that still provokes debate, and it's impossible to say the war has actually ended. As with Iran, it's difficult to think the United States would see Iraq as a target without the oil.

Countries other than the United States have also demonstrated the strategic dimensions of oil and gas. Those countries lacking deposits of these fuels are vulnerable to disruptions or may use their militaries against others. Consider two examples: Ukraine and China.

Ukraine declared its independence from the USSR in 1991, a blow that tipped the Soviet Union into dissolution. For the first fifteen years of Ukrainian independence, both the Ukrainian and Russian economies slowly recovered, and relatively little tension passed between the two. In 2006 and then again in 2009, however, tensions arose over the dependency of Ukraine on gas imports from the Russian Federation. Price disputes led Gazprom, the Russian company, to shut off supplies to Ukraine, in January, and Russia forced terms favorable to Russia on Ukraine.

Further dissension arose between Russia and Ukraine when Ukrainian unrest forced President Yanukovych from office in 2014. Many

Ukrainians wanted to move closer to the European Union and away from entanglement with the Russian Federation, but Yanukovych had moved closer to Russia. After Yanukovych fled, Russia annexed Crimea, a part of Ukraine, and aided Ukrainian separatists in eastern Ukraine. This still-active war between Russia and Ukraine involves military actions as well as Russia's political use of its oil and gas exports to Ukraine. Without self-sufficiency in petroleum and gas, Ukraine's ability to leave the Russian sphere of influence remains compromised.

China poses a different situation. For years after the victory of the Communists in 1949, China's economy grew very slowly, and not many people had automobiles. Hence China's need for oil remained minimal. The reforms of Deng Xiaoping after 1980 transformed China and created a middle class with the ability to purchase automobiles. As the number of autos grew, so, too, did the need for oil. By 2006, China, like the United States, imported over half its oil from abroad.[13]

China thus emerged as a parallel to the United States: rich, with foreign exchange to buy oil, a powerful military with nuclear weapons, and dependent on foreign oil to keep its economic machine humming. The question that now faces the world centers on possible competition between the United States and China for other countries' oil. So far both countries have procured their needs, and conflict over oil has not arisen. Nevertheless, the potential for conflict remains between these two heavily armed countries.

Uranium and Geopolitical Tensions

Geopolitical tensions around uranium began the instant the United States destroyed Hiroshima and Nagasaki in 1945. The reality of nuclear weapons held by one country immediately raised the desire of other countries to have them, and by 1964, the USSR, the United Kingdom, France, and China had joined the nuclear weapons club. The United States and the USSR built the largest inventories, and the tensions of the Cold War continually threatened to erupt into a nuclear holocaust. President Dwight D. Eisenhower tried to reorient uranium toward electricity production in 1953, but it has always been impossible to isolate civilian nuclear power from nuclear weapons.[14]

Four sources of geopolitical tensions arise from uranium used to generate electricity. First, countries that consume uranium fuel may not have good deposits of uranium ore, and foreign supplies may not be secure. Second, technology for enriching uranium fuel's content of fissionable ^{235}U for electric power is essentially identical to technology for enriching uranium for weapons. Third, as ^{235}U fissions to generate electricity, the ^{238}U present in the fuel absorbs a neutron to yield ^{239}Pu, useful for weapons. And fourth, serious accidents at nuclear reactors have dispersed radioactive debris across international borders; as a corollary, every nuclear reactor has always been a target for enemies wishing to unleash radioactive debris as a weapon.

The United States, the largest generator of nuclear electricity, illustrates the first point. Even though the United States holds 4 percent of the world's uranium reserves and produces 3 percent of the world's uranium fuel, the country imports the vast majority of its fuel for nuclear power. Between 1994 and 2014, the United States always imported at least 74 percent of its uranium fuel, and in some years the total imported exceeded 92 percent. Australia, Canada, Kazakhstan, and Russia were the main suppliers.[15] For the moment, reliance on foreign markets for uranium seems stable and reliable, but dependence on energy sources outside a country's boundaries carries the risk of lost access.

More serious tensions have arisen from the second factor, devices for enriching the ^{235}U content from about 0.7 percent in uranium ore to about 3 to 5 percent for nuclear reactors and to about 90 percent for weapons. Enrichment in the Manhattan Project used gaseous diffusion plants, requiring enormous amounts of electricity, and U.S. analysts believed in the 1940s that the expense of gaseous diffusion would prevent other countries from making nuclear weapons. That assessment proved incorrect: by 1949, the USSR had demonstrated its ability to build a nuclear weapon, based on enrichment with gaseous diffusion.[16] In the process of building an atomic bomb, however, the USSR also advanced a new technology for enriching uranium, based on ultracentrifuges, which required far less capital and electricity to operate.[17]

With either method, the ability to enrich uranium opened the door to both nuclear power and nuclear weapons. Ultracentrifuges made the process vastly easier and cheaper, and that made any country's plan to operate nuclear power plants with its own enrichment operation simultaneously a potential for building nuclear weapons. That potential sparked geopolitical tensions between, for example, the United States and Iran. Both Republican and Democratic administrations have decried the dangers of Iran possessing a nuclear weapon since 2003.[18]

Enrichment of uranium to have 90 percent or more ^{235}U, however, is only one pathway to a nuclear weapon. Fission of ^{235}U releases more neutrons, some of which turn ^{238}U into ^{239}Pu, which, if separated, can make a weapon. During the 1950s through 1970s, India, for example, produced plutonium-based bombs from the waste products of electricity production.[19] China already had nuclear weapons, and Pakistan acquired them in response to India's tests. South Asia remains an area of severe geopolitical tension and abundant nuclear weapons.

Finally, many modern nuclear reactors have about 75 metric tons of low-enriched uranium fuel inside. Operators replace about one-third, or 25 metric tons, of this fuel every year to eighteen months.[20] This fuel, generally uranium dioxide (UO_2), is modestly radioactive when first inserted in the reactor, but at the time of its removal the fuel is intensely radioactive. A person standing one meter from this used fuel would receive a lethal dose of radiation in less than a minute.[21] As a consequence, a typical reactor for generating electricity at any given moment contains many tons of radioactive materials that must have secure containment and heavy shielding to protect human health and the health of other species.

Therein lies a potent source of geopolitical tension. Well-known accidents at Three Mile Island (1979), Chernobyl (1986), and Fukushima (2011) released large amounts of radioactive debris, with health consequences for surrounding populations, as well as others. Consequently, every operating commercial nuclear reactor is a potential threat to people and economies downwind should the containment and shielding fail, for any reason. Accidents, sabotage, and terrorist attacks on the reactor can unleash debris in one country with severe consequences in another. People in one country near a reactor in another country may draw little or no benefit and still be at risk.

Taken together, these four factors make uranium's use as a fuel a source of geopolitical tension. Proliferation of nuclear weapons, and consequently increased probability of use, poses the most catastrophic threats. As with conflicts generated over access to oil, the tensions sown by reliance on uranium are intrinsic to the fuel's uses.

HEALTH AND ENVIRONMENTAL ISSUES AS "EXTERNALITIES"

The Third Energy Transition improved human health by reducing the costs of clean water, food, shelter, and medical care. At the same time, the extraction, transport, and use of the big-four fuels, plus disposal of waste products, have caused significant health problems and environmental degradation, mostly at the local and regional levels. Assessing the balances between positive and negative effects, however, remains challenging.

In the language of economists, negative effects are *externalities*, meaning that they lie outside the market transactions of using the big-four fuels. If people deciding whether or not to use the fuels don't pay the costs of the negative effects, then those costs don't affect the decision. In other words, the fuel appears to cost less and provide more benefits than it really does. The fuel may still have a surplus of benefits over costs, but inequities in bearing externalities may create situations in which most benefits accrue to users of a fuel while external costs afflict people who receive little or no benefit. If so, the externality has also become an environmental injustice.

For each fuel, externalities arise in extraction, refining, manufacturing, and transport and in disposal of wastes. Externalities often remain invisible—except to the victim—but in a few cases the mass media publicize major events, such as an explosion killing many coal miners or radioactive contamination of a wide area from a nuclear power plant. Invisibility of effects stems from their technical nature, their local scale, sometimes slow manifestation, and the victims' lack of wealth and power. Many situations involving externalities remain unexamined and uncorrected.

All of the big-four fuels have generated health and environmental externalities, but this book illustrates health and environmental externalities associated with coal in the United States (health effects summarized in Table 7.1). The two largest coal-mining areas (central Appalachia and Wyoming) and the widespread effects of coal combustion for electricity provide numerous examples.

Coal extraction methods have changed greatly since its commercial production became significant starting about five hundred years ago in England. Today, two basic methods of extraction pull coal from beneath the earth's surface.

Underground mining involves digging tunnels to reach coal seams, where miners break the coal from the seams for delivery to the surface. Surface mining involves scraping the vegetation, soil, and rocks away to extract the coal. Mountaintop removal is an extreme form of surface mining in which explosives and machines expose coal seams by removing whole mountaintops and dumping the "overburden" or "spoils" into the valleys below, obliterating streams and other nearby ecosystems.

The destruction resulting from mountaintop removal challenges comprehension, because of its scale, which a picture can only partially capture. Figure 7.1 presents before and after views of a landscape at the Hobet Mine in West Virginia. Figure 7.2 presents a closer aerial view.

TABLE 7.1.   EXAMPLES OF MAJOR HEALTH EFFECTS OF COAL

Activity	Health and Environmental Effects
Extraction of raw fuel	Deaths in U.S. mines from 18 to 48 per year, 2000–2013;[a] over 100,000 since 1900. Pneumoconiosis—over 10,000 deaths in 1990s in U.S. Increased rates of lung cancer, coke oven workers, Pittsburgh. Over 500 sites of mountaintop removal in KY, WV, VA, TN, altering 1.4 million acres: landslides, floods, damage to houses, loss of property values. Methane (CH_4) released. Mortality from lung cancer, heart, respiratory, and kidney diseases higher in mining areas. Risk of low birth weight higher in mining areas.
Transport of raw fuel	Road hazards and dust in mining areas.
Manufacture and refinement to ready-to-use fuel	Coal washed to remove clay, other rock, heavy metals. Of chemicals used or released, 19 known carcinogens, 24 linked to lung and heart damage.
Transport of ready-to-use fuel	Rail traffic is 70 percent coal with 246 fatalities due to coal in 2007.
Fuel use	Release of $PM_{2.5}$, NO_x, and SO_2 linked to respiratory and heart illnesses, low birth weights, infant deaths. Release of 48 tons per year of mercury, over 50 percent of total in U.S.
Disposal of waste at any step in fuel progression	Hundreds of impoundment ponds at mines, processing plants, and combustion sites; 53 publicized spills 1972–2008, Appalachian region; water pollution from spills. Over 1,300 fly ash impoundment sites in U.S., most poorly constructed. Abandoned mine lands (1,700 in PA alone) create mine fires and subsidence or cave-ins.

[a]United States Department of Labor, "Coal Fatalities for 1900 through 2014," www.msha.gov/stats /centurystats/coalstats.asp., May 14, 2015.

SOURCE: Unless otherwise specified, all data on coal from Paul R. Epstein et al., "Full Cost Accounting for the Life Cycle of Coal," *Annals of the New York Academy of Sciences* 1219 (2011): 73–98.

In general, surface mining is safer than underground mining for workers. Miners quickly learned the dangers of underground mining in England in the 1700s, when coal mining was still small industry, just starting to expand. All told, over 164,000 died in mines between 1700 and 2000. In the deadliest period, 1900–1950, over 84,000 died.[22]

West Virginia had been a coal-mining area before statehood in 1863, and the new state produced 445,000 tons in its first year. West Virginia's continued status as a major coal producer also means it has dramatic experience with externalities. Its mines have yielded 14 billion tons of coal since 1863 at a cost of 21,170 fatalities and 485,000 injuries. During this time, an average of 1.5 people died to produce every million tons

(a)

(b)

FIGURE 7.1. Mountaintop removal mining, before and after: Hobet Mine, near Madison, West Virginia.

Photo (a) taken June 7, 2005; (b) taken October 21, 2006. Note farmhouse and clearing at bottom left in top picture; same farmhouse and clearing at center right in bottom picture. For a series of time-lapse satellite images of the growth of the Hobet Mine between 1984 and 2015, see http://earthobservatory.nasa.gov/Features/WorldOfChange /hobet.php.

SOURCE: Vivian Stockman, Ohio Valley Environmental Coalition, used by permission. http://ohvec .org/before-and-after-hobet-mtr-operation-on-mud-river-lincoln-county-wv/, April 4, 2016.

FIGURE 7.2. Mountaintop removal coal mine. Close-up view of the Hobet Mine, with a drag-line shovel operating at upper middle.

SOURCE: Vivian Stockman, Ohio Valley Environmental Coalition, used by permission. http://ohvec .org/hobet-mountaintop-removal-operation-continues-expansion/, May 28, 2016.

of coal. The year of the most deaths, 1925, saw 686 fatalities in a workforce of almost 112,000, a death rate of about 1.2 fatalities for every 200 workers during that year. During this deadly year, the fatality rate approximated 1.5 deaths for every thousand tons of coal produced.[23]

Nationally in the United States, the year of highest fatalities was 1907, with 3,242 deaths, an average of nearly 9 deaths per day. Federal laws regulating safety in mines were enacted first in 1891, but serious declines in fatalities appeared only after other laws were enacted in the second half of the 1900s.[24]

In 2011 in the United States, about 143,000 miners produced slightly over one billion tons of coal (32 percent underground mining and 68 percent surface) with 21 fatalities. Coal mining remains a hazardous occupation, but it now has a vastly lower accidental death rate than in the first half of the 1900s. In 2013, 6 coal-mining fatalities happened in West Virginia, 5 of them in underground mining; one death occurred at a coal transport site.[25]

Serious accidents kill and injure to cause *acute* morbidity and mortality. Workers also face *chronic* hazards from inhalation of dust from coal and other rocks. Underground miners usually face the highest chronic dust hazards, but workers in other locations in and around mines also suffer exposure. These dust particles cause pneumoconiosis (black lung disease) and silicosis, lung diseases that injure and kill slowly.

Pneumoconiosis begins with shortness of breath and a chronic cough. This stage of the disease, called coal workers' pneumoconiosis, in its milder forms may not even be recognized unless the worker has chest X-rays to see if aggregates of coal dust have accumulated in the lungs. Further exposure can lead to progressive massive fibrosis, scarring and stiffening of the lung tissue as a result of inflammation. At this stage, the person will have severe shortness of breath, a chronic cough, poor lung function, high blood pressure, and heart problems.[26] The severer symptoms of pneumoconiosis disable the worker, rob him of the pleasures of daily life, and may lead to premature death.

In some areas of the United States before 1969, over 40 percent of miners who had worked thirty or more years had coal workers' pneumoconiosis. In 1969, Congress established the Coal Workers Health Surveillance Program, offering free periodic X-rays to detect early stages of the disease and permitting transfer of workers to low-dust areas of the mining operation. Between 1970 and 1974, the overall number of miners with pneumoconiosis was 11 percent, but with the new safety law the numbers with the disease dropped to 2 percent in 1995–99. Overall average dust levels dropped from 6 milligrams per cubic meter (mg/m^3) to usually less than 1 mg/m^3, in compliance with the new limit of 2 mg/m^3. Unfortunately, in 2005–6 the number of miners with the disease began to rise again, to over 3 percent. The excess cases of disease came from central Appalachia: parts of Kentucky, Virginia, and West Virginia. These regional differences suggested that more cutting of rock to obtain the coal plus working in smaller mines may have caused the increases.[27]

Progressive massive fibrosis, the most debilitating stage of pneumoconiosis, saw a resurgence in this region, to levels similar to those in 1970. This serious disease had almost disappeared by the late 1990s among miners with over twenty-five years of experience, but it reached over 3 percent of underground miners in the region by 2012.[28] Some surface miners, too, despite less confinement in dusty work settings, suffer pneumoconiosis, including its severest stage, progressive massive fibrosis.[29]

Factors other than mining, like smoking, can also cause lung diseases, but without doubt coal mining creates health hazards, both acute and chronic in both underground and surface mines. But do these health hazards constitute externalities, that is, costs that do not affect economic decisions to use coal? Economic studies from the National Research Council did not include morbidity and mortality among active miners as "externalities." Instead, the report argued that the hazardous nature of coal mining created demands for higher pay, which affected the cost of coal and integrated the cost of health hazards into decision making about the use of coal.[30] We won't try to resolve whether the pay was high enough or whether high pay alone makes use of coal economically rational. Even if economic factors make use of coal rational, important moral questions are intrinsic to the use of this fuel.

In addition to health effects on miners, considerable evidence suggests that people living near coal mines also incur health problems attributable to coal mining. Between 2008 and 2015, a series of papers focused on the central Appalachian area (Kentucky, Tennessee, Virginia, and West Virginia) found the following health effects associated with coal mining:

- increased lung cancer mortality in areas near surface and underground mining;[31]
- increased mortality in men and women from heart, respiratory, and kidney diseases in counties with coal mining;[32]
- mortality from lung cancer and "other" cancers was more highly associated with proximity to coal industry facilities and activities than to tonnage of coal mined in a county;[33]
- birth defects occurred at higher rates in areas with mountaintop removal mining;[34]
- lower birth weights occurred at higher rates in coal-mining areas of West Virginia;[35] and
- Virginia, an area where mountaintop removal was practiced, had elevated symptoms of respiratory disease compared to an area without mountaintop removal.[36]

Studies of communities living near coal mining intrinsically have a much harder time showing effects of mining, because community health statistics cannot match individuals with a health problem to exposure of those individuals to mining. The research mentioned above clearly recognized this weakness, and other researchers have emphasized this problem in research that found no significant association of mortality rates with coal mining.[37] These researchers focus instead on conditions associated with poverty as the most important factors in disease and mortality rates, which they acknowledge are higher in the areas studied. A vigorous debate surrounds these contested research findings.[38]

Human health effects dominate concerns about coal mining, but environmental impacts, especially from surface mining, increased in the decades after World War II. Surface mines in the United States produced 25 percent of the coal in 1949, 51 percent in 1971, and 68 percent in 2011. The switch resulted from the convergence of several factors in mining technology, the properties of coal from different areas, environmental regulations, and changes in rail transport costs.

Consider the rise of Wyoming as a major coal state. It had an early peak in coal production, mostly from underground mines, in 1945, but it produced then only 1.6 percent of all U.S. coal. Wyoming coal production sagged after 1945 as railroads switched from steam to diesel locomotives. Production of coal began to pick up a bit after 1958 as Wyoming's electricity production rose, but it was after 1969 that a totally new era began.

Surface mining replaced underground mining, and in 1980 the Staggers Rail Act deregulated freight rates on American railroads. Perhaps most important, however, in 1970 changes in the Clean Air Act clamped down on emissions of SO_2 to curtail acid rainfall and health effects.[39] Wyoming's coal was low in sulfur compared to eastern coal (0.35 percent vs. 1.59 percent in Kentucky). On the downside, it averages only about 8,600 Btu per pound compared to 12,000 Btu per pound in eastern coal.[40] Thus more tonnage of Wyoming coal than eastern coal must be mined and burned to produce the same amount of electricity.

Relatively inexpensive surface mining and cheaper rail transport of coal propelled low-sulfur Wyoming coal, particularly from the Powder River Basin, into the U.S. electrical generating industry (figure 7.3). The state now ranks first in the United States in coal production (2012), ahead of West Virginia, Kentucky, Pennsylvania, and Illinois. Overall the state produced 401 million tons of coal that year, approximately 40 percent of all U.S. coal.[41] Two Wyoming mines alone, North Antelope Rochelle and Black Thunder, together produced about 200 million tons, or 20 percent, of the U.S. total (figure 7.4). North Antelope Rochelle by itself nearly equaled the entire production of West Virginia, the state with the next highest production levels after Wyoming.[42]

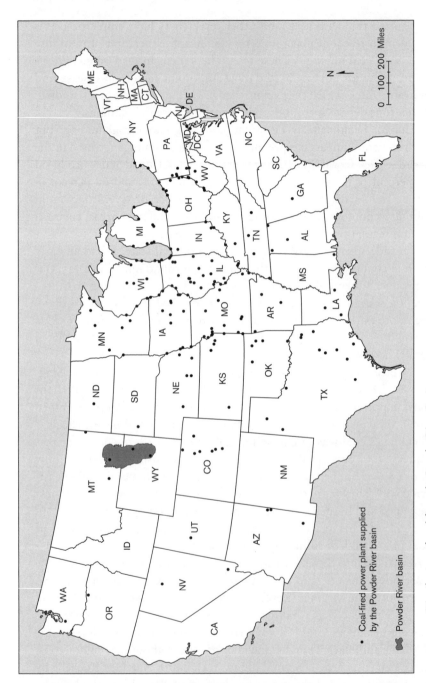

FIGURE 7.3. Distribution of coal from the Powder River Basin.

SOURCE: Kurt Menke, Birds Eye View GIS, used by permission.

FIGURE 7.4. Production technicians with a coal-haul truck at the North Antelope Rochelle coal mine, Wyoming.

SOURCE: Peabody Energy, Inc. Used by permission, https://commons.wikimedia.org/wiki/File:Miners_and_Haul_Truck.png, June 21, 2016.

The Wyoming coal story illustrates the convoluted relationships between coal and its health and environmental impacts. SO_2 from burning eastern coal has deleterious effects on health and the environment. Regulation of this pollutant proved a major factor in opening the door to Wyoming coal produced by surface mining. Control of SO_2 has been a definite success story in environmental protection, but solving the SO_2 problem with Wyoming coal created a new set of problems that offset the reduction of air pollution in the Midwest and in the eastern United States.

These problems center on whether the surface-mined areas of the western United States can be or have been successfully reclaimed. One report from the Natural Resources Defense Council and the Western Organization of Resource Councils definitively said no, exemplified by the failure, as of 2007, of essentially all Wyoming surface-mined lands to achieve the standard of fully reclaimed land demanded by the Surface Mining Control and Reclamation Act of 1977.[43]

Mountaintop removal, a highly disruptive method of surface mining, has in the past few decades increasingly affected the central Appalachian areas of Kentucky, Virginia, and West Virginia. Despite winning approval of federal and state regulators, serious questions surround whether such lands can be reclaimed at all, similar to the questions surrounding the surface mines in the arid West. Critics paint a picture of utter ecological destruction caused by the practice, with accompanying loss of habitat, biodiversity, and ecosystem functions.[44] A casual inspection of satellite images on Google maps clearly shows these mines and their complete alteration of the original Appalachian landscapes.

Combustion of coal adds to the hazards of using this mineral source of energy. When burned, coal releases carbon dioxide, methane, particulates, oxides of nitrogen and sulfur, mercury, other heavy metals, and numerous carcinogenic chemicals. The type of coal used, combustion technology, and pollution control technology can substantially affect the air pollution resulting from each of these by-products. As with attempts to measure the effects of coal mining on local communities, the effects of these pollutants from combustion become visible only with careful health and environmental studies. The complexity of the studies typically generates debate and controversy similar to that found in looking for health effects from mining.

Despite the challenges, multiple studies based on different methods indicate that a coal-fired electrical generating plant creates a health hazard. Some of the dangers fall on those relatively close to the facility, but some pollutants travel far and endanger a larger region or the entire world. The literature is vast, so here I briefly indicate only a few of the relevant dangers from a recent review.[45]

Most attention falls on emissions of particulate matter smaller than 2.5 microns ($PM_{2.5}$), nitrogen oxides (NO_x), sulfur dioxide (SO_2), carbon dioxide (CO_2), and mercury (Hg). NO_x and SO_2 can also increase particulate concentrations in reactions that occur after exhaust comes from the power plant.

Exposure to $PM_{2.5}$ is associated with mortality from all causes, including cardiovascular and cardiopulmonary as well as respiratory diseases, hospitalizations, and reduced lung function. Most of the dangers from $PM_{2.5}$ happen within a year of exposure, and it may be that no "safe" level of $PM_{2.5}$ exists. Effects arise at concentrations of 10 µg/m^3. As I write this section, $PM_{2.5}$ levels in Beijing, China, stand at about 80 µg/m^3, an amount considered "Unhealthy."[46] $PM_{2.5}$ comes largely from coal-fired power plants. China's immense increases in the use of coal in the past three decades has left many Chinese cities shrouded in a murky, particulate cloud, which sometimes reduces visibility to the point of closing airports and highways.

Studies of specific power plants have found health effects for those living nearby. For example, people living near two older power plants in Massachusetts suffered "70 deaths, tens of thousands of asthma attacks, and hundreds of thousands of upper respiratory illnesses annually."[47]

These debates highlight a recurring problem in assessing the strengths and weaknesses of all sources of energy. Scientific information on acute illnesses and fatalities generally can associate cause and effect. For example, an explosion of methane gas in an underground coal mine kills miners; no argument surrounds the fact that the explosion caused the deaths. For chronic effects, however, scientific investigation faces severe hurdles, because (a) the disease may take a long time to appear after the cause, (b) often no scientist has money to support the needed research to establish cause and effect, and (c) even if money is available, other variables such as smoking, obesity, and general poor health confound and obscure the links between coal and disease.

This conundrum troubles questions about all sources of energy. Each of the big-four fuels suffers from lack of scientific certainty on some health and environmental issues. Acute health effects, such as accidents, pose few problems, but chronic effects take great scientific sophistication to identify and quantify.

DEPLETION OF RESOURCES

Geopolitical tensions and health and environmental problems focus on damages from using fuels, not benefits. In contrast, concerns about depletion look at consequences of not having the fuel and its benefits. All three problems potentially disrupt economies and political systems, even though the mechanisms differ substantially. The first two call for reduction, regulation, or prohibition of fuel use, whereas depletion calls for finding more fuel, removing regulatory hurdles, or finding substitutes.

Depletion of nonrenewable mineral fuels involves geology, technology, and economics, connected by two questions: Does a way exist to extract the fuel from the earth and make it practically useful? If so, does it cost less to extract and prepare it than the fuel's selling price? If the answers are affirmative, technology exists for extraction, and the economic benefits of use outweigh the costs of extraction.

"Depletion" means extraction costs exceed the fuel's sale price or the mineral no longer exists at that place. A depleted fuel can "un-deplete" if new technology produces the fuel at lower cost, if customers pay more, if new ways of using the fuel require less of it to produce benefits, and/or if geologists find a new deposit successfully exploited by existing technology and sales prices.

In short, "running out of" or "depleting" a nonrenewable fuel always involves an interplay among the known geological deposits; the technologies of extraction, preparation, and use; and the costs of extraction, preparation, and use. None of the existing big-four fuels has yet reached the zone of no known geological deposits. Over the years, however, depletion fears have periodically arisen due to extraction technologies becoming ineffective or too costly compared to the prices customers could or would pay, which in turn depended on the customers' costs and benefits of using the fuel and their access to alternatives.

Perhaps the oldest and most famous alarm about the consequences of depletion came from Jevons's 1865 book, *The Coal Question*.[48] Jevons understood that the energy Britain obtained from coal in the mid-1800s far surpassed the energy that could ever conceivably be obtained from firewood harvested from the British Isles (see chapter 1). He also understood that Britain's rise to prosperity and imperial power rested on its development of coal, iron, and steam engines, with the heat of coal driving the entire economy. Jevons feared depletion of coal would ultimately undermine the entire scheme of Britain's power.

Jevons clarified that for him "depletion" did not mean that no coal would remain in geological deposits. Indeed, he saw them as inexhaustible. Instead, he saw the need—based on trends then appearing—to go deeper and deeper into the earth to find seams of coal wide enough to harvest economically. Deeper mines cost far more to dig, ventilate, prevent from exploding, keep dry, and haul coal to the surface. He also saw the enormous coal deposits of the United States, cheaper to mine, drawing the production of iron, steel, and other goods to that country. Britain, he feared and lamented, must decline due to "the probable exhaustion of our coal-mines."

Jevons articulated a deep pessimism about the fate of societies and economies based on nonrenewable resources. When the fuel ran out, as it inevitably would, darkness and poverty would reign again as modern cultures collapsed and reverted to agrarian life. This pessimism has appeared over and over since Jevons's time.

Recognition of decline has sparked an array of remedies, many incompatible with each other. Some say lifestyles must change to lower energy uses. Others tout new technology and/or fewer regulations on extracting the declining fuel. Yet others push for new fuels to substitute for the declining one. Frequently, proposed remedies pay little or no attention to the linked triad geology-technology-economics.

At least three different methods predict the future of the big-four nonrenewable fuels: business decision making, peaking, and energy return on investment. Each method has its partisan adherents vying for the attention of business and political leaders as well as the public.

Business decision making. This method dominates companies producing, refining, and selling energy. These companies survive by knowing their production costs and the selling prices of their products, so money measures factors of most importance to them. Companies recognize and accommodate depletion as a fact of business as they constantly monitor production costs and selling prices. The company may aspire to continue far into the future, but short-term profits and losses determine the ability to survive at all. Business decision making can't pay much attention to depletion in the far future.

For a business the solution to depletion lies in continued exploration for new deposits and in improving extraction technology. Each of the big-four fuels constantly faces depletion, but the oil and gas industries have grappled most seriously with it. The companies have to maintain an optimistic public face: "Not a problem; we're finding new geological deposits, developing new technologies, and our proven reserves hold steady." For economic and strategic reasons, it doesn't matter if private parties or a government owns the company. Declining proven reserves spell doom; the faster the drop, the sooner economic collapse of the company will occur.

Business decision making guides successful enterprises in the energy arena, but the inadequacies of this method quickly become apparent when assessment moves beyond relatively short-term profits and losses. Expansion of the relevant time frame beyond twenty to thirty years alone kills the utility of business decision-making methods. Realistically, no company can hope to predict future conditions of profit and losses beyond this time frame, nor will any sane lenders extend credit for longer periods.

Cultures, societies, and governments, however, must consider longer time frames, because people live longer than twenty to thirty years and they care about their children and grandchildren. Thus relevant time horizons easily stretch to a century and beyond. Similarly, people care about climate change, geopolitical tensions, and health and environmental issues. All of these relevant considerations call for different assessment methods of depletion.

Two newer methods, both developed since the 1950s, try to predict the future of energy supplies in ways that accommodate relevant time frames. In addition, they both supply insights useful to finding workable, effective solutions to climate change, geopolitical tensions, and health and environmental effects. One method estimates "peaking," the time at which the production rates of a nonrenewable fuel begin inexorably to fall due to geologic depletion. The second calculates energy return on investment (EROI), the amount of *energy* it "costs" to produce a fuel. Substantial difficulties attend both methods, but each provides unique insights.

Peaking. Discussions of peaking, in popular and in scientific media, generally refer to oil, but the concept of peaking applies to each of the big-four fuels (coal, oil, gas, uranium). Peaking means identifying the time at which the *extraction rate of the fuel per unit of time* decreases, so long as extraction technology and production costs remain unchanged. For oil, peaking means petroleum still exists physically in the ground, but existing extraction techniques can't pump it out as fast as they used to.

If customers burn the oil at a faster rate than extraction produces it, a crisis arises: some customers wanting gasoline at a specific time won't be able to buy it, and the customer loses mobility. The interruption may be for a short or long time, but it creates a problem. Past availability of oil promoted building entire lifestyles on its continued availability, and interruption threatens the lifestyle.

Serious concerns with peaking come from geologists and others focused on the physical resource, not its human dimensions including technology development and prices. Analyses of exhaustible fuel production by economists and others focused on the human dimension consider peaking theories misguided, inappropriately framed, and poor bases for natural resource policy.

The United States Geological Service (USGS) predicted that between 1909 and 1921 U.S. production of petroleum would peak within decades. The predictive abilities of geologic science, however, vastly underestimated the retrievable petroleum hidden beneath the surface. By 1931, new finds in East Texas and elsewhere had produced a glut in the oil supply, and prices crashed.[49] Nevertheless, pessimism about future oil supplies continued, as noted in the decisions of President Roosevelt to align with Saudi Arabia over worry that U.S. oil would "run out."

Marion King Hubbert (1903–89), a geologist, joined this pessimistic line of thought and developed predictive theories about future rates of oil production. Hubbert now receives credit—and, from some, derision—for developing the methods to predict peaking of oil production rates.[50] Hubbert assumed production rates over time would follow a bell-shaped curve, and the area under the curve equaled the quantity of accessible fuel in the earth. In his early work, however, he could not predict the year of peaking.[51] Later, more refined work predicted that the peak rate of oil production in the United States would arrive between 1965 and 1970,[52] a prediction included in the 1962 report, *Energy Resources*.[53]

Hubbert's predictions, contentious at the time he made them, began receiving more support in 1969–70, as supplies of oil and gas began to fall short and imports and prices began to rise. Sentiment shifted even further toward Hubbert during the Arab oil embargo against the United States in 1973–74. At the time, the United States simply could not produce oil at a faster rate to make up for the lost imports.[54]

By 1977, production statistics clearly suggested that U.S. oil production rates had peaked in 1970 and steadily declined in the next six years.[55] Hubbert's "peak" had appeared on graphs, just as he had predicted, and enthusiasm for peaking theories sparked new efforts to study the apocalyptic consequences of running out of oil and the need to move public policy away from its tight embrace of endless economic growth based on oil and other fossil fuels.[56] Hubbert doubted economists' trust in studying money and their models of unending growth.[57]

Peaking provided an uncertain guide, however, for thinking about nonrenewable resources. Hubbert spent little time on new technology for estimating reserves of oil in deposits and extracting them at a profit. Peak-oil enthusiasts similarly spent more time trying to predict the exact date of the peak rather than the dynamics of oil production driven by geology, technology, and economics.

In the United States, a new extraction technology—hydraulic fracturing—utilized horizontal drilling and fracturing shale deposits containing oil and/or gas with high-pressure water, sand, and chemicals. In 2009, U.S. producers began fracking in North Dakota for shale oil and in Texas, Ohio, West Virginia, and Pennsylvania for gas. U.S. production soared, and total oil production sailed past the levels achieved in 1966, headed for the record "peak" of 1970. The health and environmental effects from fracking stirred controversy, but the clear view of Hubbert's peak fogged over.

Energy return on investment. EROI originated not from geologic science but primarily from ecological science, supplemented with some compatible ideas from anthropology and economics. EROI focused on quite a different question from peaking, one much more akin to the business decision-making methods. Where business followed money-in/money-out in producing fuels, EROI calculated energy-in/energy-out. A simple fundamental idea underlay EROI: if extraction of a fuel requires more energy than the fuel provides, what's the point?

Many contributed to EROI, but Charles A. S. Hall (1943–) played a key role. Hall, an aquatic systems ecologist, studied fish migration and metabolism; he reasoned an animal must obtain more energy from its food than it expends finding, capturing, and consuming it. Without breaking even, the animal starves to death; without a surplus, the animal cannot reproduce or do anything else. In 1981, he began applying this principle to human society's use of energy and coined the term *energy return on investment.*[58]

Hall and his colleagues shared Hubbert's concerns about fuel supplies and, like Hubbert, wanted a thermodynamically informed measure of the efficiency of producing energy for use (see chapter 2). In contrast to Hubbert, however, Hall and colleagues focused on estimating the energy needed to find, extract, and use oil rather than predicting a future date at which production rates would decline. Like Hubbert, Hall disdained economists enthralled with unending economic growth and business leaders focused on profits alone; he believed their analyses missed vitally important features about the future of energy production and rested on a make-believe proposition, namely, that exponential growth of an economy within a finite ecosystem could continue forever.

Two closely related numbers provide insights into the prospects for surplus energy from extracted fuels. "Net energy" is the difference, measured in joules or some other energy unit, between energy obtained and energy expended: $NE = E_{out} - E_{in}$, where NE is net energy, E_{out} is energy obtained, and E_{in} is energy expended. NE will be a positive number if the extraction process produces more energy than the process requires. If NE is positive but very low or if it's negative, questions immediately arise about why bother.

EROI is a simple ratio: $EROI = E_{out}/E_{in}$. If the number is greater than 1, more energy results from the process than the process required. A number much greater than 1 suggests an efficient way to obtain energy. If EROI is low but greater than 1, the number suggests the extraction process may be of questionable worth. If EROI is less than 1, than the "source" is not really a primary source of energy as it requires a subsidy from some other source to obtain it.

Without the raw data (E_{out} and E_{in}), a value for NE indicates the amount of energy obtained but says little about the efficiency of the extraction. A value for EROI without the raw data says little about the amounts of energy extracted but indicates the efficiency of the extraction process. A low EROI, however, indicates that obtaining a large amount of net energy will require a very large input of energy (E_{in}).

Together, NE and EROI provide useful information about a purported fuel's likely usefulness. Nevertheless, calculations of NE and EROI are complex, making comparisons of different estimates invalid unless all analysts use the same assumptions. The foundation of the calculation lies in identifying the "system," that is, the components of the extraction process.

Consider the EROI of gasoline and diesel fuel produced from crude oil. Does E_{in} consist solely of the energy needed to run the machines for drilling and producing crude oil? Or does it also include the energy needed to make the machines? How about the energy to make and run the machines that made the machines? After the crude oil comes from the ground, does the calculation consider E_{out} as the energy content of the crude oil? Or does the calculation use E_{out} as the energy in the refined gasoline and diesel? What about the energy to transport crude oil to a refinery and to refine it? Ready-to-use fuel at the refinery must reach the customer before it's useful, so does EROI include the energy to transport the refined fuel to the place of actual use?

These system boundary questions reflect the reality that producing energy involves many steps, and presentations of EROI values must explicitly state the boundaries. If the ultimate aim includes comparisons of multiple EROI computations, the system boundaries and all assumptions must be the same.[59]

Despite the caveats surrounding calculating and comparing EROI (and NE), these measures provide an important insight into depletion. Estimates of EROI for oil production, for example, indicate that the values appear " . . . to be declining over time in every place we have data." Hall and colleagues estimated a fall in values from around 20–30 in the 1990s to below 20 in 2005. They speculate that production of oil by fracking in the last decade in the United States currently has a high EROI but that the value is likely to fall. A persistent decline indicates that new technology for fuel production has not outpaced the energy costs of tapping a depleted deposit.[60]

EROI analysts have also pushed the measurement into broader considerations of the quality of human life. They find that EROI and availability of energy per person correlate positively with the Human Development Index, expenditures on health, proportion of female literacy, and access to clean water. The measures correlate negatively with proportion of babies born underweight and the Gender Inequality Index. In short, they reaffirm the historical pattern that increased access to cheap energy created a more affluent way of life, and societies with expensive energy have a lower standard of living.[61]

It's important to note, however, that, like peaking, EROI calculations remain subject to the deposits-technology-economics triad. A fuel source with a low EROI value can become a source with high EROI if E_{in} drops; that is, new energy-efficient technology requires less energy inputs to extract the fuel. Nevertheless, theoretical discussions of EROI present a sobering picture of a future powered by fuels with successively lower EROI values, assuming no advances in extraction technology to reduce the energy costs of obtaining the fuel.

To what level can EROI decline before a society or culture can no longer sustain its living standards and material wealth? David Murphy, an ecologist–energy analyst, built a simple model based on the idea that, like an organism, a society needed energy to (a) procure more energy, (b) maintain its existing infrastructure, (c) consume for pleasure, and (d) grow a larger economy and/or population. He then traced what happened if EROI declined.

In the model, society produced 10,000 units of energy per unit time and that 5,000 units of energy had to be expended on "metabolism," that is, operating and maintaining the existing social infrastructure (food production, water supplies, transport, maintenance and operation of buildings). The model assumes that the society will spend its 10,000 units of energy first on procuring the next 10,000 units and on metabolism. Any remaining energy goes to consumption and growth.

At the high end with EROI = 10, E_{in} = 1,000 and E_{out} = 10,000. The first 1,000 energy units out go to producing the next 10,000 units. The allocation of the 9,000 not needed for the next batch of energy goes first to metabolism (5,000), and the remainder (4,000) supports consumption and growth.

Note what happens, however, if the EROI drops to 5. In this case E_{in} = 2,000 and E_{out} = 10,000. The first 2,000 units go to produce the next 10,000 units, leaving 8,000 units. Metabolism still requires 5,000, so only 3,000 units of energy remain for consumption and growth. In other words, when EROI dropped 50 percent from 10 to 5, the amount of energy available for consumption and growth dropped 25 percent, from 4,000 to 3,000. The society no longer has as much energy for consumption and growth as it used to have. If the drop in EROI happens rapidly, the society would face rising prices and a decline in living standards.

The situation becomes bleaker as EROI drops an additional 60 percent to 2. Here the society must spend 5,000 units of energy to produce the next 10,000 units. The remaining 5,000 units go to support metabolism, and no energy remains for consumption and growth.

If EROI drops to 1.5, the society must spend 6,667 units to procure the next 10,000 units. At this point no energy remains for consumption and growth, and, moreover, energy expenditures for metabolism drop to 3,333: below the 5,000 needed to maintain and operate existing social infrastructure. At this level of EROI, the society can no longer sustain its economy in the ways it previously could. It becomes a culture in decline. It can no longer maintain its population size and its standard of living.

This conceptual model looks bleak, but the situation in real life may have even bleaker features. The model assumes a society sufficiently egalitarian to always devote energy procured first to new energy production and to metabolism. A society with marked inequality—which includes most societies on earth—might not have so much goodwill. Powerful individuals or classes might divert energy expenditures to

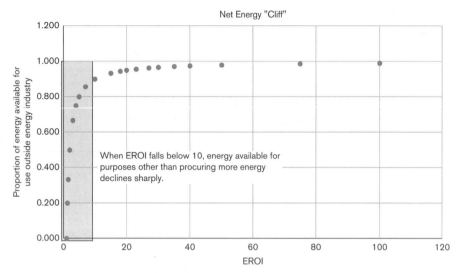

FIGURE 7.5. Net energy cliff (in shadowed area). When EROI falls below 10, net energy available to society drops sharply.

SOURCE: Figure created by author, inspired by David J. Murphy and Charles A. S. Hall, "Year in Review—EROI or Energy Return on (Energy) Invested," *Annals of the New York Academy of Sciences* 1185 (2010): figure 3.

their own consumption and growth first, thus shorting supplies to metabolism and perhaps even to future energy procurement. In such a case, inequality might well become more pronounced, as a privileged minority allows the majority's infrastructure, consumption, and growth to decline sharply while keeping themselves in comparative luxury.

The significance of falling EROI values appears starkly in a graph dubbed the "net energy cliff" (figure 7.5).[62] This graph shows EROI on the horizontal axis and proportion of energy available for energy services (other than for procuring more energy) on the vertical axis compared to total E_{out}. Note that EROI values from 50 down to about 20 don't decrease the energy available for energy services very much, from 98 percent to 95 percent. When EROI drops below 20, however, the amount of NE available to society starts to drop much more sharply. As soon as EROI drops below 10, the NE available falls precipitously: from 0.9 at EROI = 10 to 0.5 at EROI = 2.

Historically since 1700, societies embracing fossil fuels worked with EROI values above 30, which means that they used very little energy to procure vast amounts of new energy. In the EROI framework, depletion means that societies using coal, oil, and gas may be moving to a new situation with significantly lower values of EROI.

Serious challenges accompany calculations of NE and EROI, as noted earlier, and this factor alone indicates the uncertainty that accompanies any purported values obtained. Moreover, NE and EROI can't tell the whole story about the wisdom of using a prospective energy source. Specifically, EROI and NE provide no direct information about geopolitical tensions or health and environmental effects associated with extraction and use of the fuel. Nor can these calculations provide insights into how externalities (e.g., health effects), affect different groups of people. Without such information, the fuel's strengths and weaknesses remain incompletely known. Accordingly, based on NE or EROI values alone, it remains unclear whether extraction is a good public policy or whether obtaining the fuel would be a good business decision. NE and EROI inform decision making, but they cannot dictate it. We will return to EROI and the impacts of its changing values later (chapters 9 and 10).

CONCLUSION

Geopolitical tensions, health and environmental impacts, and the challenges of depletion surround the big-four fuels (coal, oil, gas, and uranium). Together these challenges create incentives to decrease or end uses of these fuels. Despite the importance of these three factors, climate change is the most important factor driving the need for the Fourth Energy Transition. Uncertainty still surrounds the seriousness of climate change's effects and when its impacts will fully arrive. Nevertheless, the risks of climate change create a powerful incentive to consider alternatives to fossil fuels. If those alternatives have attractive features, why risk the possibilities of serious effects from climate change?

In contrast, the three factors examined in this chapter supplement rather than drive the imperative to change energy economies around the world. In local areas, each of the three may be or soon become sufficiently severe to drive change, but climate change calls for worldwide action now. The benefits of the big-four fuels, of course, loom large, so stiff resistance to moving away from them comes as no surprise.

We turn to the alternatives in chapter 8. In chapters 9 and 10 we return to the big-four fuels in order to compare them with the alternatives.

Connections with Everyday Life

Up to this point, the energy story has consisted of assembling information and facts about existing sources of energy, the benefits derived from energy services, and the problems associated with the existing primary energy sources, especially the big-four fuels.

Chapter 8 takes us into a new realm, something we do every day: speculating on what might be, not what is. No one truly knows what the future will bring, but everyone has the ability to plan, embrace aspirations, set goals, and hope for things that make life better. Wisdom generally lies in the reasonableness of plans, aspirations, goals, and hope. Leaders who encourage reaching for an unattainable future by overpromising will only disappoint.

In this chapter, we begin the serious examination of renewable energy sources to ask the fundamental question, Is it reasonable to aim for an energy economy based on renewable energy used with high efficiency? Others have raised this question, and the argument here builds on earlier work to open the imagination. Does that mean humanity really can forgo the big-four fuels in favor of renewable energy used efficiently, maintain high standards of living, and end energy poverty? That is the most exciting question of the twenty-first century.

CHAPTER 8

The Fourth Energy Transition

Energy Efficiency and Renewable Energy

The modern energy economy based on the big-four fuels (coal, oil, gas, uranium) risks becoming more harmful than beneficial or simply running out of useful fuel. This chapter outlines the potential for a new energy economy based on renewable energy used with high efficiency. Amounts of renewable energy available for harvest could—in theory—provide more than enough direct and indirect solar energy to power the world at a level commensurate with today.

This is the Fourth Energy Transition, which rests on a striking irony. The appeal of the fossil fuels and uranium lay in the liberation of humanity from the annual flux of solar energy to the earth. The new energy economy, however, will return mankind to an annual harvest of direct and indirect solar energy supplemented by geothermal energy. Is this really possible? Or is it just a pipe dream? The remainder of this book explores these vital questions.

ENERGY CRISES, ENERGY EFFICIENCY, AND THE FIRST RUMBLINGS OF THE FOURTH ENERGY TRANSITION

The 1970s brought two energy crises—uranium and oil—that altered debates about energy and launched the Fourth Energy Transition. First, turmoil about uranium and nuclear power erupted in the late 1960s and very early 1970s, just at the time that nuclear power had begun to matter in the U.S. energy economy. By the end of the 1970s, the industry had ceased building new reactors in the United States, but the reasons were economic and not related to safety. Public debates over safety and industry concerns about costs together had thrown a pall over uranium as a fuel.[1]

Simultaneously, these discussions about uranium fed into larger concerns about energy that were stimulated by a crisis in petroleum supplies. In October 1973, Arab countries embargoed deliveries of oil to countries supporting Israel in the war that erupted that month with Egypt and Syria. Suddenly the United States faced gasoline shortages to which it had formerly been immune because of surplus oil production capacity in East Texas.[2]

While the nuclear power debates posed no existential threats to the country, shortages of gasoline did. Without gasoline, mobility collapsed and the American economy could not function. Arguments over nuclear power left many people unconcerned, but inability to fill the gasoline tank caught nearly everyone's attention. The U.S. Congress asked the U.S. Atomic Energy Commission for new ways to understand the dilemmas of energy economies (see chapter 5), and since 1973, reform of the energy economy has remained a perennial although still unresolved topic.

Collapse of the nuclear dream, the 1973 oil crisis, and a follow-on oil crisis occasioned by the Iranian revolution in 1978–1979 stimulated policy recommendations, including conservation, efficiency, and a search for alternative energy sources. Unfortunately, this all-of-the above strategy avoided setting priorities. All fuels, both nonrenewable and renewable, had equal value, and efficiency was a nice extra, not a key component.

In contrast to the all-of-the-above strategy, a small group of "energy intellectuals" began to forge a new vision for energy economies. Three prominent studies saw that energy extraction and use faced serious problems with stability, environmental pollution, and resource scarcity. Each advocated serious reforms, particularly the need to increase energy efficiency and energy conservation. One proposed forging a pathway to new energy sources, away from the big-four fuels.

Starting in 1971, before the 1973 oil crisis but in the middle of protests about nuclear power, the Ford Foundation published *A Time to Choose: America's Energy Future*,[3] which argued that U.S. energy policy had been too committed to expanding supply instead of curbing demand. By the time the report came out, the oil embargo of 1973 had clearly demonstrated the wisdom of emphasizing efficiency.

Energy efficiency provided enormous opportunities to continue high standards of living without the pollution, capital expenditures, and political uncertainties associated with expansion of supplies. Today this argument is so commonplace that it's hard to remember that in 1974 the explicit embrace of energy efficiency as *public policy* was both new and radical. Before 1974, energy efficiency lay strictly in the arena of private enterprise (see chapter 4).

In 1975, the American Physical Society (APS) published *Efficient Use of Energy: A Physics Perspective*, which emphasized two things. First, energy increases material wealth, but increasing gross national product (GNP) does not necessarily require increasing energy use at the same rate. In other words, GNP isn't tightly coupled with energy use. Second, physicists up to that point saw their job as helping to increase energy supply but not to help increase the efficiency of energy use.[4] Both points made efficiency of energy use an interesting problem in energy science, and technical expertise about efficiency acquired strategic legitimacy.

Arthur H. Rosenfeld (1926–2017), a particle physicist with research interests far from the nitty-gritty of practical energy issues, co-led the APS study. This experience prompted him to radically shift his research and teaching at the University of California, Berkeley, to establish efficiency as a key objective for energy policy. A graph, often called the Rosenfeld curve, showed per person uses of electrical energy in California remained approximately constant after 1980 while rising steadily in the rest of the United States (figure 8.1). A portion of the reduced use of electricity per person came from energy efficiency championed by Rosenfeld.[5]

The Ford Foundation and APS studies clearly embraced energy efficiency, but energy efficiency answered just one of two fundamental questions: How should we use energy resources? Clearly "efficiently" was the right answer. At the time, it meant energy from coal, oil, gas, uranium, and hydropower was precious and not for wasting.

The second question, From where should we obtain our energy?, remained unchanged, because for all practical purposes coal, oil, gas, uranium, and hydropower were the fuels then available. Uses of geothermal heat had importance in a few locations, but technologies for harvesting solar and wind energy in the 1970s remained in small, specialized niches only.

Nevertheless, Amory B. Lovins, a smart, somewhat brash, and strongly opinionated thinker saw the necessity of answering both questions. Yes, efficiency had to be key, but sources of energy supply also mattered. The pathway to acceptable energy could not arise solely from efficient uses of the big-four and hydropower. With a beautiful,

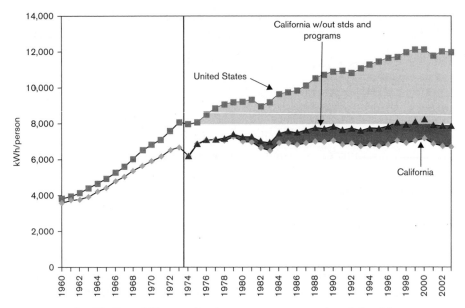

FIGURE 8.1. The Rosenfeld curve, showing per person consumption of electricity remaining approximately steady in California after 1980, in contrast to a steady rise in other states. The steadiness in California resulted in part from efficiency measures championed by Rosenfeld and others.

SOURCE: Arthur H. Rosenfeld, "Energy Efficiency in California," presentation to Climate Breakfast Group, November 18, 2008, www.energy.ca.gov/2008publications/CEC-999–2008–032/CEC-999–2008–032.pdf, May 31, 2016.

rhetorical flourish, Lovins identified the "hard" and "soft" pathways for future energy development.[6]

In the hard pathway the United States was then pursuing, he saw coal and nuclear energy eventually phasing out dependence on gas and oil, and by 2025, the country would use about 225 quads of energy per year. In contrast, the soft pathway of energy efficiency combined with existing hydropower plus new solar and wind technologies could replace coal, oil, gas, and uranium; by 2025 the United States would have a total energy economy of about 60 quads, without loss of standard of living.

It would be too much to say that Lovins, all by himself, invented the Fourth Energy Transition. His essay, however, for the first time captured a complete vision of a new way to deal with energy. It reflected the viewpoints of a new and growing group of energy analysts and encapsulated the major elements of what I have called the Fourth Energy Transition. Lovins's boldness lay in his vision that solar and wind technologies would vastly improve to the point that, combined with the drive for efficiency, they could replace the big-four fuels. Efficiency held a key role in the soft pathway, but efficiency itself was not particularly new or bold, as it had already played an important role in the Third Energy Transition (see chapter 4).

Now the question is, Can the soft pathway, or the Fourth Energy Transition, succeed? We turn first to the magnitudes of renewable energy available.

THE SUPPLIES OF RENEWABLE ENERGY RESOURCES

An assessment of the adequacy of renewable energy supplies must include not only current uses but also growth in energy use due to economic and population growth. In addition, energy poverty currently stymies the legitimate aspirations of millions.[7] Can the harvestable supplies of renewable energy satisfy existing needs, alleviate energy poverty, and accommodate population growth? Equally important, can renewable energy reach *every* location on earth, at the times needed?

The question of the adequacy of the energy supply from renewable sources logically follows from remembering the start of the Third Transition in England (see chapter 1). Renewable energy in firewood gave way to nonrenewable energy in coal, and agrarian economies changed to ones powered by energy supplies vastly larger than those based on food and firewood alone. What allows even imagining reversion back to organic, renewable energy? The short answer is new technologies.

Despite the seeming simplicity of the short answer, the question of possible imaginaries immediately distinguishes the Third and Fourth Energy Transitions. First uses of coal relied on practical knowledge of local people familiar with near-surface seams of coal, not the trained expertise of scientists and engineers, for the latter group did not exist. Over time, geologists and engineers supplanted the lore of local folks, but the initial steps of the transition required no special expertise. Also, direct uses of coal to heat buildings and drive processes such as making beer or glass closely mimicked the uses of firewood, a skill that began to develop with *Homo erectus* over a million years ago.

In contrast, the idea of using wind, solar radiation, hydropower, and geothermal heat to produce electricity—the major way of harvesting renewable energy—inherently rests on the expertise of scientists and engineers. This form of renewable energy remains opaque and unintuitive to those without scientific knowledge. Uses of solar radiation and

TABLE 8.1. ENERGY PROCURABLE FROM RENEWABLE SOURCES AS ELECTRICAL
ENERGY

Energy Source	Minimum (EJ/year)	Maximum (EJ/year)
Solar photovoltaic	1,338	14,778
Concentrating solar power	248	10,791
Wind onshore	70	450
Wind offshore	15	130
Hydropower	50	52
Geothermal	118	1,109
Ocean	7	331
Total	1,846	27,641

	CURRENT GLOBAL USE EJ/year
Electricity	61
Primary energy sources	492

SOURCE: W. Moomaw et al., Introduction to *IPCC Special Report on Renewable Energy Sources and Climate Change Mitigation* (Cambridge: Cambridge University Press, 2011), table 1.A.1, 206.

geothermal resources to heat water or buildings rest more easily on common knowledge, but renewable energy transformed into electricity does not.

Determining the amount of renewable energy procurable, therefore, requires nonexperts—most people—to trust the expertise of experts— just a few people. We will return in chapter 11 to trust as a potential barrier to the Fourth Transition, but for the moment let's examine the expert estimates of renewable energy available based on current technology.

The Intergovernmental Panel on Climate Change summarized existing literature in 2012 (table 8.1). The 2008 global primary energy use is 492 EJ/year (EJ = exajoules = 10^{18} joules). The enormity of renewable energy supplies thus leaps out: at a minimum, 1,846 EJ of electrical energy could be harvested. Current electrical energy produced is 61 EJ/year. Solar photovoltaic electrical energy alone could power the entire world with its minimum of 1,338 EJ/year. Not only could renewable sources power the entire world now, but the minimum estimates of supply suggest room for population growth and for expansion to end energy poverty.

Ecofys, a Dutch consulting firm, produced a more refined set of estimates for 2010, 2030, and 2070.[8] It used existing databases of areas as small as one square kilometer over the entire globe. For each area, they estimated both the *suitability* and *availability* for production of electricity from renewable sources. The Ecofys study also estimated potential availability from rooftops and building facades.

Designation as "technically suitable" excluded land covered in ice, land that was too elevated, urban land, many forested areas, coastal and rocky areas, excessively slopped areas, and areas of insufficient potential energy. Designation as "available" rested on an estimated "availability factor," based on current patterns of social acceptability of energy installations. Availability factors clearly depended on social norms, so the study developed Low, Medium, and High estimates. This latter factor, from Low to High, created the largest variations in estimated amounts of electricity that could be harvested from renewable energy.

The amount of renewable energy available depends on the amount of land devoted to energy production. Fossil fuels and uranium mostly avoid this issue, because they have very high *energy density* (energy per kilogram; see appendix 2). Small land areas of mines and wells suffice to produce large amounts of energy. In contrast, solar and wind have low energy densities per square meter (m^2). Compared to the currently dominant fuels, renewable energy will take more land to produce and thus potentially create land use conflicts. The Ecofys study accordingly sharply constrained the uses of forested land, agricultural land, and grasslands for energy production with the availability factor.

Onshore wind could not occupy forestland to an extent greater than 0.5 percent (Low) to 2 percent (High). Onshore wind energy could not occupy agricultural and grasslands to an extent greater than 3 percent (Low) to a High of 10 percent (industrialized countries) to 20 percent (developing countries). Wind turbines already exist on some agricultural lands, so combining agricultural uses with wind-based electricity production is feasible. Offshore wind turbines could occupy 4 percent (Low) to 5 percent (High) of suitable areas within 10 km of shore; availability rose further, to 25 percent (Low) to 80 percent (High) when turbines were 50 to 200 km from shore.

Solar energy could occupy no forested land. Use of agricultural land could be no higher than 0.1 percent (Low) to a High of 2 percent (industrialized countries) to 5 percent (developing countries). Use of grasslands could be no higher than 0.5 percent (Low) to a High of 3 percent (industrialized countries) to 5 percent (developing countries).

Based on these constraints, the land areas actually available for energy development were less than 5 million km^2 (3.5 percent of the

TABLE 8.2 ENERGY PROCURABLE FROM RENEWABLE SOURCES, AS ELECTRICAL
ENERGY, 2070

Energy source	Minimum (EJ/year)	Maximum (EJ/year)
Photovoltaic on roofs	121	121
Photovoltaic on facades	90	90
Solar & wind, land, corrected for overlaps	320	2,832
Wind, off-shore	197	652
Geothermal electricity	27	84
Hydropower	26	26
Total	781	3,805

SOURCE: Yvonne Y. Deng et al., "Quantifying a Realistic, Worldwide Wind and Solar Electricity Supply," *Global Environmental Change* 31 (2015): 239–52; data from table 9.

earth's land, excluding Antarctica). Offshore, less than 10 million km²
(about 2.2 percent) of the ocean area were available for development. In
comparison, urban areas currently occupy 0.18 percent of the land area
and agriculture occupies 11 percent.

The land areas deemed actually available, therefore, exceeded the
areas currently occupied by cities, and equaled about 32 percent of the
land currently occupied by agriculture. In some cases, agricultural land
and wind energy production could coexist. For solar technology, no
more than 5 percent of current agricultural land could change to energy
production at High availability; at the Low availability, no more than
0.1 percent would change, and for Medium no more than 0.5 percent.

How much energy could these "available" areas produce in 2070? The
total estimated production lay between about 730 (Low) and 3,700 (High)
EJ/year. Additional electrical energy procured from geothermal and hydropower (derived from other studies) brought the total, long-term electrical
energy supply to 781 EJ/year (Low) and 3,805 EJ/year (High) (table 8.2).

Based on these figures, we return to the original question: Does
enough renewable energy exist to power the world's current uses of
energy plus allow for ending energy poverty and facilitating future grow
in the world's economy and population? Based on these estimates, the
answer is yes, enough potential energy harvest exists.

Perhaps the estimates, even the low ones, are too optimistic. Perhaps
the realization of these new energy technologies will encounter cultural
and political obstacles that cannot be resolved. And perhaps the massive electrification implied by the figures will hit technological or social
snags (see chapter 10). But on the surface, the figures give every encour-

agement to a profound embrace of renewable energy at amounts up to and including 100 percent of energy used. Lack of enough renewable energy does not appear to block the Fourth Energy Transition.

PROFILES OF RENEWABLE ENERGY SOURCES AND ENERGY EFFICIENCY

Solar power, wind, and hydropower rank as the top contenders for sustainable energy sources. Other sources, such as various forms of ocean energy, biomass, and geothermal energy can play important supplementary roles to solar, wind, and hydropower. This section explains the physical functioning of each source, the resources available, and current uses of these resources.

Solar Energy

Two fundamentally different types of harvesting methods, one very old and one recognized only since the 1800s, can transform solar radiation to useful forms of energy.[9] Although they work by very different means, each method involves absorption of energy carried in electromagnetic radiation from the sun. The radiation wavelengths involved lie either in the range detected as light by human eyes or with wavelengths a bit shorter (ultraviolet) or longer (infrared).

Solar thermal energy

The first, *solar thermal energy*, absorbs solar radiation and transforms it directly to heat. The absorber rises in temperature, and different kinds of devices can raise temperatures either small amounts (to less than 100°C, the boiling point of water) or large amounts (greater than 100°C).

The former can heat water and building interiors to comfortable temperatures. This form of energy capture may seem unspectacular, but space heating and hot water generally represent the majority of a building's energy costs. Special, dark-colored solar collectors for water heating already play important roles in Greece, Turkey, Israel, Australia, Japan, Austria, and China. In years past solar hot-water systems were common in the United States, and they could be again.

The potential for using solar space heating depends on the building's appropriate design, construction, orientation to the sun, and location, so optimal investment in this form of solar energy begins before construction. South-facing walls and windows capture solar radiation and convert it directly to heat. Roofs and other shading devices limit solar inputs in summer but maximize them in winter. Properly insulated, such buildings require relatively little or no additional energy, compared to traditional construction, to be comfortable year-round.

Appropriate building design also maximizes sunlight to reach the interior and thus decreases the need for artificial lighting. Natural light inexpensively replaces artificial light without transformation first to heat or electricity.

Large temperature increases in ovens or special solar collectors can cook food, produce electricity by making steam to spin turbines, or drive other processes. In countries still using biomass (wood and dung primarily) for cooking, solar stoves may provide a less expensive way to cook without the smoke that creates unhealthy indoor air pollution.

Concentrating solar power, an industrial-scale technology, uses solar radiation to make steam, which spins a turbo-generator to make electricity. Arrays of mirrors concentrate solar radiation on a boiler sitting at the top of a centrally located power tower. The immense amount of radiation focused on the boiler heats water directly or indirectly to produce steam. Alternatively, parabolic-shaped troughs focus solar radiation on pipes with water or other fluids. The heated liquids then make steam for electrical generation (figure 8.2).

In both arrangements of concentrating solar power, the heat can be stored for periods of cloudiness or darkness and then used to generate electricity. This form of solar-generated electricity thus provides one method for handling intermittency in solar radiation.

Solar photovoltaic energy

Photovoltaics is a most elegant energy source. Light shines on a crystal and produces electricity. It's as simple as that. There are no moving parts. The fuel source (sunlight) is free, abundant and widely distributed, available to every country and person in the world.[10]

The second and much newer method of using solar energy produces *photovoltaic (PV) electricity*, a process in semiconductors that transforms light energy directly into electrical energy. As noted in the quotation above, photovoltaic electricity is an elegant energy source, but making it happen appears "simple" only in hindsight. Development of photovoltaic electricity rested on Becquerel's discovery of the photovoltaic effect in 1839 (see chapter 2).

FIGURE 8.2. PS10 concentrating solar power plant (foreground), Seville, Spain. This was the first commercial-scale concentrating solar power station in the world. Construction started in 2005, and production of electricity started in 2007. Area receives 2,012 kWh/yr/m². Planned production from PS10 is 23,400 MWh/yr. PS20 is in background. NREL, www.nrel.gov/csp/solarpaces/project_detail.cfm/projectID=38, June 21, 2016.

SOURCE: Koza 1983, licensed under the Creative Commons Attribution 3.0 Unported license, Wikimedia https://commons.wikimedia.org/wiki/File:PS20andPS10.jpg, 24 June 2016.

Perhaps the most important reason for the delay in developing practical applications of photovoltaic electricity was discerning the subtle distinctions between conductors and insulators. Faraday recognized the easily observed differences between conductors and nonconductors based on their ability to transmit electric forces. His work on retardation of electrical conduction on undersea telegraph cables formed his ideas about what happened in material substances as electric currents moved through a conductor wrapped in an insulator or nonconductor.

The term *semiconductor*, first used in 1838, simply meant a material that was intermediate between a good conductor and an insulator. In the 1930s, it acquired a very new theoretical and conceptual meaning.[11]

Semiconductors were materials with a very particular intermediate status between conductors and insulators. All materials consisted of atoms composed of nuclei and electrons. In conductors, the electrons moved from nucleus to nucleus on the application of comparatively little force or energy. Insulators had electrons that took extraordinarily high levels of force or energy to make them move. Semiconductors were intermediate in the amount of force or energy required to make its electrons move.

In the 1900s, work on semiconductors incorporated *quantum mechanics*, which postulated that electrons attached to nuclei had specific energy levels. If they moved at all, they moved only to new, higher or lower, specific energy levels but not to intermediate energy levels.

The term *band gap* denoted the difference between the original energy level and the new one. Conductors had low band gaps, semiconductors had band gaps of intermediate sizes, and insulators had extraordinarily high band gaps. The semiconductors of most use in turning light into electricity consist of materials in which the band gap equals the energy carried by a *quantum*, or particle, of visible light. If light quanta have the energy needed to increase the energy levels of electrons in a semiconductor from their low energy position across the band gap to higher energies, they make an electric current flow.

A second finding by the 1930s was that the electrical properties of semiconductors were extremely sensitive to minute impurities. Once recognized and controlled, the impurities became the trick that catapulted semiconductors from objects of curiosity to the foundation of the many industries now built on the manipulation of semiconductors, including among others the photovoltaic and computing industries.

Some impurities make a semiconductor richer in electrons, and some make it richer in "holes," the positive charge of a nucleus that has lost an electron. Semiconductors are doped with different, specific impurities to make them rich in electrons (*n-semiconductors*) or holes (*p-semiconductors*). When a layer of n-semiconductor interfaces with a layer of p-semiconductor, an *n-p junction*, the electrical interactions between the two layers, triggered by light energy, produce an *electromotive force*, or voltage, that forces the electrons to flow.

A *photovoltaic cell* is a device in which a p-semiconductor rests under an n-semiconductor. When light energy from the sun strikes the device, the electrons able to flow become vastly more numerous and the electrical interactions between the two layers create the voltage to make the electric current flow (figure 8.3). Today, silicon, a component of sand and thus highly abundant on earth, is the semiconductor, when appropriately doped, most relied on to make photovoltaic or solar panels.

Solar photovoltaic cells, whether made of silicon in several possible forms or from other kinds of semiconductors, thus embody a multitude of developments in science and engineering since Becquerel's first observation

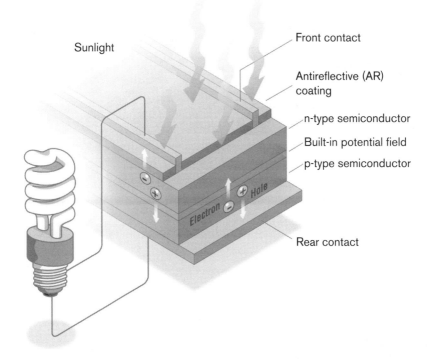

Sunlight

Front contact

Antireflective (AR) coating

n-type semiconductor

Built-in potential field

p-type semiconductor

Electron

Hole

Rear contact

FIGURE 8.3. Diagram of photovoltaic cell, showing the *p*- and *n*-silicon layers. When sunlight is absorbed by the cell, electrons flow from the *n* layer through the circuit to light the bulb, and then continue back to the *p* layer.

SOURCE: U.S. Department of Energy, National Renewable Energy Laboratory, courtesy of Wayne Hicks.

that light could produce electrical effects. In practice, cells are collected into modules, which in turn are connected as panels. Multiple panels form an array.[12] It takes far more than solar panels and arrays, however, to create a system capable of powering a civilization with photovoltaic electricity. Moreover, compared to fossil fuels and uranium, practical uses of photovoltaic electricity depend heavily on site-specific issues over significant areas of land. Human organizations and behavior must mesh with the physical capabilities of solar cells to produce usable electricity delivered at the right times in the right amounts to the places where needed.

Consider the most important other factors impinging on the project of electrification by means of photovoltaic panels. First, the rate of electrical energy producible faces a real limit: the rate of sunlight energy falling on

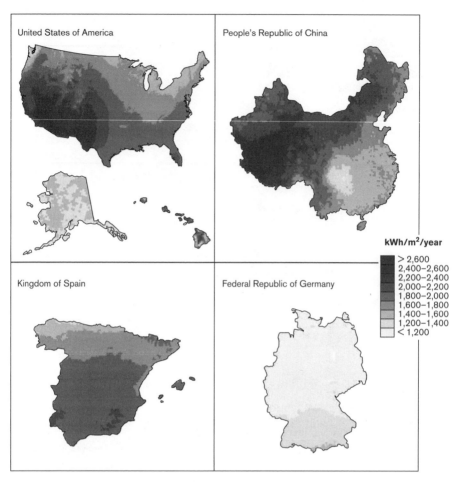

FIGURE 8.4. Solar resources, United States, China, Spain, and Germany
(in kWh/m²/year)

SOURCE: U.S. Department of Energy, National Renewable Energy Laboratory.

a particular spot. It is true that the fuel is free, but it varies in every place
throughout each day and throughout each year. Some parts of earth have
average annual solar fluxes of over 2,600 kWh/m² per year, for example,
southwestern China. The southwestern United States has up to 2,600
kWh/m² per year, but other areas have much less; for example, Alaska
and the U.S. Northwest have less than 1,200 kWh/m² per year. Spain is
like many areas of the United States, at 1,200 to 2,000 kWh/m² per year,
but Germany has mostly below 1,200 kWh/m² per year (figure 8.4).

FIGURE 8.5. Roof-top solar panels, on two sections of house roof, California, facing southeast

SOURCE: John H. Perkins, 2016.

The amount of solar energy falling on the land, however, does not entirely dictate the willingness of individuals and countries to harvest it. For example, Germany and Spain have lower amounts of incident solar radiation than the United States, but in 2008 they had more installed solar panels. Spain and Germany have no areas receiving up to 2,500 kWh/m^2 per year (6.8 kWh/m^2 per day) like the American Southwest. Germany, in fact, has resources similar to Washington State and Alaska, the least well endowed areas of the United States.

A separate set of factors affecting photovoltaic electricity production centers on the size of the panel array, the array's ownership-management, and its location relative to customers' electric meters. Individual home owners and small businesses have installed relatively small arrays; for example, a maximum production of 2 to 5 kW on a home. Much of the electricity generated by small arrays goes directly to the building on which they are installed, with excesses going to the transmission and distribution (T&D) grid (figure 8.5).

Utility-scale installations, in contrast, typically generate 1,000 kW or more at their maximum. These installations are managed to sell electricity to the transmission grid; they may be owned by traditional electric utility companies or by independent companies selling to the grid. Small

FIGURE 8.6. Utility-scale photovoltaic power plant. "Nine million cadmium telluride solar modules now cover part of Carrizo Plain in southern California. The modules are part of Topaz Solar Farm, one of the largest photovoltaic power plants in the world. At 9.5 square miles (25.6 square kilometers), the facility is about one-third the size of Manhattan island, or the equivalent of 4,600 football fields. Construction at Topaz began in 2011. The plant was mostly complete by November 2014, when it was turned on and began to generate electricity. By February 2015, all construction activity ended. . . . When operating at full capacity, the 550-megawatt plant produces enough electricity to power about 180,000 homes."

SOURCE: U.S. National Aeronautics and Space Administration (NASA), Earth Observatory image by Jesse Allen, using EO-1 ALI data provided courtesy of the NASA EO-1 team. Caption quotation by Adam Voiland, http://earthobservatory.nasa.gov/IOTD/view.php?id=85403&src=eoa-iotd, October 31, 2016.

arrays typically are located on the customer side of the electric meter while utility-scale installations are located before the customers' meters on the grid (figure 8.6).

Differences in size, ownership-management, and location strongly affect the costs and ways of managing the variable amounts of electricity produced. At the time of writing, the costs of generating solar PV electricity have fallen to make small- and utility-scale installations more economically attractive.[13] The ultimate amounts of PV electricity produced in the future and the characteristics of the arrays that will produce it remain speculative today. Whatever the ultimate amount produced, the characteristics of the producing arrays will depend on the laws and policies governing their installation and ownership.

Chapters 9 and 10 delve into further aspects of PV electricity, but an important point about solar PV energy has already emerged. In certain parts of the world, for example, Germany and California, solar PV had already risen to the level of a "significant source"; that is, it is no longer such a tiny part of electricity generation that it hardly registered in statistical summaries.

Consider these figures. In the United States as a whole in 2014, solar electricity contributed only about 0.4 percent of the total electricity. This was above the amount of a decade earlier, but for the country as a whole solar electricity was not a significant source of energy. In California during 2014, solar electricity contributed 5.0 percent of the total. During the first six months of 2015, solar electricity was 8.3 percent of the total electricity used in the state. On September 7, 2015, solar PV generated 6.6 percent of the total electricity used in California. At about noon on that day, over 20 percent of California's electricity came from solar electricity.[14] Germany in 2014 generated, over the entire year, 6.9 percent of its electricity from solar PV. On some sunny weekdays, PV provides up to 35 percent of the total and up to 50 percent on weekends.[15] These numbers, already significant in California and Germany, continue to rise in these places and others with new PV installations. In a book, it is impossible to have "current" numbers, because the rate of increase in solar PV far exceeds the speed with which a book can be produced.

Without a doubt, solar electricity is no longer a curiosity. It has begun to be a major player in powering industrial economies. The question remaining is how much it will ultimately contribute.

Not only has solar electricity arrived; research and development on solar technologies continue to produce new developments promising lower costs and higher outputs. One of the potentially most interesting developments comes from combining PV with solar thermal installations to produce both electricity and hot water from the same panel. The panel may be able to turn 15 to 20 percent of the light energy into electricity, but the unconverted light energy heats the panels and diminishes the efficiency of conversion to electricity. A solar thermal absorber can put the excess light energy into hot water, thus cooling the PV panel and making use of a higher proportion of the incident light energy.[16]

Wind Energy

The energy in moving air masses—wind—has been harvested since very ancient times by sailing ships and windmills for pumping water and grinding grain.[17] Adaptation of windmills to make electricity began in 1887 in Scotland and the United States, just five years after Edison's first coal-fired power station in New York City.

By the 1930s, small-scale efforts appeared in many countries in Europe, including feeding wind-generated electricity into the grid. After 1973, engineers in the United States began to see wind power as potentially a major source of electricity. Wind energy resides in the movement of masses of air; this mechanical energy turns the blades of a turbine, which turns a generator to induce electrical energy in the wires, much as Michael Faraday did in the 1830s (figure 8.7).

As the number of wind installations increased, engineers learned three important points. First, the total kinetic energy in moving air masses increased as the *cube of the velocity* (v^3) of air intercepted by the blades of the windmill; higher wind velocities carried a great deal more energy than lower ones. When wind speed goes up 2 times, for example, 4m/s to 8 m/s, the kinetic energy in the wind goes up 8 times ($2^3 = 2 \times 2 \times 2 = 8$).

Second, the circular area defined by the sweep of the blades sets the maximum amount of energy that a single wind turbine can harvest. The larger the area, that is, the longer the turbine's blades, the more harvestable energy intercepted. Third, higher velocities of wind occurred at greater heights above the ground. Taller wind turbines reach wind with greater amounts of energy.

These three factors together govern the theoretical amount of energy obtainable from wind, but additional details based on the turbine-wind interactions further limit the actual amounts of energy obtainable. Two

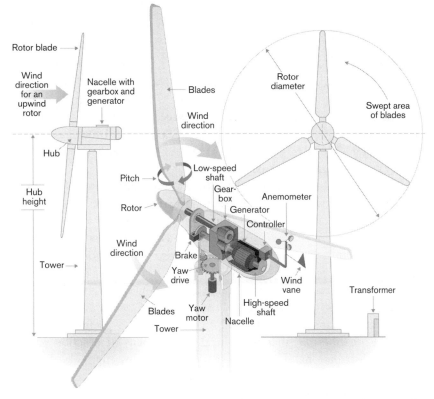

FIGURE 8.7. Diagram of a wind turbine nacelle, which sits atop the tower and contains the gearbox, low- and high-speed shafts, generator, controller, and brake; some nacelles are large enough for a helicopter to land on.

Anemometer: Measures wind speed and transmits wind speed data to the controller.

Blades: Lift and rotate when wind is blown over them, causing the rotor to spin. Most turbines have two or three blades.

Brake: Stops the rotor mechanically, electrically, or hydraulically, in emergencies.

Controller: Starts up the machine at wind speeds of about 8 to 16 mph and shuts off the machine at about 55 mph. Turbines do not operate at wind speeds above about 55 mph because they may be damaged by the high winds.

Gearbox: Connects the low-speed shaft to the high-speed shaft and increases the rotational speeds from about 30 to 60 rpm to about 1,000 to 1,800 rpm; this is the rotational speed required by most generators to produce electricity. The gearbox is a costly (and heavy) part of the wind turbine and engineers are exploring "direct-drive" generators that operate at lower rotational speeds and don't need gearboxes.

Generator: Produces 60-cycle AC electricity; it is usually an off-the-shelf induction generator.

High-speed shaft: Drives the generator.

Low-speed shaft: Turns the low-speed shaft at about 30 to 60 rpm.

Pitch: Turns (pitches) blades out of the wind to control the rotor speed, and to keep the rotor from turning in winds that are too high or too low to produce electricity.

Rotor: Blades and hub together form the rotor.

Tower: Made from tubular steel (shown here), concrete, or steel lattice. Supports the structure of the turbine. Because wind speed increases with height, taller towers enable turbines to capture more energy and generate more electricity.

Wind direction: Determines the design of the turbine. Upwind turbines—like the one shown here—face into the wind, while downwind turbines face away.

Wind vane: Measures wind direction and communicates with the yaw drive to orient the turbine properly with respect to the wind.

Yaw drive: Orients upwind turbines to keep them facing the wind when the direction changes. Downwind turbines don't require a yaw drive because the wind blows the rotor away from it.

Yaw motor: Powers the yaw drive.

SOURCE: Diagram, National Renewable Energy Laboratory, courtesy of Wayne Hicks; text explanation: U.S. Department of Energy, http://energy.gov/eere/wind/inside-wind-turbine-0, April 2, 2016.

stand out. The Betz limit, sometimes called the Lanchester-Betz limit, derived from calculations based on the fall of wind velocity as it passes the turbine blades, holds that the maximum efficiency of turning wind kinetic energy to electrical energy can't exceed 59.3 percent. Practical measurements of the *power coefficient* indicate modern turbines operate between 40 and 50 percent efficiency, but theory indicates this efficiency cannot surpass 59.3 percent.

The other practical limitation of utmost importance comes from the simple fact that wind velocities are quite variable. Even a windy area can have times of dead calm or light breezes. Similarly, a place with relatively low winds may at times have gusts or storms with much higher velocities. The *average velocity* conveys information about the average kinetic energy of the wind, but the *distribution of velocities* generates important constraints on the installation of a practical system of wind turbines.

Mathematicians provided wind engineers with a useful tool, the *Rayleigh distribution,* that by coincidence happens to describe wind velocity distributions. This mathematical tool helps engineers design wind turbines for specific geographic sites based on each site's characteristic wind patterns.

The distribution of velocities means that wind has a *distribution of kinetic energies* associated with the distribution of velocities. The fact that the kinetic energy goes up as the cube of the velocity (v^3) means that the vast majority of the energy in wind comes at the higher velocities, or in the engineer's language, the energy distribution is higher than the velocity distribution.

This offset drives compromises in the design of wind turbines and the electrical transmission system. Use of electrical energy goes through a steady rise and fall each day: low overnight, rising in the morning, peak use in the afternoon and evening, and falling as the night progresses. The design of wind farms must recognize this daily dance of rising and falling power usage.

Wind turbines can be designed to work in low velocity wind sites or ones with high velocities. They can also be designed to maximize total energy production or maximize the steadiness of energy production. Depending on the criteria specified by the turbine owner or the manager of the transmission grid, the engineer will vary the height of the turbine, the area swept by the blades, the adjustability of the blades, the lowest wind velocity that will turn the blades, and the maximum electrical energy that the turbine will generate.

After installation of a specific wind turbine, the *capacity factor* indicates the percent of its actual production compared to its maximum designed production. High capacity factors enable a turbine to produce a steadier power output, which becomes more important as the amount of wind-generated electricity rises.

Wind energy depends indirectly on solar radiation. Air masses heat unevenly, because the highest solar radiation falls at the equator, with lower amounts toward the north and south. In addition, solar radiation at a particular place varies constantly during each day and over the course of the year. Air receiving greater radiation heats more, expands, increases in pressure, and moves toward the poles where pressure is lower. Uneven heating thus puts the atmosphere into constant motion (i.e. wind). Oceans and land heating at different rates also result in movements of air and water to create the patterns of wind on earth.

Figure 8.8 displays the geographic variation in average wind velocities in and around the United States.[18] Clearly the middle sections of the country have rich wind energy supplies, as do offshore areas. As rich as these areas are, the middle portion of the continent lies far from the coasts, the areas of highest electricity consumption. The costs of transmission increase the costs of electricity from these areas. Similarly, offshore areas have higher costs to install, operate, and maintain wind turbines, plus higher costs of transmission. By 2010, wind produced between 4 percent and 19 percent of electricity in the Netherlands, Germany, Ireland, Spain, Portugal, and Denmark. Texas and California, the two most populous states, generated over 10 percent and over 6 percent of electricity, respectively, from wind in 2014.[19]

These figures show that wind is a minor source of electricity today. Like solar energy, however, during some times of the day wind is already a significant actor on the energy stage. In the United States, Texas has more capacity to generate wind power than any other state. In the winter, when electricity demand is low and the wind velocity high, wind contributes a significant amount of the total electricity. For example, on February 11, 2015, wind at 5:00 p.m. comprised 30.7 percent of the total electricity. At the peak, wind electricity equaled 10,398 MW. In summer heat, electricity demand soars, but wind velocities diminish. For example, at midnight on August 12, 2015, wind generated 9.6 percent of the total electricity. At its peak, wind electricity was 4,534 MW. Operators of the grid serving most of Texas, the Electric Reliability Council of Texas (ERCOT), are now experienced in managing the fluctuating and sometimes highly significant production rates of wind electricity.[20]

FIGURE 8.8. Wind resources in the United States.

SOURCE: U.S. Department of Energy, National Renewable Energy Laboratory, courtesy of Wayne Hicks.

Wind speed
m/s

> 10.5
10.0
9.5
9.0
8.5
7.5
7.0
6.5
6.0
5.0
4.5
4.0
< 4.0

Hydropower and Ocean Energy

The use of water to turn wheels to run pumps and machines, like wind power, dates to ancient times.[21] At the end of the 1700s in Britain, manufacturers often preferred the steady power delivery from waterwheels to run machines instead of direct reliance on steam engines. Some of the first steam engines in manufacturing pumped water from downstream to upstream to run over the waterwheel once again (see chapter 1).

Water power today consists of two broad types. *Hydropower* uses falling river water to turn a turbine that turns a generator to make electricity. This source of renewable energy has developed on a major scale for over a century and in 2012 produced 16.2 percent of the world's electricity (see chapter 3).[22]

Ocean energy refers to a mixed collection of technologies, mostly still in development, to extract energy from the kinetic energy of moving water in waves, tides, and ocean currents, and from energy derivable from heat gradients and salinity gradients in the oceans. A few small installations currently transform tidal currents to electricity, but the other devices have not yet been deployed commercially. Nevertheless, ocean energy potentially could provide a great deal of renewable energy.

Hydropower

Hydropower—making a turbine spin a generator to move metal wires through a magnetic field—changes kinetic energy of falling water to electrical energy. The amounts of energy obtainable from hydropower depend primarily on only two things: the mass of water available and the height through which it falls. The conversion efficiency of hydropower, that is, the amount of kinetic energy turned into electrical energy, can be as high as 95 percent for large installations and from 80 to 85 percent for small ones.[23] This high conversion efficiency makes falling water the most efficiently used fuel of all primary energy sources.

Indirectly, hydropower ultimately derives from solar energy. Solar radiation evaporates water from oceans, rivers, and lakes. Water vapor moves with air masses. When the air masses cool, the water condenses and falls as rain or snow. From higher elevations, the water runs downhill, acquiring kinetic energy from the earth's gravitational field. A large river flowing over a waterfall, like Niagara Falls, or held in an artificial lake behind a dam hundreds of meters high may be capable of producing thousands of megawatts in a huge power plant (see figure 3.6).

Alternatively, a small stream may produce anywhere from a few hundred watts to a few tens of megawatts.

The designs of turbines differ substantially. *Impulse turbines* operate by the force of water hitting them, and *reaction turbines* have the turbine blades completely submerged in the water, where they move in reaction to changing water pressures as water flows over them. Turbines must fit with the characteristics of the water resource as well as the transmission and distribution system.

On a global level, hydropower contributes significantly to the world's electricity supplies, about 16 percent, or 3,672 TWh of the 22,668 TWh total (2012). Hydropower ranks third after coal at 40 percent and gas at 23 percent. Hydropower produces more than nuclear power (11 percent), oil (5 percent), and other renewables (5 percent).[24] Hydropower's production came from installed capacity of 926 GW (2009 figures), which produced at 44 percent of capacity.[25]

In a few areas, hydropower now provides huge proportions of electricity. For example, in 2012, Norway generated 97 percent of its electricity from hydropower and the United States generated 7 percent.[26] Washington State in 2013 generated 68 percent of its electricity from hydropower, the highest percentage of any state in the United States.[27] Washington State generally has the lowest prices for electricity in the United States, reflecting its number one rank in hydropower.[28]

Hydropower could become a larger source of electricity, even in areas with large existing capacities. Estimated total production of hydropower could reach 14,576 TWh per year based on a total capacity of 3,721 GW, about four times as much as at present. This expansion would include upgrading existing hydropower facilities, adding hydropower to existing structures currently without it, and instream installations to produce small amounts. Europe has already developed 53 percent of its maximum, North America 39 percent, Asia and Australia-Oceana 20 percent each, Latin America 26 percent, and Africa 8 percent.[29] Thus even in well-developed Europe, there is substantial room for additional hydropower.

Hydroelectricity is one of the lowest cost sources of electric power. Multiple challenges, however, complicate a full assessment of the costs of hydroelectricity. On one side, hydropower installations, especially dams, often have multiple functions producing economic value: irrigation, flood control, drought buffering, and recreation. Dams are expensive to build, and estimating the relative values of multiple benefits is often controversial. On the other side, impoundments of rivers for hydropower causes damages and losses. They displace people and destroy communities, obliterate fish runs and other aquatic life, destroy recreational values of rivers, and significantly alter the landscape by flooding beautiful areas. We will further assess the strengths and weaknesses hydropower and other energy sources in chapters 9 and 10.

Ocean energy

Of all the renewable energy sources, *ocean energy* has the largest theoretical flux (flow of energy per year) of any save solar energy: 7,400 EJ/y compared to 3.9 million EJ/y.[30] Within ocean energy, *ocean thermal energy conversion* (OTEC) and *wave power* probably have the largest potential for practical extraction, 300 EJ/y and 20 EJ/y, respectively.[31] Neither of these forms has achieved commercial success yet, but other technologies have.

Tidal range means the daily fluctuation in the height of the ocean resulting from the gravitational attraction of the moon and sun for ocean water plus factors associated with the earth's rotation. The known changes range from 17 meters at the Bay of Fundy in Canada to less than 1 meter. The kinetic energy of rising and falling water can be transformed to electrical energy. Daily cycles of the tides mean intermittency in production of electricity, but their predictability exceeds the predictability of intermittent solar and wind power.

Tapping the tidal range involves a barrage or dam in an estuary that traps incoming water behind or upstream from the dam. As the water level falls, turbines in the dam activate generators. As much as 1,000,000 MW of capacity might be installed, but that may be too optimistic. As of 2015, only three projects exist: La Rance (France, 240 MW), Sihwa Lake (South Korea, 254 MW), and Annapolis Royal (Canada, 20 MW). Two additional plants are in construction: Swansea Bay (Wales, 240 MW) and MeyGen Tidal Energy (Scotland, 398 MW).[32]

Tidal currents refer to ocean water mass flowing downhill, associated with the rise and fall of tides, as modified by the shape of the tidal basin. No dam is involved, which reduces costs, but only a portion of the falling water moves past the turbine, limiting its output. Promising sites exist in various parts of Europe, Asia, and South America, but this form of ocean energy is not yet commercial. As with wind power, the kinetic energy of the flowing water will vary as the cube of the water's velocity (v^3), so fast-flowing tidal currents have particular value.

Ocean currents result from wind, temperature and salinity variations in ocean water, and gravity. Some, such as the Gulf Stream off eastern North America, have regular steady flow patterns and move at rates sufficient to turn a turbine. Devices to harvest their energy as electricity would be similar to those used to harvest tidal currents, but no commercial project currently exists.

Wave power results from wind blowing across ocean water. Some of the wind's kinetic energy transfers to the water in the form of potential energy (water above average sea level) and kinetic energy (water molecules in motion). Oceans between 30 and 60 degrees, the mid-latitudes, have the most wave energy. The quantities of energy in waves vary seasonally and daily as wind velocities vary. All devices to convert wave energy to electricity involve moving water turning a turbine attached to a generator, but test results with over fifty kinds of devices have not yet firmly identified optimal designs. The total wave energy may be as large as 32,000 TWh/year, about twice the amount of the world's current production of electricity. Until successful commercialization, however, the amount practically harvestable remains unknown. One estimate puts it at about 20 EJ/year (5,556 TWh/y), about 25 percent of the world's total (2012).[33]

Ocean thermal energy conversion (OTEC) recognizes that the differences in temperature between the surface of the ocean and deeper waters can make the ocean a heat engine, analogous to a steam engine. The temperature difference comes from solar radiation warming the surface waters, while the depths remain cool. As with the steam engine, the temperature difference can expand a gas in the warm waters and condense it with cooler water, thus forcing a piston to move back and forth to turn a generator.

Practical exploitation of these temperature differences will likely be far smaller than the annual flux. One estimate predicts that 300 EJ/year (83,333 TWh/y) might be made available, 3.7 times as much as the global electricity generated (2012). Despite the huge potential of OTEC, successful operation of commercial installations has not yet happened.[34]

Salinity gradient as a source of electricity is among the newest and least developed possibilities for renewable energy, but it originates from a physical process—*osmosis*—first recognized in 1748.[35] Osmosis is a physical process in which a special membrane separates two batches of water, which differ in, for example, salinity. Water will naturally move across the membrane from low salinity to high, but the salt does not cross the membrane. This movement of water could generate electrical energy. Over two hundred years elapsed before a practical machine produced a voltage and current.[36]

Current research focuses on the development of membranes to allow selective passage of ions, atoms, or molecules with a positive or negative charge. If membranes allowed selective passage of ions, the resulting voltages could force a current to flow. Small-scale experiments in the Netherlands and Norway have explored using salinity gradients and different kinds of membranes to allow selective passage of water or ions.[37] Both OTEC and salinity gradient power could produce power steadily; they are not as intermittent as solar or wind power.

Biomass Energy

Biomass includes a wide range of materials storing solar energy.[38] Photosynthesis transforms the energy in light into energy of chemical bonds holding the molecules of the material together. Biomass is plant or animal material, whether still living or not, or a material derived from living organisms, and it comprises the renewable energy resource best known to humanity.

Biomass fueled both the First and Second Energy Transitions. Firewood and other plant and animal materials first gave human beings heat and light and forever distinguished us from all other species. Agriculture, the human-directed production of biomass, stabilized and increased food production and changed us from hunter-gatherers into civilized, settled communities (see chapter 1).

The fossil fuels of the Third Energy Transition could also be called a form of biomass, because coal, oil, and gas all had their origins in the remains of plants and animals. The fossil fuels, however, don't renew themselves on an annual basis, and the term *biomass* refers to annually renewable materials. From biomass come *biofuels* and *bioenergy*.[39]

Biomass energy is released by combustion, a chemical process that changes the original biomass into mostly carbon dioxide (CO_2) and water (H_2O) (see appendix 2). Combustion of biomass in oxygen results in more stable products, that is, products holding less potential chemical energy. Fuel at the high ignition temperature has chemical bonds with high potential energy ("unstable bonds"), which change into chemical bonds with low potential energy ("stable bonds"). Combustion releases excess potential energy as heat and light, and, with only a few exceptions, heat is the most valuable product.

Despite the antiquity and familiarity of biomass energy, it, of all the renewable sources of energy, remains the most difficult to characterize and project its contributions to the Fourth Transition. It did not appear in table 8.1 or 8.2 above because of uncertainties about (a) the amount of energy that biomass can provide, (b) the best way to harvest its energy, (c) its potential conflict with raising food, feed, and fiber essential for humans, and (d) its sustainability (see chapter 10). Nevertheless, biomass will retain important uses in the Fourth Transition. In some local areas, it may long remain the best source of energy for tasks like cooking and home heating.

It would take an entire book to explain all the intricacies related to bioenergy, so here we will just indicate the major considerations and provide estimates of the energy biomass will likely provide. Table 8.3 summarizes the various likely uses of biomass in the next few decades.

At the household level, in both industrialized and less industrialized countries, some people will find various forms of biomass the cheapest, easiest, or only fuel choice open to them. Unprocessed biofuels include firewood and dung. Physically altered wood pellets and chemically altered charcoal may be available and preferable to raw wood or dung. These uses today are common, especially in rural areas and in the agrarian economies of countries without much industry. The International Energy Agency estimates the use of about 50 EJ per year of energy from biofuels, 70 to 80 percent of which (35–40 EJ) is in traditional, noncommercial use.[40]

Reliance on biomass, particularly unprocessed materials, may indicate energy poverty (i.e., lack of access to modern energy sources, particularly electricity). Moreover, biomass used indoors for cooking produces extremely unhealthy indoor air quality, damaging the health of women and children especially. Biomass collection requires a great deal of labor and prevents other activities.[41] Eliminating energy poverty for people who desire integration with the modern world represents a major challenge in the Fourth Energy Transition (see chapter 11).

Biomass energy can power many commercial uses with both processed and unprocessed fuels. Cooking, for example, in restaurants, can utilize various forms of biomass energy, either because it's the cheapest or only fuel available or because, in high-end restaurants, it provides a specialty status.[42] Beyond commercial cooking, from an engineering perspective biomass can provide many energy services: water and space heating, generating electricity, and powering vehicles. Except for mobility, unprocessed firewood and physically or chemically processed materials will work.

TABLE 8.3. LIKELY CONTINUING USES OF BIOMASS

PROCESSING STATE	UNPROCESSED	PHYSICALLY PROCESSED	CHEMICALLY PROCESSED
Uses			
Household Level			
Cooking, heat, light	Firewood, dung	Wood pellets	Charcoal; biogas from agricultural, municipal, and forestry wastes
Commercial Level			
Heat for commercial cooking	Firewood	Wood pellets	Charcoal; biogas from agricultural, municipal, and forestry wastes
Space and water heating	Forest, municipal, and agricultural wastes	Wood pellets	Biogas from agricultural, municipal, and forestry wastes
Electricity generation	Forest, municipal, and agricultural wastes	Wood pellets, bagasse, stover, straw	Biogas from agricultural, municipal, and forestry wastes
Mobility			Biogas from agricultural, municipal, and forestry wastes; liquid fuels from biomass

FIGURE 8.9. Corn ethanol plant, West Burlington, Iowa.

SOURCE: Wikimedia, Steven Vaughn, https://commons.wikimedia.org/wiki/File:Ethanol_plant_cropped.jpg, from U.S. Department of Agriculture, Agricultural Research Service.

Electricity generation in the United States, for example, currently uses wood and wood-derived products, landfill gas, municipal solid waste, and other (unspecified) forms of biomass. In 2014, these forms of biomass generated 64,319 GWh of electricity out of a total of 4,092,935 GWh, or 2 percent of all U.S. electricity. Seven states (California, Florida, Georgia, Virginia, Maine, Alabama, and Louisiana) generate about 47 percent of this biomass electricity.[43]

Preparing firewood and dung for uses as fuel requires no elaborate processes or tools other than perhaps a saw and an ax. In contrast, processing biomass on an industrial scale involves machinery and energy inputs. Installations to prepare wood pellets or agricultural wastes, for example, generally are sited in central locations surrounded by a catchment area feeding raw agricultural or forestry wastes into the plants for baling or more elaborate processing. Chemical transformations to make biogas or liquid fuels such as ethanol require even more elaborate equipment and significant engineering expertise.

As noted above, biomass currently contributes about 50 EJ/year, or about 10 percent of the total world energy economy of almost 500 EJ/year. Various estimates of future potential range from 100 to 500 EJ/year.[44] Significant controversies, however, may stymie many proposals for using biomass on industrial scales. For example, use of wood pellets for electricity generation and use of corn (maize) for fuel-grade ethanol have raised accusations of destruction of forest habitat and competition with food needs, respectively (figure 8.9).[45]

It is beyond the scope of this book to resolve the conflicting views about biomass energy, but suffice it to say that biomass probably provokes as many or more disagreements than any of the other sources of renewable energy. These opposing views may limit development of these energy sources and explain why biomass does not appear in tables 8.1 and 8.2. Chapter 10 has more details on the strengths and weaknesses of biomass energy. The other sources will also generate conflicts, perhaps more easily resolved and in any case with greater potential to replace fossil fuels and uranium.

Geothermal Energy

The amount of *geothermal energy* (heat) in the earth's crust (up to about 15,000 meters in depth) vastly surpasses the size of the world's energy economy.[46] Estimates in the first 10,000 meters vary from 110 million to 403 million EJ, compared to the world's current energy use of about 500 EJ per year. In other words, each year people use about 1/200,000 or less of the amount of heat in the crust. In the first 5,000 meters, the amount of heat may be between 56 million and 140 million EJ. And in the first 3,000 meters, one estimate puts the heat content at 34 million EJ.

Strictly speaking, geothermal energy is not really "renewable" in the sense of being perpetual. Heat left from earth's formation and from radioactive decay dissipates slowly to the surface and then to space as infrared radiation. This resource is declining over time and does not replenish; tapping it hastens natural heat transfer.[47]

Geothermal energy also may not be renewable because of water shortages. Without water, it's costly and difficult to bring the heat to the earth's surface for use. If tapping the resource exhausts preexisting water in the deposit, then further harvest of the heat depends on bringing new water sources to the site.[48]

Geothermal energy, however, generally appears in lists of renewable energy resources, because the amounts of heat in the earth are stupendous in comparison with the approximately 500 EJ used each year by people. For all practical purposes, we will continue to treat geothermal energy as renewable, with the caveat that water shortages may be an Achilles' heel of this energy source.

Geothermal energy exists in two forms. The largest but deepest deposits of heat have temperatures that easily generate electricity. The second form involves tapping the lower but steady temperatures very near the surface.

Large, deep, hot deposits. The interior of the earth has a temperature perhaps as high as 7,000 K, or about 6,700°C. This enormous amount of heat, left from the earth's first formation and from radioactive decay of mostly uranium and thorium, keeps part of the earth's interior melted or very hot. The temperature increases as one goes deeper, a *heat gradient.* The best sites for tapping this heat lie in areas with high temperatures near the surface, that is, those with a steep heat gradient.

Hydrothermal systems use steam or other vapor to turn electric generators and occur in three forms: *vapor dominated, liquid dominated,* and *enhanced geothermal systems* (EGS, also called *hot dry rock*). Vapor-dominated deposits have steam that comes to the surface without pumping to operate generators. These are the rarest but most easily harvested deposits of geothermal energy.

Liquid-dominated deposits have pumps (a) at the surface to force water into the deposit and (b) below the surface to deliver heated water either as steam (flash system) or as hot water to heat a liquid with lower boiling point to make a vapor to operate a generator (binary system). These deposits are more common than vapor-dominated deposits.

Enhanced geothermal systems, or hot dry rocks, include most of the exploitable heat deposits, but water cannot move easily through the rocks to gain heat for transfer to the surface. High-pressure water can *hydrofracture* this rock and enhance the ability of water pumped from the surface to become hot enough for flash- or binary-system generation of electricity. Enhanced geothermal systems have higher production expenses than vapor- or liquid-dominated systems.

Typically, production of electricity works best with temperatures above 150°C. Lower temperatures can also spin a turbine, but the Carnot efficiency of the geothermal heat engine will be lower (see chapter 2). Once generated, the electricity can move long distances. The heat can also directly provide space or water heating in nearby areas without transformation to electricity.

Hotter temperatures for electricity occur at many locations. Harvest requires drilling to great depths (as much as 12,000 meters) to access a large enough temperature difference between the surface and the depths. In practice, engineers have installed geothermal electric plants at sites where tectonic plates meet and access to high temperatures occurs at relatively shallow depths.

The amounts of electrical energy that engineers might succeed in generating from the first 3,000 meters is about 89 EJ per year. The highest estimate of heat transformable to electricity from the crust is 364 EJ per

FIGURE 8.10. The Geysers geothermal power plant, near San Francisco.

Steam pipelines snake from production wells to a power plant at The Geysers geothermal field. The plant consists of a turbine room in front and a row of cooling towers behind. In cold weather, the cooling towers emit tall columns of cloud. (*Andrew Alden/KQED.*)

SOURCE: Andrew Alden. Used by permission.

year.[49] These estimates, although far less than the theoretical amounts, still dwarf the current global electricity production, about 61 EJ/year (see table 8.1).

Electric plants first tapped steam from vapor-dominated deposits at a depth of only 250 meters in 1904 in Larderello, Italy, which became commercial in 1913. Larderello generated highly significant levels of electricity after the 1930s, and the amounts generated increased steadily after 1945. The largest vapor-dominated facility is The Geysers, 75 miles north of San Francisco, California, which began production in 1960 and now yields 725 MW_e.[50] The deepest wells at The Geysers plant reach about 3,700 meters; the average depth at the plant is about 2,600 meters (figure 8.10).[51]

Low, near-surface deposits with steady temperature. The second form of geothermal energy utilizes the temperature of the earth's surface, from about 1 to 2 meters down to about 100 meters, which varies little with annual seasons. In temperate latitudes, for example, temperatures will often vary over the course of a year by only a few degrees Centigrade, between about 7° and 21°.[52] Near-surface temperatures result from a balance between heat transmitted from the earth's interior to the surface, solar radiation warming the near-surface, and heat lost by infrared radiation to space.

In locations where the ground is warmer than the air in winter and cooler than the air in summer, a *ground source heat pump* (GSHP), powered by electricity from an external source, can provide highly efficient heating in winter and cooling in summer. The GSHP acts like a two-way refrigerator that can cool a building in summer and heat it in winter. Use of steady ground temperature with a ground source heat pump can happen almost anywhere on earth.

CONCLUSION

The profiles of renewable energy answer a critical question: does enough physical supply exist to contemplate a global energy economy based on energy efficiency and renewable energy? Could such an energy economy really replace the big-four fuels of coal, oil, gas, and uranium? Based on the information summarized here, yes.

No one, however, yet knows for sure if a renewable energy economy can power modern life completely, because this energy economy doesn't yet exist. Serious analysts think it's possible,[53] but barriers and challenges lie ahead (see chapters 10 and 11). Nevertheless, the weaknesses of the big-four fuels justify this strategic goal: *use energy efficiency to lower the amounts of energy needed and derive that energy from 100 percent renewable sources or as close to 100 percent as possible.* We must open up our imaginations to new possibilities.

Does that mean the transition away from the big-four fuels will involve no difficult choices or sacrifices? Will the transition happen without promotion or insistence by government policies? Here the questions become more difficult. Just as the big-four fuels have enormous benefits along with serious drawbacks, energy efficiency and renewable energy, too, come with benefits, costs, and risks. As a result, choices remain, and decisions about those choices entail criteria, to which we turn in chapter 9, and standards for the criteria (chapter 10).

Connections with Everyday Life

Most children don't have to make complicated decisions about life because adults make those decisions for them. As soon as people enter adolescence, however, and then for the rest of their lives, they face a constant barrage of complicated decisions. Should I stay in high school? Should I get a job immediately after high school? Should I go to college? What should I major in? What do I want to do with my life? Should I marry? If so, whom should I marry? Should I take this job offer? Where do I want to live? Should I have children? How should I care for them? What should I do if my child misbehaves or has problems? And the list goes on and on. Only death provides respite.

One way of coping with life's complexities is to throw up your hands. Life will go on, and factors in the cultural context will inevitably dictate events. Alternatively, people can actively try to shape their futures. They can assess cultural contexts and adjust personal plans and aspirations to increase the chances of events working out as they wish. The most active planners of the future will identify specific criteria they think important and then make decisions in line with them.

Energy sources used by societies can greatly affect people's lives, and choices among energy sources can pose complicated trade-offs. Where are the biggest risks, and what pathway looks best for now and for future generations? This chapter urges societies to take an active role in planning a new energy future. Throwing up one's hands is the ultimate cop-out.

To help with assessment and planning, it proposes the criteria most likely to be useful. Energy is too important to be left to the existing cultural context, because that context will unnecessarily and damagingly prolong the unsustainable reliance on the big-four fuels to provide nearly all energy services.

CHAPTER 9

Energy Sources

Criteria for Acceptability

Chapter 8 demonstrated that the physical size of renewable energy sources could—in theory—power the world, and the excess of potential supply over current demand could—in theory—end energy poverty and provide for economic growth as population increases. On the surface, therefore, the possibility exists for the complete elimination of fossil fuels and uranium.

Theoretical possibilities, however, do not equal engineering and practical reality. *All* energy sources—big-four fuels, renewable sources, and efficiency—have strengths and weaknesses. The political ecology framework directed attention to the need for ongoing investments in energy infrastructure (see chapter 5), but what criteria can be used to make the best investment decisions in the future? How should societies decide whether to embrace the benefits of an energy source and tolerate its weaknesses?

This chapter explains the criteria that governed investments in the Third Energy Transition. These criteria will remain important for future investments, supplemented by new criteria recognizing the downsides made prominent by the big-four energy sources. Chapter 10 utilizes all criteria to characterize the strengths and weaknesses of each primary energy source.

CRITERIA GOVERNING THE THIRD ENERGY TRANSITION

The five criteria traditionally used for investments in energy infrastructure are profits and risks, security of access to the resource and any profits earned, size of the resource, matching availability of energy to time of energy use, and geographic distributions.

Profits and Risks

Investing money to make a return, or profit, formed the very heart of commerce in both the premodern and modern industrialized states operating with abundant energy supplies. Before the Third Energy Transition, investments in energy focused on human and animal muscle power, firewood, small amounts of coal, wind power for sailing and pumping water, and water power for running machines.

As the Third Energy Transition unfolded, over the seventeenth to twentieth century, investment opportunities to produce energy expanded in scope and scale. Coal competed with muscle power, firewood, and water power as it entered commerce, but this competition vanished in the 1800s, with coal the victor. After petroleum began to find increasing markets in the mid-1800s, followed later by hydropower, natural gas, and uranium, competition among the fuels continually affected decision making about investments in energy production.

Inevitably, capital invested precedes any possible returns, and every financial investment risks the future not working out as predicted. Political and business leaders of modern states encouraged investments and minimized risks by emphasizing (a) rule of law; (b) a stable, well-regulated credit currency; (c) the expectation of unending, exponential, economic growth; (d) policies for taxes and subsidies; and (e) trade policies.

Rule of law allows the investor to put securely owned money into securely owned property to produce something that without question will belong to the investor. In addition, any regulations or taxes on the project must reasonably balance the needs of the investor with the needs of the public. Insecurity of property rights and unreasonable or unequal taxes and regulations diminish the willingness of investors to take risks. (There are, however, many arguments in democracies about what is "reasonable" and the definition of "equality." One person's reasonable is another's disaster.)

The original investment and realization of a return requires confidence that money—the medium in which commercial transactions occur—will be in ample supply to facilitate commerce and have a reasonably stable value. If insufficient money is available or if the value of money fluctuates unpredictably, especially by deflation or too rapid inflation, investors will avoid the risk.

Embrace of the mineral fuels (and hydropower) created economic growth far beyond that enabled by organic fuels (see chapter 1). In developed energy-rich nations, growth has, during many periods, been exponential; that is, it has grown at a certain percentage rate, year after year. Modern states embrace the idea of unending exponential economic growth to foster investment (but the downsides of the big-four fuels threaten it).

Successful investments made the Third Energy Transition possible by building infrastructure to produce, refine, transport, and use the big-four fuels and hydropower. Inside the industries, these different phases of energy investments are referred to as upstream, midstream, and downstream.

Governments increasingly recognized the key role played by energy, and the energy industries sought subsidies for their projects and products delivered. In the case of fossil fuels, the International Monetary Fund estimates the total global subsidies for coal, oil, and gas at $5.3 trillion in 2015, or 6.5 percent of global GDP.[1] The subsidies promote more use of these fuels than would otherwise occur, and much of the value of the subsidy consists of not paying the costs of externalities: damage to health and the environment. Instead these costs are borne only by the individuals affected.

The upstream, midstream, and downstream components of energy production have generally occurred in the same country, but increasingly after 1900 globalization of trade meant that these steps occurred in different countries. Accordingly, trade and other policies profoundly affected profits and risks in energy investments. Coal, oil, gas, and uranium were first to confront global supply chains, but after 2000 these matters also affected solar, wind, biomass, and other forms of energy.[2]

The perception and the reality of costs, risks, and potential returns, bolstered by subsidy, have therefore always depended on government policy, so decisions about energy investments never have and never will occur in the nonexistent "free market," a market free of governmental rules and subsidies. Costs, risks, and potential returns will always be contingent on government policy, and investors in energy will always be at risk of political events that suddenly alter the rules and threaten their estimations of profits and risks.

The above considerations made investment decision making on energy production contingent on two overarching contexts: competition with other energy sources and the rules of the market established by government policies. These two contextual factors heavily influenced the course of the Third Energy Transition, and they will continue to guide the Fourth. The criterion Profits and Risks assesses the status of competition and markets; without a favorable outlook on this criterion, investment in an energy source won't occur.

Security of Access and Profits

Investors may prefer investment opportunities in their home countries, because they perceive more security of access to an energy resource. For home-country investments, investors will also want stability of the policies governing access. Environmental regulations, designations of protected areas by government, and revisions of tax codes, for example, will affect investors' decisions about the risks they face. Political leaders may also want government investments to generate jobs and local economic activity.

Investments abroad may appear less secure than investments in the home country. The foreign government may seize the project or alter the policy environment governing returns and risks. Investors may have little opportunity to protect their interests when foreign governments change the rules.

For over a century, for example, private British, French, Italian, Spanish, German, and American investors have sought to develop foreign petroleum deposits. China has more recently joined these other countries in making substantial foreign investments. Changes in the policies of foreign governments, however, frequently generated requests from investors to their governments for diplomatic and military actions to secure the investments (see chapter 7). Such security concerns will also attend foreign investments in efficiency and renewable energy.

The criterion Security of Access and Profits assesses the abilities of investors to reap financial rewards that an investment may generate. A low rating on this criterion will discourage investments.

Size of the Resource

In general, a project to develop an energy source that offers only a small supply will not attract investors. Projects have *fixed costs* that an investor incurs regardless of the size of the supply and *variable costs* that depend on the amount of energy produced. Thus investors tend to prefer projects that can harvest large supplies.

For example, geologists may discover a 12-inch seam of coal 300 feet below the surface in one place but a 120-inch seam at the same depth in another place. All other things being equal, the investor will prefer the thicker seam. Digging a pit 300 feet deep costs the same for the smaller seam as for the larger, but the amount of coal harvestable will be ten times higher from the larger deposit.

The same principle operates as depletion exhausts the large, easy to reach deposits of a mineral fuel. Subsequent exploitation of smaller deposits costs more, and the investor will want to see projections, before investing, that the energy produced will sell at adequately high prices.

Investors in energy efficiency and renewable energy will have the same preferences. Developers of a utility-scale solar energy project, for example, may choose a large sunny site with one owner rather than an equally sunny and equally large site with many owners. Transaction costs to buy or lease the area are fixed, and the site with one owner will have lower total production costs.

The criterion Size of Resource assesses the strengths and weaknesses of an energy source based on how much energy an investment might produce above fixed costs.

Time of Use

From Thomas Edison's first electric power plant in 1882, engineers recognized the need to burn coal at rates that made just enough heat to make just enough steam to make just enough electricity to match the demand for electrical energy by consumers. Later power plants using hydropower, oil, gas, geothermal heat, and biomass operated under similar constraints. Successful managers of nuclear power plants also learned to do the same with uranium by the mid-1950s: control rods governed the rate of fission to make heat, steam, and electricity in just the amounts needed.

As electricity demand rises and falls each day, power plants ramp up and down to exactly match the production of electricity with its uses. Some plants, such as those powered by coal and uranium, do not ramp up and down easily, but hydropower, geothermal, oil, and gas do. Accordingly, coal and uranium supply predominantly *base load* power, or the minimum amount of electricity used 24 hours per day every day. The other fuels can produce base load but often produce mostly *peak load*, which typically rises and falls each day.

Solar and wind are *intermittent* sources. Solar energy varies quite predictably with time of day and time of year; it varies more irregularly with cloudiness. Wind energy also has daily and annual variation patterns, but its exact velocity at any given moment and place varies. Variations in solar and wind energy do not easily match the regular daily rise and fall of electricity demand.

Intermittency poses a challenge to using solar and wind energy to generate electricity. The future of the Fourth Energy Transition will depend heavily on the success of engineers, government policies, and individuals learning to use these intermittent sources in ways that match their production of electricity, and this learning is already well under way. Nevertheless, some critics have declared the impossibility of matching electricity from intermittent sources with use and thus have rejected the possibility of an energy economy based on 100 percent renewable energy.

Time of use concerns also affect energy for mobility and other purposes. Gasoline demand, for example, rises in the summer and declines in the cooler and cold parts of the year. Operators of oil refineries know this pattern and routinely adjust their output to match the changing seasonal demands for gasoline. Gasoline stores easily and inexpensively, however, in contrast to electrical energy, where battery storage remains expensive. Thus time of use adjustments have been easier for petroleum than for fuels generating electricity.

Electrification of transport will bring the constraints of managing electricity production and demand to mobility, another adjustment that will require mutually reinforcing innovations by engineers, companies, government, and individuals. Production of electricity for transport will be a component of mastering the uses of intermittent energy sources.

The criterion Time of Use assesses the flexibility of a primary energy source to easily match the shifting patterns of its use over time.

Geographic Distributions

In most cases, the phrase "nearby energy source" has meant "cheapest energy source." Accordingly, different geographic areas developed distinct differences in their respective uses of energy sources. A source good for one area might not be good for another. "Nearby," of course, was just one criterion, and it did not always determine use patterns. If, however, a desirable fuel source was not "nearby," the cost of this fuel would likely be higher due to transport expenses.

The deposits of the big-four fuels have spotty distributions around the globe, which has generated geopolitical tensions (see chapter 7). Traditional fuels, however, have high *energy densities*, or joules per kilogram, which makes it economical to move them physically. Their energy can also move as electricity.

Solar and wind energy are essentially everywhere, although people living at high latitudes near the poles have many months of each year with little or no solar radiation. Hydropower resources are found where rivers run downhill, which is common but does not occur everywhere. Solar, wind, and hydropower can move only as electrical energy, but that is a relatively cheap mode of transport.

The criterion Geographic Distribution assesses the breadth and evenness of the deposits or supplies of a primary energy source.

NEW CRITERIA

Each of the above criteria has long affected energy investments, and each will continue to affect decision making on investments in the Fourth Energy Transition. To these traditional criteria we must add new ones, which follow from the importance of sustainability and democratic choices.

Sustainability

Sustainability has become a meaningless buzzword.[3] A proponent of a project can seek support for a project or idea by claiming sustainability. Similarly, critics of a project can usefully allege unsustainability, because favoring an *unsustainable* idea is silly. Adding specifics, however, makes the word useful.

The concept of sustainability gained prominence in 1986 with the publication of *Our Common Future* by the World Commission on Environment and Development, organized by the United Nations and chaired by Gro Harlem Brundtland, former prime minister of Norway. The "Brundtland Report" offered the essential definition of sustainable development as "development that meets the needs of the present without compromising the ability of future generations to meet their own needs."[4]

For the Commission, sustainability involved an inextricably intertwined set of concerns: economic growth, elimination of poverty, needed constraints on deployment of technology, reductions of inequality, and the protection and nurturing of ecosystems and resources. The aim was to enable future generations to have as good a life as the current one, which meant that sustainability rests on a time line of *forever*. People have never been able to predict the future to "forever," however, so practical predictions seldom go beyond a century. The Commission believed it was impossible to have a sustainable economy without meeting all the intertwined concerns.

To put it bluntly, sustainability simply did not count as a criterion for decision making on investments in the Third Energy Transition.

- The very concept of nonrenewable resources meant the impossibility of a time line of forever. The time line clearly stretched over centuries, but concerns about ultimate exhaustion of supply first arose in the 1800s and have appeared sporadically ever since.
- Equality in the distribution of costs, risks, and benefits mattered seldom and only when citizens wielded power over energy companies.
- The risks posed by the big-four fuels to climate, geopolitical stability, health, and environment threaten the well-being of the current generation and all generations to follow.
- Depletion of nonrenewable resources will at some point confront future generations with EROI values signaling that deposits no longer produce net energy.

More recent work has emphasized that sustainability questions reside within a cluster of "nexus questions." For example, production of usable energy, such as electric power from coal and uranium, requires immense quantities of water. Similarly, production of energy and food, feed, and fiber is interlinked for several reasons. High agricultural yields depend on fertilizers—chemical nutrients requiring energy to manufacture—and on water for irrigation—a physical input requiring energy to move it to agricultural fields. Some of the most contentious debates about the sustainability of energy sources will focus on these types of nexus questions: will, for example, supplies of water or food, feed, and fiber be adequate or adversely affected under specific energy strategies?[5]

Reversal of the Third Transition's indifference to sustainability requires new outlooks. Protection of common property resources, such as the atmosphere, ecosystem services, and visual landscapes, draws attention to preserving well-being for future generations. Similarly, seeking equality in distribution of costs, benefits, and risks points to the need for equity among people.

Democratic Choices

A sustainable energy project must minimize disparities and conflicts among affected communities, and wide public participation in decision making provides the best method to ensure incorporation of all concerns. Democratic choices about energy development, however, have been rare during the Third Energy Transition.

Private investors based investment decisions on projections of profit. Governments invested in energy projects such as oil field development and hydropower dams based on political processes that may or may not have involved extensive input from the public. All too often, small groups with political clout made decisions reflecting their own narrow interests.

Decision making during the Fourth Energy Transition risks repeating disparities and conflicts, and it can be truly sustainable only if energy development involves broader input of public preferences. The most important strategic decisions center on the overall objectives of the new energy economy. For example, should it aim for 100 percent renewable resources within a certain time frame? If yes, then investments directed away from energy efficiency and renewable energy become a barrier to reaching the goal. If no, then which nonrenewable sources should remain, and what plans will deal with nonsustainability?

The mix of technologies also has immense strategic importance. For example, what relative roles should energy efficiency, solar, wind, and other renewable resources play? Regardless of the ultimate goal, what short-term roles should nuclear power and fracking for gas play? Are they bridges or barriers to the long-term goal? For mitigation of climate change, what roles do carbon capture and storage and geoengineering play? Do these technologies (neither of which is developed at the moment) help or hinder the strategy?

The need for specialized expertise will of necessity leave most tactical decisions at the project level to engineering, financial, and management experts. But for sustainability the experts must work within a strategic framework created with broad public input. Without such a framework, decision making about energy will continue to look like it has during the Third Transition: property owners will want to maximize the resources on their property, and companies based on particular fuels will forever argue that their expertise is sufficient.

Public engagement guiding decision making sounds wholesome but trivial. After all, in democracies doesn't the public already guide decision making? Don't elected officials rely on public input to regulate energy markets and set standards of conduct for energy producers?

Yes, of course, democracies already have avenues for the public to express choices about energy. All too often, however, the public suffers when pitted against individuals and companies that pay for lobbyists, make large campaign contributions, organize into trade associations, and hire articulate lawyers and engineers to make the case for their products. With vastly uneven resources, energy debates quickly become David-Goliath face-offs, except that in this case David usually loses.

Sustainability is a prime criterion for guiding investments in new energy infrastructure. We further specify the meaning of sustainability by examining protection of common resources; distribution of benefits, costs, and risks; ethics; and aesthetics.

Protection of Common Resources

Protection of common-property resources centers on protection of ecosystems, the atmosphere, and the oceans. Terrestrial ecosystems have mixed public and private ownership, but no country and no individual owns the atmosphere. Countries can control economic activities in the ocean within 200 miles of their shores but cannot own the oceans. Air and oceans belong in common to all humanity, and all people survive courtesy of the global ecosystem.

Global energy economies have treated both the atmosphere and ecosystems as dumping grounds for garbage. They have dumped carbon dioxide, methane, and other greenhouse gases (GHGs) into the atmosphere with little or no constraint or charge. Increased GHGs are raising global mean surface temperatures, with multiple effects on terrestrial and aquatic ecosystems. Carbon dioxide dissolved in the ocean has changed its acidity and abilities to support life. Sea levels are rising, threatening cities and agroecosystems. The potential for "tipping points" risks sending ecosystems into degradation spirals too rapid to manage or respond to (see chapter 6).

In essence, individuals and companies appropriated dumping rights, but dumping without constraint must not continue if future generations are to have a chance to meet their own needs. Future investments in energy infrastructure must consider GHG emissions as a factor affecting choices of energy sources.

Investments in renewable energy and efficiency also affect common property resources. Birds, bats, and other wildlife may be harmed or killed by wind turbines and concentrating solar power plants. Wind farms may create shadows and noise, with adverse effects on human health and wildlife. Replacement of natural landscapes with high biodiversity by plantations for biomass energy damages common resources and the livelihoods of people in the area. Seismic activity can follow hydrofracturing, whether the aim is to produce oil, gas, or geothermal energy. Similarly, dams for hydropower have profound effects on the landscape, ecosystems, and seismic stability. Simply because energy is renewable does not automatically mean it has no adverse effects.

Distribution of Benefits, Costs, and Risks

Monetary and nonmonetary costs, benefits, and risks have always been part of energy development. During the early stages of the Third Transition, from 1700 into the mid-1800s, monetary concerns reigned supreme. Investment in coal, oil, and gas tracked property rights. Either property owners developed their own resources, or entrepreneurs leased rights to explore and extract. Investors bore the risks and reaped returns. The *distribution* of costs, risks, and benefits didn't play much of a role, because decision making focused only on profits.

In each country, energy development enabled agrarian states to become fully modern ones, and issues of distribution increasingly came to the forefront. Distribution of both monetary and nonmonetary costs, risks, and benefits sparked debates among disparate and conflicting interests of investors, property owners, labor, consumers, and citizens.

Highly inequitable distributions of costs, risks, and benefits have frequently fired political rancor and, sometimes, violent conflict. Discord hindered investments and the change to a new, sustainable energy economy based on efficiency and renewable energy; equity, in contrast, produces a more favorable environment for investment in the Fourth Energy Transition.

Ethics

Ethics points to the need to consider equity among people immediately affected by energy investments and effects on unborn generations. Ethical concerns reflect the strong moral overtones of recent public debates about energy, climate change, health and environmental effects, geopolitical tensions, and the exhaustion of resources. Effectiveness and costs—the issues dominating energy investments up to this time—will continue to be important, but questions of morality add new dimensions previously unrecognized, avoided, or ignored.

Ethicists raise questions about the existence of moral truths and the psychology of moral reasoning. They ask, Which normative standards should prevail? Does the ethical decision maker focus on rules of virtue, duties, or consequences? What happens when ethicists disagree? If the moral pathway for one group of people differs from that of another, then what?[6]

In short, ethical assessment raises thorny but important questions about the right things to do. Decision making on energy investments must include issues of ethics.

Aesthetics

Visual landscapes are a particular kind of common property resource, but it's important to draw specific attention to them.[7] Unlike protection of ecosystems, the atmosphere, and oceans, which rests on quantifiable physical-chemical properties, aesthetics entails inherently subjective perceptions. Subjectivity, however, does not mean unimportant.

Investments in energy infrastructure have always altered landscapes and thus the visual appearances of rural and urban areas. After Newcomen's engine launched the Third Energy Transition in 1712, buildings housing steam engines began to dot the landscapes around mines, and later factories belching smoke from the engines became the new urban look. Judgments about the beauty or ugliness of these new visual features was largely in the eyes of respective beholders and their differing links to benefits and risks of the facilities.

Over time people with access to political power increasingly insisted that energy installations such as mines, wells, power plants, and electric transmission lines be located out of sight of their neighborhoods. Low-income people and minorities have often lacked sufficient political power to keep new installations from their neighborhoods, or they have had to move close to existing facilities because of lower prices.

Battles have erupted over the aesthetic impacts of wind-turbine farms and the installation of utility-scale solar power plants. Similarly, complaints that hydropower dams destroy the beauty of previously wild areas have affected dam installations and motivated dam removals. Other types of environmental and health impacts have also attended wind, solar, and hydropower installations, but the aesthetic impacts merit serious attention as a distinct issue.

Are such facilities "ugly" and therefore unacceptable? Or are they the symbols of a "beautiful" new sustainable energy economy? Who should decide? Must the "loser" receive compensation? From whom? Of what kind? Voluntary or mandatory? What is the ethical pathway for decision-making on aesthetics? Investors and regulators must consider aesthetics to avoid or resolve disputes before they inevitably erupt.

METHODS PROVIDING INSIGHTS TO THE CRITERIA

Chapter 10 applies both the traditional and new criteria—based on *qualitative* indicators of strengths and weaknesses—to the big-four fuels and to renewable sources of energy. Three *quantitative* indicators provide additional useful insights to both groups of criteria. This section briefly explains these quantitative indicators.

Life Cycle Analysis

Life cycle analysis (LCA) originated in the late 1960s in the United States and Europe to assess the costs and environmental impacts of a product over its entire life cycle. The Coca-Cola Company in the United States spurred its invention as it studied the feasibility of producing the packaging of its product, including the possibility of moving from glass to plastic. The term *life cycle analysis* was first used in 1990.[8]

LCA was one of the first methods seeking to define, on a quantitative and scientific basis, the increasingly serious environmental impacts of modern industry. The idea focused on quantitative estimates of all inputs and outputs related to a product, from its origins as raw materials through to the end of its functional life and disposal as waste. The method can focus on any input or output of interest. Over time, LCA developed great utility in estimating energy inputs and outputs and the evolution of greenhouse gases like carbon dioxide and methane.

In the field of energy, LCA now includes estimates of costs, raw material inputs, energy inputs, waste products like carbon dioxide, and ancillary materials such as methane. High outputs of carbon dioxide and methane over the life cycle of a product imply a process detrimental to climate stability, a signal that increasingly will deter investment. Levelized cost of electricity (LCOE) and energy return on investment (EROI) are specialized LCA estimates.

Levelized Cost of Electricity

The *levelized cost of electricity* estimates costs of electricity from a power plant by calculating the even stream of revenues needed from sales to exactly balance the costs of building, operating, and maintaining the plant over its entire life. LCOE estimates typically play a role in a utility's decisions about what kinds of new plants to build, including selection of fuel. The measurement has increasingly played a role in strategic choices about fuels to be used for decades into the future over the life expectancy of the plant.

LCOE uses assumptions about the prices of the plant's fuel over many years, operation and maintenance costs based on the durability and reliability of the plant, and other known or assumed costs. Investors find LCOE estimates useful in estimating likely profits to be made by investing in the plant, especially when the plant must sell its electricity into a deregulated or competitive market. All other things being equal (almost never the case!), a low LCOE attracts investment while a high value deters it.

Energy Return on Investment

Energy return on investment (EROI) and net energy (NE) (see chapter 7) played little explicit role in the early periods of the Third Energy Transition, not because they didn't matter, but because high values of EROI made obvious the advantages of investments in coal, oil, gas, and hydropower. Costs and returns measured with money alone, not energy units, adequately guided investments in these mineral fuels.

EROI and NE provide useful information about processes of depletion, the factor that intrinsically makes energy economies based on nonrenewable resources unsustainable. Renewable resources by definition have an inexhaustible supply, so reliance on them can in theory last forever.

EROI and NE may not be as high for renewable energy as those in the early 1900s were for fossil fuels and hydropower. In contrast, EROI and NE for energy efficiency investments may be stunningly large, perhaps rivaling or surpassing those for fossil fuels and hydropower. In addition, improvements in technology for harvesting renewable energy and for energy efficiency will likely improve EROI values over time.

EROI and NE, however, remain key for renewable energy technologies. If, for example, energy from solar, wind, and hydropower cannot power the production, reproduction, operation, and maintenance of the facility, plus provide ample power off-site, then these particular fuel-technology combinations cannot sustainably power an industrial economy. Too low a value for EROI for renewable energy indicates the need for an energy subsidy from "somewhere else" to succeed, and only the big-four fuels can provide that energy subsidy. It is conceivable that low EROIs may indicate the impossibility of an energy economy based on 100 percent renewable energy. Energy efficiency mitigates low EROI values, and better harvesting technology for renewable energy raises them.

Chapters 10 and 11 return to this point, because knowledge of EROI values for renewable energy systems are just now beginning to appear. Closer margins between energy costs to harvest renewable energy make this criterion especially important to monitor. Moving as close as possible to the strategic goal of a 100 percent renewable energy economy will require efficient use of every joule harvested and continued technology innovation.

CONCLUSION

One of the most difficult challenges in the future of energy economies lies in organizing information in ways that people other than engineers and technical experts can understand the issues at stake and the strengths and weaknesses of energy sources in addressing those issues. The criteria identified here, supplemented by the three quantitative measures, LCA, LCOE, and EROI, and aided by visualizing energy flows in Sankey diagrams within a framework of political ecology (see chapter 5) can help organize productive discussions about energy sources. They will have served their purpose if they help all people, not just engineers and businesses, make strategic choices about future investments in energy.

Criteria by themselves, however, are only half of the story. The other half lies in *scoring* each energy source by each criterion and then deciding the *standards* needed on each criterion for a source to rank as *strong* or *weak*. Chapter 10 turns to the latter task by ranking each primary energy source and energy efficiency against each criterion with qualitative indicators of strengths and weaknesses. The three quantitative measures help inform the ranking, but by themselves they do not constitute either a strength or a weakness.

Connections with Everyday Life

When a person sets a goal to achieve some objective and devises a plan to reach the goal, two processes immediately take place to judge their likely effectiveness. First, the person needs criteria to assess the actions. Second, the person needs to assess the strengths and weaknesses of each action based on the criteria.

This book promotes the search for a workable strategy to achieve the goal of a sustainable energy economy. The actions focus on adoption of energy efficiency and on selection of primary energy sources for needed energy services.

Chapter 9 laid out the criteria for judging the actions. Chapter 10 now assesses each potential action for its strengths and weaknesses according to those criteria. In other words, we will produce a scorecard to determine which actions look like winners and which look like they won't work. The narrative here, therefore, will be familiar to anybody who has sought to establish and reach a challenging goal.

Strengths and Weaknesses of Primary Energy Sources

Productive debates about energy must focus on the strengths and weaknesses of energy sources, choices made by democratic societies, and investments in sources judged most acceptable for the future. Debates on whether an energy source now in use should continue in use for the immediate future may also be important, but they are secondary to the issues addressed here. Current energy sources reflect investment choices of the past, and, with some exceptions, those decisions will last until the facilities and machines involved require replacement. The scorecard focuses on the future.

THE SCORECARD

Table 10.1 summarizes the rankings of different energy sources in terms of the criteria established in chapter 9. Inspired by Likert scales used in survey research, it uses an *ordinal* scale. In other words, this table holds my answers to a survey in which I rated energy efficiency and the different primary energy sources on each of the criteria developed in chapter 9. The rating +++ shows a strong, well-established strength; ++ is strong and tending stronger; + indicates a strength unlikely to grow or with qualifications. On the negative side, – – – indicates a debilitating, fatal flaw; – – is a serious flaw either tending worse or very difficult to mitigate; – indicates a weakness that is remediable or tolerable. A rating with both + and – indicators reflects a volatile, mixed, or geographically diverse situation.

TABLE 10.1.   COMPARISON OF PRIMARY ENERGY SOURCES AND EFFICIENCY WITH CRITERIA OF CHAPTER 9 (GLOBAL VIEW)

CRITERIA SOURCE	Profits and Risks	Security of Access and Profits	Size of Resource	Time of Use	Geographic Distributions	Protection of Common Resources	Distribution of Benefits, Costs and Risks	Ethics	Aesthetics
	Traditional Criteria					New Criteria for Sustainability			
Coal	+++ / --	+++	+++	+	++ / -	---	+ / ---	---	---
Petroleum	+++ / --	+ / --	+	+++	---	---	+ / ---	---	- / ---
Gas	+++ / -	+ / --	+	+++	--	---	---	---	- / ---
Uranium	+ / ---	+++	+	+	-	+ / --	+ / ---	---	-
Efficiency	+++	+++	+++	+++	+++	+++	+ / -	+++	+++
Solar	++ / -	+++	++	-	++ / -	+++	+++	+++	+ / -
Wind	++ / -	+++	++	-	++ / -	+++	+++	+++	+ / -
Hydropower	+ / -	+++	++	+++	+++	+++ / -	+++ / -	+++ / -	+ / -
Biomass	+ / --	+++	--	+++	+++	+ / --	+ / --	+ / --	+
Geothermal	+ / -	+++	+ / +++	+++	+++	+++	+++ / -	+++	-

NOTE: Symbols: + means scores well on this criterion; ++ means strength with trend toward stronger; +++ means exceptional strength; – means does not score well; –– means weak and trend to weaker; ––– means exceptional weakness. Mix of + and – means contradictory indicators of strength and weakness.

Table 10.2 summarizes the quantitative attributes of each primary energy source in studies of life cycle assessment (LCA), levelized cost of electricity (LCOE), and energy return on investment (EROI) (see chapters 7 and 9). In its summary of comparative emissions of carbon dioxide and other greenhouse gases, the Intergovernmental Panel on Climate Change showed that renewable energy, except possibly biomass, emits far lower amounts of greenhouse gases than the fossil fuels (coal, oil, natural gas). Most estimates of greenhouse gas emissions for uranium also are lower (figure 10.1).[1]

Analysts have frequently turned to numbers as objective assessments of important dimensions, especially when factional disputes and mistrust cloud social debates.[2] Despite their usefulness, numerical measures rely on assumptions, which in turn reflect subjective and often contested inputs. Full appreciation of the information contained in the numbers, therefore, requires an assessment of the methods underlying their calculations.

Tables 10.1 and 10.2 focus on critical issues surrounding each fuel, but no ranking system will ever have complete objectivity or permanence. Rankings of necessity reflect value and ethical considerations, and they also reflect conditions at the time of ranking; technological and other changes will affect strengths and weaknesses. Finally, rankings of energy sources for particular geographic areas, not global, may well vary from global rankings. Nevertheless, these rankings are my judgments about the primary energy sources best suited to build a new, sustainable energy economy over the next thirty to forty years.

The industries involved in each primary energy source have three components: production, transport, and sale to users. In some cases, one company integrates all three functions. In other cases, different groups of companies may be involved. Competition among companies associated with a particular energy source varies from weak to strong and affects rankings. Table 10.1 emphasizes the strengths and weaknesses of the primary energy source at the site and time of production.

Neither table presents a decision matrix. Ordinal rankings and numbers inform but should not "determine" decisions. Good decisions about choices for energy investments must reflect public debates, which will usually involve multiple issues. Choices about energy will generate winners and losers, and ethics or pragmatism may require the majority to mitigate or compensate losses.

The discussions below about respective energy sources follow a standard format. Most important, they summarize prospects for

TABLE 10.2. COMPARISON OF RANGES OF NUMERICAL ESTIMATES FOR PRIMARY ENERGY SOURCES (LCA FOR CO₂, EROI, LCOE)

Primary Energy Source	EROI (E_{out} / E_{in})		LCA for CO_2 emissions (g CO_2-eq / kWh)			LCOE ($ / MWh, 2020)	Energy Payback Times for Electricity Generation (years)	
	Min	Max	Min	Median	Max		Min	Max
Coal	27	80	675	1,001	1,689	95.1	0.5	3.7
Petroleum	10	45	510	840	1,170	75.1	1.2	3.9
Gas	20	67	290	469	930	95.2	0.8	3.0
Uranium	5	15	1	16	220	125.3	0.2	8.0
Solar PV	6	12	5	46	217	239.7	0.7	7.5
Solar CSP	1.6		7	22	89	73.6	0.1	1.5
Wind – onshore	18		2	12	81			
Wind – offshore					196.9			
Hydropower	>100		0	4	43	83.5	0.1	3.5
Biomass	0.8	10	−633	18	75	100.5	0.6	3.6
Geothermal	32.4		6	45	79	47.8		

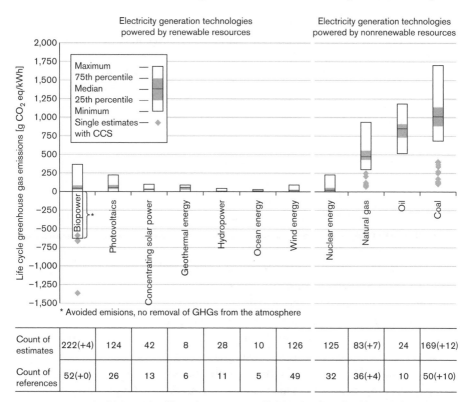

FIGURE 10.1. Comparative life cycle assessments (LCA) of carbon dioxide emissions from different primary energy sources.

Estimates of life cycle GHG emissions (g CO_2eq/kWh) for broad categories of electricity generation technologies, plus some technologies integrated with CCS. Land use–related net changes in carbon stocks (mainly applicable to biopower and hydropower from reservoirs) and land management impacts are excluded; negative estimates1 for biopower are based on assumptions about avoided emissions from residues and wastes in landfill disposals and co-products.

SOURCE: Figure SPM.8 from IPCC, 2011: Summary for Policymakers. In *IPCC Special Report on Renewable Energy Sources and Climate Change Mitigation* [O. Edenhofer, R. Pichs-Madruga, Y. Sokona, K. Seyboth, P. Matschoss, S. Kadner, T. Zwickel, P. Eickemeier, G. Hansen, S. Schlomer, C. von Stechow (eds.)]. Cambridge University Press, Cambridge, UK, and New York, NY, USA. Used by permission.

industries to make a profit from capturing energy. That is, they rank the energy source on the Profits and Risks criterion, which is key to that source's future.

This criterion assesses current technology and market conditions. Evidence included the economic size and stability of industries, the role of policies and subsidies, and prospects for the future. The criterion does not assess the financial health of any particular company, as a company producing a strong primary energy source may find itself facing losses or bankruptcy for reasons outside the general prospects of its products. These rankings reflect the global situation in 2016.

THE TRADITIONAL BIG-FOUR FUELS

Coal

Coal investments have generated great profits at low risk for well over three centuries. In two critical industries alone, generating electricity and making iron and steel, coal remains a fuel in high demand. In 2012, coal ranked second only to oil and accounted for 29 percent of the joules powering the world.[3]

Coal production and consumption grew rapidly over the past forty years, from 3 billion tons in 1972 to over 8 billion in 2013. Growth after 2000 followed China's march to become the largest producer and consumer. Once mined, coal is divided into three major streams. The largest in 2014 (77 percent by weight) makes steam for electricity, the second (13 percent) goes into coking for the iron industry, and the third (10 percent) is lignite, a low-heat material used mostly for making steam, manufactured gas, and ammonia for fertilizer.[4] Coal has many markets and customers, which is good for investors and producers.

A special strength of coal is coking: heating coal in the absence of oxygen to drive off volatile materials and leave nearly pure carbon. In eighteenth-century England coke replaced charcoal (and liberated iron-making from the constrained supply of firewood. Making iron, a critical industry, produces about 7 percent of the total carbon dioxide emissions each year. Iron smelting requires about 24 EJ per year, about 5 percent of world energy use. Coke made from coal is not the only way to smelt iron from its ore, but the utility of coke will be a challenge to replace.[5] These strengths earn coal a rank of +++ under Profits and Risks.

Given the immense amounts of coal used in key industries and the immense amounts remaining (see chapter 4), the question arises, what could possibly detract from the attractiveness of coal investments? There are two factors. First, competition with natural gas and renewable energy has replaced coal in the United States and other places. For example, in 2007, coal supplied 49 percent of U.S. electricity, natural gas 22 percent, and renewable energy other than hydropower 3 percent. Just seven years later, in 2014, coal had dropped to 39 percent, while gas rose to 27 percent and renewables rose to 7 percent.[6]

Second, coal produces more carbon dioxide per joule than the other fossil fuels, oil and gas. No global scheme, other than the Paris climate agreement of 2015, exists to eliminate fossil fuels yet, but concerns about climate change continue to mount. The future of coal looks bleak as movements to curtail carbon dioxide solidify (see chapter 6). This, plus increasing competition with other fuels, earns coal -- under Profits and Risks for an overall score of +++ / --.

Coal production does not currently suffer risks from geopolitical conflicts. Companies invested in coal production have access to the land for mining, and stable rule of law governs profits earned. In a few areas, for example, eastern Ukraine, political turmoil and war have hindered coal production and made both access and money earned unsafe. But few such instabilities currently affect coal. Coal earns +++ for Security of Access and Profits.

Coal reserves stretch to over a century globally (see chapter 4). This extended period of readily available coal gives it a +++ for Size of Resource.

Coal's main use, electricity generation, has a constraint. Coal-fired power plants work best when they run steadily. Electricity demand, however, rises in the morning and declines in the afternoon and evening. Coal-fired plants easily meet base load, the steady minimum always needed. Coal power plants can ramp up and down but not as well as, for example, gas, solar, or hydropower. For other uses, coal does not present a problem on the Time of Use criterion. Overall, coal merits a + for this criterion.

Coal, like oil and gas, has uneven distribution around the world. Three countries (United States, Russian Federation, China) hold over half the reserves, and over 90 percent of reserves lie in just ten countries. A total of forty-six countries use coal but produce none domestically (see chapter 4). Despite disparities between have and have-not nations, coal trade has generally been reliable, without disruptions from political strife. The size of trade has more than doubled since 2000 and currently amounts to 23 percent of total consumption, compared to only 11 percent in 1980.[7] Overall, coal merits a ++ / - for Geographic Distribution.

Protection of Common Resources, one of the criteria for sustainability, puts coal in quite a different light. Coal burning produces many pollutants that enter the atmosphere and oceans, including sulfur oxides, nitrogen oxides, mercury, radioactive materials, particulates, and carbon dioxide. All of these materials damage health and disrupt ecosystems (see chapter 7). Carbon dioxide's association with climate change, sea level rise, and ocean acidification poses serious risks (see chapter 6). On this criterion, coal merits ---, a deadly menace counter-balancing its benefits.

Coal has a decidedly mixed record in terms of Distribution of Benefits, Costs, and Risks. Industrialized countries with extensive use of coal for generating electricity and making iron have generally spread the benefits of these two commodities fairly broadly through government policies. In coal mining, labor often fared poorly until it received protection from strong unions and government policies. Labor struggles in coal mining began in England and continue today.[8] All countries suffer the risks of climate change, sea level rise, and / or ocean acidification. Industrialized countries share the benefits and the costs and risks of coal, not equally but not with profound gaps.

In contrast, less industrialized countries have burned little coal, received few benefits, and yet will suffer the full force of coal's damages to common property resources. The gaps between beneficiaries and losers are unjust, as climate change will affect all people. Overall coal merits a ranking of + / --- on Distribution of Benefits, Costs, and Risks.

The low rankings of coal on Protection of Common Resources and on Distribution of Benefits, Costs, and Risks generates moral concerns. Climate change, sea level rise, and ocean acidification will affect unborn generations. The gaps between industrialized and less industrialized countries indicates inequality without moral justification. Overall, coal ranks --- on Ethics.

Coal mining has left great scars on the earth. Huge slag heaps mar the landscapes of Britain. Coal mine fires smolder in Pennsylvania and China. Strip mines and mountaintop removal, even if "reclaimed," have destroyed the beauty and aquatic resources of Appalachia and lowered the biological productivity of the plains of Wyoming. Perhaps these profoundly altered landscapes have intrinsic beauty to those who have profited from the mining, but most people would prefer the original landscapes. On Aesthetics, coal merits a rank of ---.

Petroleum

In terms of Profits and Risks, investments in petroleum, like coal, have a mixed ranking. Profits from petroleum have generated some of the largest personal fortunes ever amassed, and countries with national oil companies earn billions of dollars in revenue each year. Mobility on the land, sea, and air creates a continuing market belonging, at the moment, almost 100 percent to refined petroleum products. Petroleum also goes into plastics, lubricants, and other products. Production costs in the Middle East remain the lowest on earth, and money invested to bring that oil to market will continue to generate profits. On the surface, therefore, investments in petroleum production merit a ranking of +++ on this criterion.

Despite vital uses and a product worth over a trillion (10^{12}) dollars per year, various factors have upset old expectations for returns on investment, as well as international flows of money and geopolitics. First, more efficient automobiles in North America and Europe have required less fuel to provide mobility. At the same time, increased numbers of autos in China and India have increased these countries' imports of oil and altered the flows of money. In the future, the share of the world's oil going to these countries will increase, with potential geopolitical implications.

Second, significant production of oil from shale by hydrofracturing (fracking) after 2010 in the United States reduced the dependence of the country on imported oil. In addition, fracked oil by relatively small producers introduced a kind of manufacturing process into oil production that responds rapidly to changes in oil prices, a factor that stabilizes global market prices. At the same time, fracking producers rely more on bank financing than their own cash reserves, which can destabilize prices. Thus fracking for oil mixes expectations about returns on investment for other producers.

For conventional oil production, deposits now under development lie very deep, offshore, or in difficult places like the Arctic. Investments now and in the future will likely face higher drilling expenses with greater risks. For example, Royal Dutch Shell in 2015 suspended its search for offshore oil in the Chukchi Sea north of Alaska. A $7 billion investment produced no returns.[9]

Uncertainties roil the oil markets, but more fundamental changes may alter the future even more. Electrified automobiles and regulations of carbon emissions to mitigate climate change hang like dark clouds over investment in oil. These trends merit a -- ranking on Profits and Risks and an overall ranking of +++ / --.[10]

On Security of Access and Profits, petroleum, unlike coal, is mixed and volatile. For the first century of the "petroleum age"—roughly 1860 to 1960—petroleum assets fell to imperial capture and military adventurism, primarily to Britain, France, the Netherlands, Japan, the United States, the Russian Empire, and Germany (see chapter 7). After World War II, the formation of OPEC and national petroleum companies solidified national ownership of petroleum reserves previously captured by empires. With one exception, therefore, political strife over access to oil diminished. Where strife ended, investment in petroleum became safe as well as lucrative, meriting a ranking of +.

Middle Eastern petroleum remained the overwhelming exception, despite the formation of national oil companies. In 1953, the United States and the United Kingdom conspired to overthrow the Iranian government and reinstalled the shah. Iraqi nationalists overthrew the British-installed monarchy in 1958. Iranian Islamists overthrew the shah in 1979 and held American hostages for over a year. A year later, Iraq attacked Iran, aided by the United States. In 1990, Iraq attacked Kuwait but was repelled by an international coalition led by the United States. In the 1980s through the 2000s, the United States and Iran constantly traded accusations of dire crimes. In 2001, a group consisting mostly of citizens of Saudi Arabia, led by Osama bin Laden, attacked the United States, with the loss of over three thousand lives. This attack provided an alleged but false reason for the United States and Britain to invade Iraq in 2003 and overthrow its government, followed by the execution of its president. The invasion and conquest of Iraq destabilized the entire Middle East, a situation that remains unresolved.

Regardless of claims to the contrary, it is clear that petroleum was a major underlying cause of these multiple conflicts. The region holds over 800 billion barrels of proven petroleum reserves, about 49 percent of the world's total.[11] Most of this oil is inexpensive to produce, thus earning high profits even when prices dip. Since the first British incursion into Persia in 1901, the region's oil reserves have provoked greed and aggression around the world.

If one factors in high consumption in countries with scanty reserves and high imports (e.g. the United States, Japan, EU, and China), the constant tension around the world's petroleum markets grows. Although not about oil, U.S. support of Israel and Israel's continued settlements of Palestinian lands further inflame the entire region.

An optimist would summarize these varied geopolitical factors as -- for petroleum's score on Security of Access and Profits; a pessimist might conclude that only --- captures the essence of the situation. I have settled on optimism, with a + as noted above, for an overall ranking of + / --.

The Size of Resource criterion for petroleum merits a +. Proven reserves provide at least fifty years of consumption at current rates (see chapter 4). This is less than the reserves of coal, but it is more than enough to justify current investments, despite the likely increased costs of accessing more difficult deposits.

Time of Use for petroleum merits +++. Once extracted, the resource can be stored as unrefined petroleum or as refined fuels. Petroleum's high energy density, easy transport, and stable storage make it an ideal fuel for use at any time in any place.

On Geographic Distribution, petroleum earns a ---. Over 165 countries must import all or some of the petroleum they use. Petroleum's lopsided distribution patterns greatly weaken its desirability.

Petroleum also merits --- for Protection of Common Resources. Most of the problems stem from inevitable emissions of carbon dioxide, leading to climate change, rising sea levels, and ocean acidification. Particulate emissions and nitrogen oxides, both damaging to health, also are concerns.

On Distribution of Benefits, Costs, and Risks, the ranking + / --- captures petroleum's mixed attributes. Benefits of mobility have penetrated even into poor, less industrialized countries. Motor vehicles, often fueled by imported gasoline, traverse many roads where electricity has not arrived. Trade in goods has benefited some of the world's poorest people. It defies logic to not recognize these benefits, even where the sharing of them remains highly unequal. At the same time, the rich world burns by far the biggest share of oil products, which exacerbates climate change, sea level rise, and ocean acidification.

Multiple issues in the Ethics criterion mandate a ranking of --- for petroleum. The geopolitical barbarity incited by the desire to capture oil reserves never had or will have moral justification. In addition, inequities in the distribution of benefits and inequities in the suffering from climate change and ocean acidification create fatal weaknesses.

On Aesthetics, conventional petroleum earns -. An engineer or resource owner may think an oil derrick, pipeline, oil refinery, or corner gasoline station is beautiful. After all, the technology is clever and lucrative. But most people don't see much beauty. Given today's means of mobility, these things are of course necessary and practical. Some of the ugliness can be mitigated by design. Fracked petroleum covers the landscape much more densely with wells, plus wastewater ponds, bright lights, and noise. Fracked petroleum merits -- on Aesthetics.

Gas

Uncertainty best describes the past four decades of the natural gas industry. New technologies, changing public policy, and intrinsic physical properties of natural gas (primarily methane, or CH_4) sent prices and uses of natural gas in the United States on a roller coaster ride.

Large-scale use of gas depended on pipelines to move it from well to consumer. Citizens early on called on government to regulate gas prices to prevent abuses by pipeline companies with monopoly powers. Pipelines locked commerce in gas to places with pipelines and/or facilities for liquefied natural gas (LNG), the latter developed commercially after 1950. Gas markets therefore remained regional while a global market for oil developed early in the 1900s. LNG made gas markets somewhat more global but only in places with expensive LNG facilities. The necessity for pipelines also created geopolitical issues for transporting gas across national boundaries.

From the 1950s into the 1980s, the U.S. government and gas producers often had contrasting views about the price level appropriate for a regulated commodity. This divergence coincided with the supply and price turmoil around gas and oil after the Arab oil embargo in 1973. Where Congress had seen shortage, the industry had seen prices too low to support exploration and discovery. Congress thoroughly reworked gas regulations in 1978 to increase some prices and then phase out regulation of most prices at the well by 1987.[12]

Gas regulation continued to have many facets,[13] but the shortages seen in the late 1970s abated, and by the mid-1990s production in the United States again equaled that of the early 1970s. After hydrofracturing technology began to deliver substantial quantities of gas to market in 2008, production rose steadily and prices of U.S. gas dropped compared to prices in Europe and Japan. Cheap gas replaced coal for electricity generation, a use that had been prohibited for new U.S. electrical plants between 1978 and 1987.[14]

Once delivered, gas burned cleanly and had great versatility in industry, generation of electricity, commercial establishments, homes, and nonfuel uses. Based on revised regulatory schemes and fracking, investors seeking a return will find natural gas an attractive investment, and it ranks +++ for these reasons. The heavy dependence of gas on changeable regulatory policies, however, adds uncertainty to these investments, thus ranking gas also as – on Profits and Risks, for an overall rank of +++ / –.

For Security of Access and Profits, gas ranks a mixed + / ––, for much the same reason that oil did. A great deal of the world's gas accompanies oil production, much of which occurs in the volatile Middle East. This, plus the need to move gas in pipelines across national boundaries, means there is great uncertainty in investments in and reliance on gas from the Middle East and elsewhere. Yet the markets for gas have become much more global with LNG facilities, and this trading makes money that is reasonably secure.

On Size of Resource, gas, like petroleum, ranks +. Global reserves at current rates of use are fifty-seven years, significant but not as strong as coal. If fracking technology moves successfully out of the United States, these reserves may expand, which might change the ranking on this criterion to ++. At the moment, however, + seems more prudent.

Gas stores easily, and devices to burn it can be easily turned up or down as demand for heat changes. Gas ranks +++ on Time of Use.

The top ten countries hold 79 percent of gas reserves (see chapter 4). Such disparity across different countries weakens gas to −− on Geographic Distribution, a weakness difficult to mitigate. Given the uncertainty about the use of fracking to open up new reserves in other countries, including those without significant reserves of conventional gas, the ranking could move either to −−− or to −.

Gas, despite its release of less carbon dioxide per joule of heat than coal or oil, nonetheless still releases it. The only possible virtue of gas as a fuel is that it may not degrade these common resources as fast as coal or oil. On the more negative side, however, CH_4 is itself a more powerful greenhouse gas than CO_2 (see chapter 6). Some gas inevitably escapes into the atmosphere at every stage of moving it from well to final use, thus exacerbating climate change. On Protection of Common Resources, therefore, gas ranks −−−.

On Distribution of Benefits, Costs, and Risks, gas ranks −−−. Widespread availability of coal and gasoline gave those two fossil fuels a + ranking, but gas has more limited distribution, due to its complete dependence on pipelines and LNG facilities. Thus it can't merit a + ranking. Like the other fossil fuels, however, gas fully merits a −−− rank. Its downsides spread everywhere by contamination of the atmosphere and ocean acidification. Without widespread benefits, its downside dominates.

On Ethics, gas ranks −−−. No moral justification attaches to the degradation by gas of the atmosphere and oceans. Others who see gas as a "bridge" to a future of energy efficiency and renewable energy might be inclined to rank gas much higher. The rankings shown in table 10.1, however, focus on the investment processes for the future. If gas has a bridge function, past investments will provide gas to "bridge" to the future. Going forward, new investments in gas will produce only gas; they will not build the new energy economy.

Conventional gas wells and the transmission-distribution networks don't have great beauty, despite their clever engineering. Nevertheless, they don't dominate landscapes, and proper design can mitigate their stark visual appearances. Thus conventional gas has a rank of –, tolerable. Fracking has proven quite different. To reach as much shale gas as possible, fracking wells must be so close that they dominate the landscape with huge industrial facilities and great ponds of waste water. Fracking wells generate conflicts between neighbors. This is a serious weakness, possibly trending worse and thus meriting a –– for fracked gas and an overall rank of – / –– for Aesthetics.

Uranium

In terms of Profits and Risks, production of uranium fuel by mining has been volatile for many decades. Similarly, construction of nuclear reactors—the task of utility companies, public and private—has required national government support or ownership.

Ux Consulting, which follows and tries to predict market trends for uranium, noted three prominent uranium production booms, followed by busts. From the 1940s into the mid-1960s, the federal government was the sole purchaser of uranium in the United States, and to stimulate production it paid a high price to build U.S. inventories, essentially all of which went to military use. Once inventories contained all the material needed, the government lowered the price and stopped buying. Uranium mining dwindled.

As utilities began to construct nuclear power plants in the late 1960s, more purchasers of uranium appeared. The fuel, however, needed enrichment of ^{235}U to about 3 to 5 percent for use as a fuel, and the U.S. government had the only enrichment facilities. To make the enrichment plants more economical for the government, nuclear power plant operators had to sign enrichment contracts that inflated demand above the utilities' needs, thus inflating demand for uranium and its price. This phase, too, passed, and prices and mining activity declined.

Starting in 1993, agreements between the United States and the Russian Federation made fuel for reactors from highly enriched, weapons-grade material as the two countries dismantled some of their nuclear weapons. This arrangement lasted until the end of 2013. As a result, inventories of uranium for power plants other than that derived from weapons declined through the late 1990s and early 2000s, but the uranium supplies from weapons hid the decline behind continued low prices. Prices began to climb again after 2005, moving from $14.36 per pound of U_3O_8 to $55.64 in 2011.

Japan shut its nuclear power plants after the Fukushima catastrophe, which reduced demand. Prices in the United States dropped to $46.16 per pound in 2014.[15]

Today, the futures market for uranium projects uncertainties. Older nuclear power plants, by far the largest consumer of uranium, are either closing or having their useful lives extended. New construction is under way, but no certainty attaches to how many or how soon new plants will appear. Uranium is also in steep competition with other fuels for generating electricity, especially natural gas and wind. Solar photo-voltaic is not too far from competing head to head with uranium. All three of these fuels have far fewer dangers of catastrophic accidents and political controversies than uranium, which dampens uranium markets.

Further complications for this fuel come from the risks and costs of building nuclear power plants. These devices, without which uranium is merely a heavy metal, currently have cost estimates running to $8 billion and more for a 1,000 MW facility, with construction times averaging ten years.[16] Such a huge amount of money over such a long construction period has utterly blocked private investment; only government support or guarantees against losses will bring private capital into this industry.

Even more complications stem from what to do with the wastes of used fuel. They remain dangerous for well over 100,000 years, and no politically acceptable disposal method exists in the United States. Most wastes remain at the plant site, and work on a government depository ground to a halt after the 2008 presidential elections. Utilities currently have the unenviable job of curating the wastes generated by their plants.

As Peter Bradford, a former commissioner of the U.S. Nuclear Regulatory Commission, noted, nuclear power has always been a ward of government. This has been true even in the United States, the country that tried the hardest to privatize nuclear power. Other countries generally kept nuclear power in the hands of government.[17]

Money can be made in the business, maybe. Severe price competition, combined with public fears, makes investments to produce and use this fuel risky and dependent on government policies. The weight of these various considerations puts uranium at + / −−− in terms of Profits and Risks.

Despite the severe risks of losing money on uranium and nuclear power, any profits made will be quite safe in most places. Governments dominate the uranium-electricity domain, and the profits, if it makes money, are safe. On these grounds, for Security of Access and Profits, uranium earns +++.

On Size of Resource, uranium ranks +. The World Nuclear Association, an industry group, estimates consumption at 66,000 tons per year, with 5.9 million tons of reserves extractable at a price about 1.5 times current prices. This translates to approximately ninety years of known reserves, clearly a strength. At higher prices, miners could produce even more uranium at a profit.[18]

Uranium, like coal, works best as a fuel for base load electricity generation. Nuclear power plants cannot ramp up and down rapidly in their power output. The fuel, however, stores well once made. In contrast to oil and gas, therefore, uranium merits a + for Time of Use.

On Geographic Distribution, uranium, like the fossil fuels, has a few countries with most of the known reserves and best deposits (see chapter 4). In contrast to oil and gas, however, uranium trade has not generated tensions and disputes based on unequal geographic distribution. Uranium lies squarely in the middle of very serious disputes, but these disputes occur for other reasons, primarily military ones. In addition, low grade uranium ores occur quite widely. On Geographic Distribution, therefore, uranium merits a –, not a strong ranking but not fatally flawed.

For Protection of Common Resources, uranium has a mixed record. This criterion focuses primarily on emissions of CO_2, the major greenhouse gas behind climate change and ocean acidification. In life cycle assessments of different fuels in terms of the CO_2 emissions per kWh (see table 10.2 and figure 10.1), uranium looked like the other renewable, low-carbon fuels. Most LCA estimates fell ten times or more below estimates for natural gas.[19] The nuclear industry has celebrated these findings as a major reason to consider nuclear power as essential to mitigate climate change.[20]

Critics of nuclear energy generally base their opposition to uranium on other grounds, primarily safety and costs. Some emphasize, however, the uncertainties of the claims for uranium as a low-carbon fuel.[21] If the system analyzed in LCA includes only operations of a nuclear reactor in generating electricity, then nuclear power has low CO_2 emissions. Uranium, however, has multiple other steps that produce significant carbon emissions, such as mining, ore processing, enrichment, transport, reactor decommissioning, and waste disposal. The values reported in table 10.2 included two or more phases of the fuel's life cycle, but the cycle has more than two. An LCA analysis omitting these steps would underestimate CO_2 emissions. The values in table 10.2 need deeper analysis.

Moreover, uranium is not a renewable resource. Uranium is abundant and widely distributed over the earth, but most deposits have low concentrations. Any serious expansion of nuclear power would force mining into ores with progressively lower uranium concentrations. As ores declined in quality, carbon emissions associated with mining will increase; alternatively, electrified mining will consume a steadily increasing proportion of any electricity generated. The power could not serve other energy services. In addition, increased uses of nuclear power will inevitably increase the amount of mining and spent-fuel wastes, with resulting increases in either CO_2 emissions or consumption of electricity generated to manage these radioactive wastes. Either common resources will suffer or EROI values will decline.

Wastes from the nuclear industry and nuclear reactors have also posed serious pollution problems in oceans near the shore. The catastrophic melting and explosions of the Fukushima Dai-ichi plants following an earthquake and tsunami and the operations of the Sellafield nuclear complex in Britain, for example, have polluted the ocean environment with radioactive materials.

Based on these multiple considerations, the current rank of uranium on Protection of Common Resources is + / --. Low carbon emissions from power generation constitute a strength, but the uncertainties about weaknesses, with trends looking more negative, warrant a ranking of --. With expansion of nuclear power, this ranking will move to ---, as dependence on ever lower grade ores grows, accompanied by increasing accumulations of wastes from mining and spent fuel. Expansion of nuclear power at a scale large enough to mitigate climate change and energy poverty would hasten the arrival of the rank of ---.

On Distribution of Benefits, Costs, and Risks, the current nuclear power industry ranks + / ---. Nuclear power plants developed in the 1960s and 1970s have electrical outputs in the hundreds to over 1,000 MW. Countries that adopted this technology have abundant supplies of electricity shared broadly over wide geographic areas. By this measure, nuclear power clearly has a strength or + rating for Benefits.

At the same time, smaller subsets of populations have just as clearly borne disproportionate exposures from different parts of the nuclear fuel cycle. Mining and milling of uranium ores, for example, have polluted and sickened thousands of people in the Navajo Nation in the United States and the Deline First Nation in Canada. Collapse of a dam holding wastewater and processed ore at the United Nuclear Corporation in 1979 in Church Rock, New Mexico, caused dangerous exposures to radiation and chemical poisoning over a wide area. Various accidents in the former Soviet Union caused widespread contamination and health problems.[22]

"Normal" radiation exposures to workers in fuel fabrication and reprocessing facilities and to people living around nuclear power plants have shown higher levels of cancer than populations that are less exposed. Detecting these low-level radiation effects has caused some controversy over whether they are "real" or not, but a 2013 review article by many public health specialists found increased risks of solid cancers and leukemia in nuclear power workers that were consistent with low-dose radiation effects. According to many studies, children living near nuclear power plants may acquire sufficient radiation exposure, perhaps in utero, to cause increased incidence of cancer.[23]

Catastrophic accidents at nuclear power plants have contaminated significant areas and populations. Three Mile Island (Pennsylvania, 1979), Chernobyl (USSR, now Ukraine, 1986) and Fukushima (Japan, 2011), each led to evacuations and health effects. Health studies after the events lacked thoroughness and adequate support and, predictably, led to contested findings. The explosion and widespread radioactive contamination from Chernobyl, for example, led to published estimates ranging from 4,000 fatal cancers,[24] 9,000 deaths from cancer,[25] and 30,000 to 60,000 cancer deaths,[26] to nearly 1,000,000 deaths from various causes.[27] The published numbers differ due to differences in the affected populations studied, different methods, and different background literature searched. Regardless of the cause of disagreements, however, the death toll from that accident will probably be larger than the minimum estimate of 4,000 fatal cancers.

The negative health consequences imposed on some people compared to others creates high inequity in the distribution of risks with uranium and nuclear power. Both the health consequences and the inequities create a serious problem for the sustainability of nuclear power, with no easy way to mitigate either dimension. On these grounds, uranium also ranks --- for Distribution of Benefits, Risks, and Costs.

Despite the genuine benefits of electricity from nuclear power, uranium has morally fatal attributes. While benefits may be widespread, inequities without moral justification raise serious objections. Uranium threatens the oceans and atmosphere, with a likely trend of those threats worsening if uranium use expands. Radiation exposures pose particularly serious threats to the unborn fetus and young children, hardly a way to protect future generations. To top it off, production of plutonium and radioactive wastes seems impossible to justify on ethical grounds. For these reasons, uranium merits --- on Ethics.

On Aesthetics, uranium has weaknesses that are real but not fatal. The scars from mining and mine tailings undoubtedly raise the most concerns. Nuclear power plants themselves are no uglier or more beautiful than other heavy industry facilities. Communities tolerate such installations, even if they are not admired. On these grounds, uranium merits a – ranking on Aesthetics.

EFFICIENCY: THE FOUNDATION OF THE THIRD AND FOURTH ENERGY TRANSITIONS

Efficiency played key roles in building the Third Energy Transition, but many lessons of energy efficiency dropped by the wayside in the 1950s and 1960s. The Fourth Energy Transition refocused attention on this vital concept (see chapter 8). Efficiency's pivotal roles in both transitions places it between the big-four fuels and renewable energy sources (see table 10.1).

For Profits and Risks, efficiency ranks +++. Efficiency, for example, provides the best investments for saving electricity and natural gas. Based on data from the period 2009–12, it cost $0.028 per kWh to save electricity, about one-half to one-third the cost of building new electricity production facilities. For natural gas, it cost $0.39 to save a therm (100,000 Btu) of gas compared to $0.49 to buy a new therm.[28] An earlier report indicated a comprehensive, national effort on efficiency could save $1,200 billion with an investment of $520 billion; demand for energy would fall 9.1 quads, nearly 10 percent of U.S. energy use per year.[29]

Energy efficiency comes from design of motors, light fixtures, insulation, and other devices or equipment that is built into an energy consuming process. Once built, the investment is quite safe so long as rule of law prevails. Accordingly, efficiency ranks +++ on Security of Access and Profits.

Efficiency has already demonstrated its ability to replace substantial infrastructure for generating energy (see chapter 4), and the above figures indicate that substantial amounts of energy remain to be saved. On Size of Resource, efficiency merits a rank of +++.

Time of Use for efficiency ranks +++. Once designed and built, the efficiency measures operate at any time.

Similarly, efficiency measures can operate in any geographic area. Therefore, on Geographic Distribution, efficiency ranks +++.

Energy efficiency essentially means energy not generated. Nongeneration produces no damage to the atmosphere or oceans, so efficiency earns +++ on Protection of Common Resources.

For Distribution of Benefits, Costs, and Risks, efficiency ranks + / –. The – score results from the requirement of investment to achieve efficiency, and low-income people may have trouble making the investment. This problem vanishes if policies are put in place to subsidize or otherwise reduce the costs. The + ranking reflects efficiency's genuine strengths, but the effects may not spread beyond the person who made the investment.

On Ethics, efficiency ranks +++. Lack of harm from energy not generated raises no ethical problems.

Efficiency technology usually resides invisibly in buildings and machines. No Aesthetic issues arise, so efficiency scores +++ on this criterion.

A casual glance at table 10.1 above indicates that of all energy sources, big-four as well as renewable, efficiency of energy use ranks the best, without question.

RENEWABLE SOURCES OF ENERGY

Companies large and small, some owned partly or entirely by national governments, produce the big-four fuels. Revenues range to over one billion dollars *per day* in the case of Saudi Aramco, the world's largest petroleum producer. Many of the companies have survived, often under different names, for over a century. Stable technologies produce, refine, transport, and use fossil fuels and have spread across the world; they continue to improve incrementally and from time to time upset established commercial operations. Similarly, changing government policies can quickly alter prices and returns. Nevertheless, investors know a great deal about the Profits and Risks of investing in them.

In contrast, investment in the renewable energy sources, aside from hydropower, means investment in technologies with far thinner track records and revenues that at best may top one to two billion dollars *per year*. Electricity, the major product of renewable energy, however, has an extensive history that mitigates uncertainty about renewable sources.

At the same time, rapid technological change still greatly affects solar, wind, and biomass. The harvesting and production of energy from these sources is still constantly changing, and transmission and distribution of electricity from renewable sources is still on a learning curve to be reliable and of high quality. Government policy changes, too, can quickly alter investment landscapes, and the renewable energy industries lack the political clout to affect policy compared to the big-four. Investment in renewable energy faces far more uncertainty than in the big-four.

Solar

Solar radiation provides heat and light directly, and various devices can transform it into electricity. Tremendous value lies in directly using as much heat and light as possible to make buildings and homes energy efficient. Passive utilization of heat and light rests mostly on the design and orientation of new buildings or relatively simple retrofits to old structures. Similarly, relatively simple devices can directly heat water for everyday use. In less industrialized countries, solar ovens for cooking may provide many advantages at very low costs. These investments have few or no risks and controversies, and every indicator supports their adoption.

Potentially more lucrative but also riskier technologies transform sunlight into electricity for transmission across wide geographic areas. *Central power plants* produce large amounts of electricity compared to *distributed power*, which produces small amounts for local consumption. Central power plants, also called *utility-scale* facilities, can utilize either *photovoltaic* (PV) panels or thermal methods in *concentrating solar power* (CSP). Distributed power uses PV only. PV turns sunlight directly into electricity; CSP plants concentrate sunlight to heat a working fluid to make steam for generating electricity (see chapter 8).

Sunlight is free, but PV and CSP require investment, and high production costs of solar electricity, all other things being equal, will diminish sales, lower potentials for profits, and increase risks. Unfortunately, LCOE estimates for PV and CSP in 2020 (see table 10.2) indicate that solar electricity does not yet compete well on prices with coal, gas, nuclear, wind (onshore), biomass, geothermal, or hydropower. PV comes in at lower costs than CSP, but both remain high. Based on price alone, solar electricity is still the last source a customer would choose, if s/he had access to power from other fuels and price the only issue. Based on this measure alone, solar electricity merits a – on Profits and Risks; this is a weakness but not a fatal one.

Government policies, motivated by other attributes of solar energy, have enhanced the interest in PV and CSP far beyond what it would have attracted based on prices. Consider first the following partial list of government policies supporting development of solar power.[30]

- Feed-in tariffs: require electric utilities to purchase solar electric power produced by independent generators at a price that covers full costs of production plus a profit
- Investment tax credits: allow an investor to deduct a percentage of the investment cost from taxes otherwise owed
- Direct subsidies: provide direct government grants to investors to offset investment costs
- Renewable portfolio standards (RPS): require sellers of electric power to include a specified percentage of that power from renewable energy, such as solar. Some utilities satisfy a RPS by purchasing solar renewable energy credits (SRECs). SRECs originate from private generators of PV electricity that generate more than they need. The price paid for an SREC varies.[31]
- Public investment: enables governments to invest directly in solar or other renewable energy in order to build the industry or to serve particular populations
- Net metering: requires electric utilities to buy excess electricity produced but not used by a customer, at rates lower than feed-in tariffs and sometimes with a cap on amounts sold
- Mandates: require electric utilities to buy power from or to allow transmission of electricity produced by independent generators.

Without one or more of these government-support policies, solar electricity would not exist anywhere in the world, except in remote places far from grids or in specialty applications such as space vehicles or calculators. Not to put too fine a point on it, but solar electricity is as much a creation of governments as is nuclear power. And the reasons were similar—qualities of the energy source perceived by political leaders as valuable or essential to the public interest.

For solar electricity, the main qualities motivating governments included low CO_2 emissions and thus mitigation of climate change, plentiful domestic supplies for security purposes, low emissions of toxic chemicals for health protection, and a production rate each day that goes up and down at about the same time as peak electricity consumption. Peak electricity, regardless of fuel, is the most expensive to produce, so higher production costs with solar matter less at peak times. Jump-starting the solar electric industries with policy support also spurred technological innovation and learning to steadily reduce the costs and LCOE estimates.

Rapid rates of growth of solar have proven extraordinarily exciting in the years after 2000. In the United States, for example, the amount of electricity provided by PV rose from 16 GWh in 2005 to 15,874 GWh in 2014, an increase of 992 times. Just in one year, from 2013 to 2014, PV electricity jumped nearly 100 percent, from 8,121 GWh to 15,874 GWh. Solar thermal had a less stunning but still quite amazing increase in this same ten-year period, from 535 GWh in 2005 to 2,447 GWh in 2014, an increase of 4.6 times. Despite rapid growth rates, total solar electricity was only 0.4 percent of all electrical energy in 2014.[32]

These figures take into account only the solar power generated before electricity arrives at the meter of a consumer; that is, they don't count power generated on rooftops lying "behind" the meter. One estimate for March 2014–February 2015 concluded that the total PV power was 30,400 GWh rather than the 20,200 GWh reported by the U.S. Energy Information Administration above.[33] Even with this addition, however, solar hardly counts as a source on the national scale, even though it has achieved significant status in places like California and Germany (see chapter 8).

Based on figures for 2014 in the United States, the solar electric industries have lowered their production costs and may soon reach a stage of competing with other fuels with lowered and maybe even no policy support. In 2014, a total of 3,934 MW of solar electric installations brought the total installed capacity of solar electric to about 20,000 MW. By 2020, the estimated capacity of solar electric installations will be nearly 70,000 MW.[34]

By themselves, these capacity numbers seem large, but in comparison to total capacities of the United States, they remain small. Utility-scale solar electric capacity in August 2015 totaled 11,545 MW, or 1.1 percent of the total U.S. capacity of 1.07 million MW. The estimated total of 20,000 MW solar capacity was only 1.9 percent of the total. If the total rises to 70,000 MW in 2020, that will be a large increase but still only 6.6 percent of the capacity in August 2015.[35]

For utility-scale facilities, installation costs have dropped more than 50 percent since 2007–9, and these costs now hover at about $3 per watt. At these prices, which include the policy supports of governments, utility-scale generators will now sign long-term supply contracts to utilities in the $40–$60/MWh range, which competes with natural gas (table 10.2).[36]

Despite lower prices now achieved, solar electric power nonetheless poses real challenges. For long stretches of each year, the high latitudes with long winter nights challenge PV. Nevertheless, recent research indicates that even in such areas PV can play a significant, useful role when combined with battery storage and other energy sources.[37] Zero solar energy overnight at the midlatitudes makes PV unsuitable for base load power during parts of each day, although CSP can store heat energy to operate at high power during cloudy weather and into the evening hours.

Manufacture of PV panels involves environmental and occupational safety and health issues, which require careful management and regulation.[38] Moreover, the ultimate amount of solar energy that can be harvested remains unknown. Chapter 8 provided optimistic, high estimates, but others have been more pessimistic.[39] Storage technology and other management practices for grid operations show promise for alleviating the intermittent production,[40] but the scale and costs of this new technology are still in flux.

Based on all these factors, does investment in solar electricity look like a good way to spend money? Home owners, businesses, and utility-scale investors have purchased PV panels and built CSP plants, thus indicating that producers and consumers have already found the technology attractive. Based on current trends and tremendous opportunity to expand into electricity production markets, the future looks good for solar investments. Even with supportive government policies, however, companies that make PV panels and install them on rooftops must contend with rapidly changing technology, resulting in volatility and sparking trade disputes between China and the EU and United States.[41] Investments in the stocks of manufacturers of PV panels may be risky, even if personally buying and installing PV panels is not.

This mixed profile merits an overall score on Profits and Risks of ++ / – for solar electricity. It merits ++ mostly for existing strengths and favorable trends. It also merits a – because of the weaknesses and challenges mentioned above, real but not fatal. Solar electricity is a good idea, but that does not by itself make every investment lucrative, at least at the moment.

On the second criterion, Security of Access and Profits, solar electricity merits a +++ rating. Current policies in countries with strong rule of law will protect property rights in investments made. Regulations will need to include protection of insolation through, for example, building codes and rules for management of trees. But if the area has sunlight, access to it, for all practical purposes, extends forever, and the price will always be free.

Size of Resource merits a rank of ++. On the most optimistic estimates, the size of solar resources vastly exceed the energy needs of the world. Ultimate amounts of solar energy that can be harvested for electricity, however, remain unknown. Trends suggest the optimism is correct, but at the moment the ultimate limits remain uncertain.

One of solar electricity's weaknesses lies in Time of Use. Direct production takes place in the daylight only. For solar to become fully functional requires storage technology for dark periods and grid management processes to mix with other electricity sources. For the moment, a rank of − captures the problem. It's a real weakness, but trends suggest solutions will emerge.

Virtually the entire world has significant solar radiation, but prolonged nights in the winter at high latitudes discourage its uses in those areas. The radiation itself cannot be transported or traded over long distances, but high voltage transmission can deliver significant amounts of solar electricity from areas with high resources to those with low resources. These considerations merit a mixed rank of ++ / − for solar energy on Geographic Distributions.

Solar electricity promises relief from the stark weaknesses of the big-four fuels on Protection of Common Resources. It merits a +++ rating, but this strong showing remains dependent on proper regulation of toxic materials used in manufacturing PV panels and the disruptions of land use that may affect vulnerable wildlife or disrupt scenic landscapes. In addition, recycling of PV panels beyond their economic life span must prevent solid waste and in some cases toxic waste pollution.

For Distribution of Benefits, Costs, and Risks, solar electricity merits +++. Again, this strong ranking depends on regulations and management that create universal access to electric power and avoids any dangers to exposure to toxic materials used in the manufacture of PV panels or in their disposal.

Solar power merits +++ on Ethics. As with the preceding two criteria, this score depends on effective regulation and management of toxic materials and waste. In contrast to the big-four fuels, these problems have solutions in the case of solar power.

On Aesthetics, solar power ranks + / −. This ranking depends on the location of facilities. When PV panels operate from rooftops, they are nearly invisible. Utility-scale installations, whether of PV panels or CSP, have the same weaknesses as any industrial installation. They aren't pretty, but design and regulated location can minimize their visual disruptions of the landscape. It's a weakness but not fatal.

Wind

Wind turbines transform the kinetic energy of moving air into electrical energy, either for local use at the generation site (distributed generation) or for transmission to more distant consumers (centralized or utility-scale generation). Distributed generation comes entirely from on-shore turbines, and utility-scale generation comes from both onshore and offshore sites.

As with solar, the fuel for wind electricity is free, but substantial investments must build turbine-generators, blades, and towers. In the United States in 2014, for example, the largest four turbine makers were GE Wind (United States), Siemens (Germany), Vestas (Denmark), and Gamesa (Spain).[42] Vestas is the world's largest supplier, with over 13 percent of the global market. Siemens (8 percent), GE (4.9 percent), and Gamesa (4.6 percent) are fourth, sixth, and seventh, respectively.[43]

Installation and/or operation of wind turbines involves different companies. For example, Terra-gen Power owns and operates the Alta Wind Energy Center (AWEC) in Tehachapi, California. AWEC in 2015 had a capacity of 1,020 MW, scheduled to grow to 1,550 MW, and ranked as the world's largest wind farm in 2013.[44] Terra-gen in 2015 listed twenty-seven different wind projects in its portfolio in California, Wyoming, Minnesota, Colorado, and Texas.[45] They operated as an affiliate of ArcLight Capital Partners and Global Infrastructure Partners, private equity and venture capital firms with holdings in traditional and renewable fuels.[46] Terra-gen had acquired AWEC from a bankrupt Australian developer in the early 2000s.[47]

Willingness of firms to invest in the costs of production, installation, and operation of wind turbines depends on the revenues from the electricity they produce. Prices of turbines from the manufacturer depend on multiple factors, including turbine characteristics, the inclusion or not of towers, installation, service warranties, and timing of delivery. Prices of installed wind farms also depend on many factors, including size of the turbines, blades, and towers; production capacity of the project; prices agreed on for turbines; and area of installation.

Costs in the United States for turbines dropped after 1980, rose in the period 2002–8, and dropped steadily after 2009. Installed prices for wind farms dropped after the 1980s through the early 2000s, rose to a peak in 2009–10, and have since declined again. In 2014, installation costs averaged $1,710 per kW, down $580/kW from 2009–10. Costs were highest for projects with turbines capable of producing 1 MW of power and lower for projects with turbines ranging from 1 MW to greater than 3 MW. In 2014, the vast majority of U.S. installations had turbines ranging in capacity from 1.5 MW to 2.5 MW, a move away from turbines larger than 2.5 MW; this trend reflects installations with higher capacity factors, that is, with turbines that produce steadier but lower amounts of power over longer periods instead of maximum amounts of power over limited periods.[48]

Estimated costs of electricity produced reflect the costs of turbines and installed wind farms. LCOE estimates for onshore wind in 2020 of $73.6/MWh beat the estimates for all other fuels except geothermal (see table 10.2). Higher expenses, especially for installation, operation, and maintenance, of offshore wind increase its LCOE estimate to $196.9/MWh. The low and therefore competitive estimate of LCOE in 2020 suggests that wind will compete on its own merits (i.e. without subsidy) by that time.

Up to 2015, however, the links between wind installations and government policies strongly indicated that most facilities producing electricity from wind owed their existence to the policies. For example, in the United States a number of federal and state policies drove installations:[49]

- Federal Production Tax Credit (PTC), which since 1992, with some interruptions, allowed producers selling electricity to deduct for ten years an amount per kWh from their taxes; as of 2014 this PTC equaled $0.023/kWh. The PTC lapsed four times in the 2000s, each time accompanied by a substantial dip in new installations, until installations resumed after Congress renewed the PTC.

- In December 2014, Congress allowed installer-producers to elect either an Investment Tax Credit (ITC) of 30 percent or a PTC for projects under way by the end of 2014.

- Accelerated depreciation schedules allow installers to depreciate the investment faster than warranted by the projected lifetime of the facility.

- The Clean Power Plan released in 2015 requires states to reduce carbon emissions from power plants and may stimulate installations of wind power.

- Renewable Portfolio Standards (RPS) in twenty-nine states and the District of Columbia have driven over half of the wind installations by requiring utilities to utilize a stated proportion of renewable electricity.

- RPS standards may drive installation of 4,000 to 5,000 MW of wind capacity per year through 2025, somewhat below the average of 7,000 MW per year installed from 2007 to 2014.

Like solar, therefore, wind power owes its emergence to government policies. And like solar, governments chose to support development of wind energy due to its characteristics, especially its low emissions of CO_2 and other pollutants. Landowners responded by leasing their land to producers for installations. Farmers saw the steady market for electric power as a good way to create a stable cash flow from leases. Throughout many farming areas today, wind turbines rise from fields still producing crops or livestock, and Iowa ranks as the state with the highest penetration of wind-generated electricity (28.5 percent in 2014).[50]

Wind, too, like solar, grew rapidly in the United States during the 2000s. Total amount of wind-generated electricity increased 10.2 times, from 17,811 GWh in 2005 to 181,791 GWh in 2014. Wind remained a small fraction of total electricity, 4.4 percent in 2014, but that still represented solid growth from 0.4 percent of U.S. electricity in 2005. National-scale statistics, however, hide the highly significant contribution to total power in Iowa, Texas, and California.

Most wind power capacity in the United States resides in utility-scale wind farms with relatively little distributed capacity "behind the meter" on homes and businesses. In August 2015, wind capacity equaled 66.7 GW out of a total 1,064 GW, or 6.3 percent of the total.[51] Distributed wind capacity amounted to 0.9 GW at the end of 2014, or 1.4 percent of all wind installations. Despite the small fraction of distributed wind capacity, it, too, grew rapidly after 2000.[52]

In the United States as of late 2015, no offshore sites produced power, but the Block Island project off the coast of Rhode Island began production in late 2016 with a capacity of 30 MW, a very small proportion of total U.S. wind capacity. At the end of 2014 globally, offshore wind generation capacity had reached 7,700 MW, mostly in Europe. Total wind capacity in the world stood at 372,112 MW at that time, with nearly 98 percent installed onshore.[53]

Regardless of the excitement generated by the wind industry after 2000, based on conditions in 2015 the future of wind power remained uncertain. Prices of wind technology have steadily decreased, and voluntary purchases by individuals and businesses have increased; both these trends bolster the fortunes of wind as an investment. Expiration of the PTC, limited new RPS mandates, continued low cost of natural gas, projected low growth in demand for electricity, and lack of a federal RPS, however, dampen the future prospects of wind.[54]

Based on these multiple considerations, how does wind rank on the criterion of Profits and Risks? Rapid growth in the immediate past and the continued lowering of costs of production suggest a ++ rating. Wind investments have successfully sold electricity, and this fuel choice has attractive features. At the same time, its dependence on uncertain government policies remains a weakness and merits a ranking of –. Thus the overall rank on Profits and Risks merits a mixed ++ / –.

In jurisdictions with strong structures of rule of law, investments made in wind technology will be safe. Thus on the criterion of Security of Access and Profits, wind merits a ranking of +++.

Solar energy resources vastly outrank wind, but harvestable wind resources alone are substantial, possibly even greater than the total current world energy use (see tables 8.1 and 8.2). Uncertainty about the ultimate amounts of energy procurable from wind makes its ranking on Size of Resource ++. It has already achieved significant status in some areas, and the trends to higher penetration look robust.

On Time of Use, wind energy suffers the same intermittency as solar. This creates a challenge to its use. Either it must be supplemented with other energy sources or energy storage must make up for periods of low wind. Innovations in storage technologies of multiple sorts strongly suggest grounds for optimism, and wind merits a rank of – on this criterion.

Wind has very widespread distribution and can serve as an energy resource in most areas. Some areas have much stronger average wind velocities than others, and the electricity made in those areas can be transmitted long distances at high voltage. Prolonged periods of dark in winter at high latitudes may make offshore wind difficult to maintain in those areas. Based on these considerations, wind merits a ranking of ++ / – on Geographic Distributions: the resource has substantial advantages, trending higher, with weaknesses that are not fatal.

Like solar, wind energy promises relief from the unsustainability of traditional fuels. Thus on Protection of Common Resources, wind ranks +++. This high rating, however, requires proper policies to avoid placement of turbines, especially large ones, in places not acceptable to local people. In addition, policies must require turbine operators to use designs, placements, and operating schedules that avoid or eliminate damage to bird and bat populations.[55]

On the criterion of Distribution of Benefits, Costs, and Risks, wind merits a rank of +++. The benefit, electricity, already has wide distribution, and government policies can bring power to those currently without it. The advantageous LCOE estimates for wind make it attractive for widespread use. Risks to health and safety require policies to ensure proper placement and safe working conditions for those who maintain equipment at the top of the towers and elsewhere.

Improper locations and operations of wind farms can create strong moral problems for wind, but policy remedies can eliminate these issues. In other respects, on Ethics wind merits a rank of +++.

Wind turbines on the landscape have a mixed rank of + / − on Aesthetics. Individual turbines have a gracefulness as they spin in the distance. But not everyone agrees, and large clusters of turbines and their towers, especially in undeveloped areas, have already aroused great opposition to installation of wind farms in particular locations. Placement of wind energy facilities requires public participation, democratic choices, policies to protect beautiful areas, and comparisons to the alternatives for relative rankings on Aesthetics.

Hydropower

With few exceptions, hydropower means the generation of electricity by falling water, either over a waterfall or behind a dam; engineers build a penstock to divert water from the river or allow it to fall through a dam to turn a turbine generator. Other forms of falling water to derive electrical energy are still experimental or have insufficient deployment to assess their profiles, and this section focuses on electrical production from rivers or reservoirs (see chapter 8).

Water and wind generated electricity very early in the development of electrification, but only hydropower quickly became established, before 1900, as a significant source of revolutionary electricity (see chapters 2, 4, and 8). In the first half of the 1900s, hydropower attained its greatest prominence, particularly in the western United States, because of massive investments by the federal government to develop water and energy resources. Today the U.S. government owns 49 percent of the hydropower capacity produced by 8 percent of the hydropower plants.[56]

Does a heavy federal presence mean that future investments in hydropower are, with only relatively minor exceptions, likely to result from political decisions reached by Congress? And what about in other countries? Will national governments be the likely builders and operators of hydropower, making this part of the Fourth Energy Transition primarily a matter of political decision making? We return to these questions shortly.

Two major industries specifically support the production of hydropower. First, construction industries design and build dams, penstocks, and other structures to channel water from a height through a turbo-generator to a lower elevation. A great deal of this work involves building concrete structures and tunnels; frequently auxiliary services, irrigation, flood control, and recreation, accompany electricity production. Second, different firms make turbo-generators to generate the electricity. Large firms dominate both industries, and commerce in turbo-generators involves international trade.

As with other fuels, hydroelectric facilities must connect to transmission and/or distribution systems. Some small hydroelectric facilities distribute power locally, but large installations feed their electric power to transmission lines that may cross regional and national borders.

As with solar and wind, the fuel for hydropower—falling water—is free. Solar energy evaporates water, which falls as snow or rain and then runs downhill courtesy of gravity. An investor in hydropower need never fear price volatility of the fuel, but variations in precipitation from year to year mean variations in the amounts of falling water available and thus potential revenues.

Once constructed, hydropower price, measured by LCOE, competes well with gas and onshore wind as one of the cheaper sources of electricity (see table 10.2), but assigning values to auxiliary services complicates the assumptions needed to calculate LCOE. Despite low electricity prices from hydropower, the capital expenditures to build the plant are significant. A U.S. Department of Energy survey of dam construction projects between 1980 and 2015 estimated costs of $1,000 to $9,000 per kW of capacity.[57]

Hydropower investments over time have shown striking differences. From the 1890s to the 1920s, construction of about three hundred facilities, mostly between 100 kW and 100 MW in capacity, established hydroelectric power. From the 1920s to the 1960s, the U.S. government dominated construction, and facilities with capacities of greater than 100 MW and often greater than 500 MW began to produce massive amounts of power. The 1970s and 1980s saw a return to building facilities with less than 10 MW capacity, and after the 1990s construction nearly ceased.[58]

Solar and wind power began to surge in the 2000s, but hydropower remained static until enthusiasm for increasing it revived after 2010, stimulated especially by the Hydropower Regulatory Efficiency Act in 2013. This new law considerably simplified and in many cases exempted small projects from regulation by the Federal Energy Regulatory Commission and launched a new wave of construction.[59]

The United States may never again build a massive hydropower facility such as Hoover Dam, the Grand Coulee Dam, and others built between the 1920s and 1960s. Nevertheless, significant potential for new hydropower exists in three areas: upgrading existing hydropower facilities, adding hydropower to existing dams and conduits, and locating small plants on previously undeveloped streams. The easiest additions are to existing hydropower dams. From 2005 to 2013, engineers added 1,563 MW of capacity to existing hydropower dams, small but locally significant additions to the existing hydropower capacity of almost 80,000 MW.[60]

The largest potential for new hydropower, however, lies in dams and conduits without hydropower facilities and on streams not yet dammed. About 2,500 dams currently have hydropower capacity, but over 80,000 dams do not. A detailed study of over 54,000 of these dams indicated a potential hydropower capacity of 12,100 MW. About two-thirds of this capacity, that is, about 8,000 MW, exists at just one hundred sites, mostly dams operated on large rivers such as the Mississippi and Ohio. Even more potential, over 65,000 MW, exists in streams not yet dammed but outside of national parks, wild and scenic rivers, and wilderness areas. Development of these resources could nearly double U.S. hydropower potential.[61]

Hydropower has not benefited as much as solar and wind power from federal subsidies, but in the past five years the subsidies have increased, and as of the end of 2014 a total of 331 projects were proposed for construction. If all the projects are completed, they will add 4,370 MW of new hydropower capacity.[62] Among the incentives provided by the United States are the following:[63]

- grants for research & development
- production tax credit of $0.018 per kWh, up to $750,000 per year
- investment tax credit of 30 percent of the cost of the facility
- loan guarantees

- ability of nonprofit agencies to issue Qualified Energy Conservation Bonds to finance projects
- grants for rural energy development projects

Based on the above considerations, hydropower has the potential to expand, in the United States and elsewhere. U.S. projects under way will add significant renewable energy to local areas, and the federal government provides incentives and other forms of financial help to install new projects.

At the same time, challenges remain. Despite the 2013 Hydropower Regulator Efficiency Act, there are complicated regulations, driven mostly by the fact that hydropower of necessity modifies rivers and streams, potentially a controversial action. Larger projects (over 10 MW) always require a license, and in some cases so do smaller plants.[64] Variability in precipitation has always caused fluctuations in output of electricity, and the unfolding events of climate change may alter precipitation patterns even more (see chapter 6).

Is hydropower a good investment for the Fourth Energy Transition? It clearly has strengths in terms of cost of electricity measured by LCOE. The technology for hydropower from rivers and reservoirs has matured over the past century, so startling new innovations will probably not vastly increase the ability of falling water to produce electricity. Thus the trends of hydropower's strength will probably not change much. At the same time the technology has elaborate licensing procedures, and government incentive policies may change. Based on these factors, hydropower merits a + / − rank for Profits and Risks. Strengths will remain as they are, and the weaknesses are both stable and tolerable.

Once investments launch new hydropower facilities, those investments will in all likelihood be safe. Countries with rule of law will protect access and profits. Hydropower thus merits +++ on Security of Access and Profits.

On Size of Resource, hydropower merits a ++ rank. It falls far short of the potential energy that solar can probably provide, but hydropower is a known technology. Europe and North America have already developed a significant portion of their hydropower potential, but even these areas can add significant capacity. Asia, Australia-Oceana, Latin America, and Africa have an even greater proportion of their potential still untapped (see chapter 8). And, as mentioned below, other factors make hydropower an attractive source of energy in many parts of the world.

Most rivers tapped for hydropower have enough sufficiently steady amounts of water coursing through every year to make hydropower generally reliable. Once the water is stored behind a dam or ready to course through a penstock, it can be used at any time. Thus, when compared to solar and wind, hydropower has greater flexibility, a trait it shares with biomass and geothermal. Thus on Time of Use, hydropower merits a rank of +++.

By definition, hydropower occurs only in areas with sufficient precipitation to create reliable runoff every year. These conditions occur in many areas of the earth, and the electricity can be transmitted great distances. Therefore, hydropower ranks +++ for Geographic Distributions.

Hydropower ranks high, +++, on Protection of Common Resources. Its emissions of CO_2 are low (see table 10.2 and figure 10.1), so it causes minimal damage to climate and the oceans. At the same time, hydropower disrupts aquatic ecosystems, and many hydropower projects have displaced millions of people from their homes and livelihoods, in the United States and in other countries. For example, the Narmada dam in India and Three Gorges dam in China both created severe social dislocations and controversy. These, too, are common resources, and hydropower also merits a rank of –. Projects must have a widely acceptable regulatory regime to avoid unacceptable changes to ecosystems and places of living. Together these considerations make its rank a mixed +++ / – on Protection of Common Resources.

For the same reasons, on Distribution of Benefits, Costs, and Risks, hydropower has a rank of +++ / –. The electricity in theory can serve many people, even those far away. At the same time, projects can disproportionately and adversely affect people in a facility's location. Strong regulations must protect those damaged by projects.

On Ethics, the same reasoning prevails, giving hydropower a mixed rank of +++ / –. Compared to the big-four fuels, hydropower ranks much higher on this criterion, but the potential for damage to ecosystems and for displacement of people requires recognition and mitigation. Hence the – rating; it's a weakness but not fatal so long as those disrupted receive justice.

Aesthetics also comes in with a mixed ranking of + / –. Many people find large hydropower facilities awesome and beautiful, thus the + rank. At the same time a wild and beautiful natural river may be rendered into a placid, dull pond. But people receiving irrigation water from the reservoir or running their boats around it may see nothing but a beautiful lake. Thus the – ranking, a weakness but not fatal and on which opinions will differ.

Biomass

On the surface, biomass looks like the perfect primary energy source for the human economy. Numerous new technologies have resurrected interest in firewood and other biological materials, collectively referred to as *biomass*, as significant players in the Fourth Energy Transition. Photosynthesis stores energy in biomass at a rate faster than humans use energy in the global economy,[65] suggesting the availability of at least a portion of biomass energy. And indeed it has potential for some contributions, but a number of factors will dampen prospects for more than token roles.

Three well-developed industries, each based on new technology, currently use biomass energy at commercial scales. First, conversion to liquid biofuels for transport—fuel ethanol, biodiesel, and other materials—developed on a large scale after 1975 and especially after 2000, particularly in the United States and Brazil. Second, numerous facilities around the world burn wastes from forestry and agriculture to make thermal electricity. Finally, both small and large facilities in many countries turn agricultural and municipal waste into gas for use in transport or for generating electricity.

Biomass has the greatest chance of serving as a locally useful primary source of energy for generating electricity. Production of liquid biofuels and gas for transport has utility, but constraints of energy, pollution, and geography diminish the attractiveness of biofuels.

Energy constraints in developing biomass energy. Use of biomass energy, particularly for fuel ethanol, excites farmers growing crops used in ethanol production.[66] Nevertheless, fundamental constraints based on energy science insert a note of sobriety (see chapter 2). The story begins with photosynthesis, through which green plants absorb the energy of sunlight to make carbon dioxide and water into stored chemical energy.

The basic problem lies in the low percentage of energy in sunlight captured by green plants. On a per square meter basis, about 0.04 percent of the sunlight energy hitting earth turns into biomass energy.[67] If sunlight is absorbed by a green plant, the plant captures about 3.9 percent of the energy.[68] Thus reliance on biomass energy requires using a great deal of the earth's land area for production. Farmers will always use the most productive land, so biomass for energy inevitably competes for land already producing food, feed, and fiber.

Humanity already harvests between 13 and 39 percent of the theoretical maximum produced by photosynthesis. Pushing the amount harvested to higher levels might render ecosystems vulnerable to collapse and imperil both existing levels of harvest and the possibilities of harvesting biomass energy.[69] Harvests of significant amounts of biomass energy may be unsustainable.

Pollution constraints in developing biomass energy. Plants produced for biomass energy require water for growth, enhanced by fertilizer and protected by pesticides. Rainfall alone may provide sufficient water, but some biomass production may require irrigation and thus compete with food crops for water. Fertilizer and pesticides may increase the need for water to obtain maximum yields.

Fertilizer and pesticides also move from fields into waterways, lakes, and the oceans. Fertilizers may promote growth of algae and other plants that die, decompose, and deplete oxygen in the water. Fish and other animals may die, and pesticides in the air and water may also kill plants and animals.

Land currently not farmed tends not to be the best farmland. Expansion of biomass energy production may thus bring these lower-quality lands into production, with increased levels of soil erosion. Water turned muddy by eroding soils, possibly contaminated with fertilizers and pesticides, may not support aquatic ecosystems or supply drinking water.

These agronomic facts of life mean that biomass energy will change the environmental impacts of farming. Government programs to induce farmers to considerably expand production of biomass energy will, therefore, have environmental consequences that become ever larger as more land changes to energy farming.

Yet another type of pollution constraint on biomass energy centers on the concept of low-carbon fuel standards (LCFS), currently under development and implementation in California, Oregon, British Columbia, Washington, and the European Union. LCFS create incentives for and mandate producers of transportation fuels to reduce the amount of carbon dioxide released per joule of energy supplied by the fuel. Life cycle assessment calculates the needed ratio of CO_2/MJ—the carbon intensity of the fuel—and the LCA calculation in California must specify the exact pathway used to produce the fuel. California's aim is to lower the carbon intensity of the state's transportation fuels by 10 percent compared to petroleum-based fuels by 2020. Biofuels in theory can lower the carbon intensity, but results of calculations of specific pathways indicate that some biofuels have a higher carbon intensity than gasoline and diesel fuels. Further adoptions of LCFS will therefore not necessarily lead to increased adoption of all biofuels.[70]

Geographic constraints on biomass energy. Biomass energy in the field has little energy per square meter, meaning that its energy density is low. Food and lumber, two of the most valuable biomass harvests, are so valuable for uses other than fuel that the costs in dollars and energy to consolidate and distribute the biomass appear "normal." Consolidating, processing, and distributing biomass for energy, however, places biomass in competition with wind, solar, and hydropower, as well as the big-four energy-dense fuels. And biomass will not win.

To put it simply, biomass for energy is heavy and bulky to transport. At the same time, the facilities for making it into liquid biofuels or electricity have to be of large scale to be economically viable. In addition, the expensive processing plants, once built, must have a steady feedstock supply of the biomass to make a profit for investors. To produce large amounts of biomass, these expensive plants need to be located at regular intervals over wide areas.

The upshot of these factors is twofold. First, the processing plant must have an adequate, sustainable feedstock supply within relatively small distances, generally not more than 30 to 50 miles. Second, once the facility is built, the plant owner will want continued production of enough feedstock as long as the plant operates. Good practices in agriculture and forestry, such as crop rotation and letting trees grow long enough to make forests with biodiversity, may drop by the wayside due to economic incentives and imperatives to "feed the processing plant" without interruption.

To produce electricity requires transport of heavy biomass from its place of harvest to the power plant; from there, the energy moves to customers as electrical energy, which is done much more cheaply than hauling biomass. Biomass processed into liquid fuels must move heavy biomass to the plant and then move relatively heavy but more energy-dense fuel to customers. The second step is akin to moving gasoline from refinery to customer, but for biomass this is the second step of moving heavy products long distances. The need to move heavy products with low energy densities at least once intrinsically lowers the EROI of biomass energy and makes it less useful to society.

Net results of constraints on biomass energy. The extensive land requirements, the need for many expensive processing plants spread over the landscape, and the need to move heavy raw materials and, for fuels, to move heavy products combine to limit the prospects of biomass energy. Three types of circumstances, however, may provide a rationale that justifies minor uses.

First, biomass energy may be a way to offset costs of doing something that is useful but otherwise too expensive to justify. Use of forest products, such as small trees from thinning new growth and the branches left from harvesting lumber, can be burned to produce electricity.[71] Similarly, processing agricultural crops like rice may leave a concentrated mass of rice husks suitable for burning. In these cases, selling the electricity offsets the costs of thinning, clearing debris from lumber harvests, and disposing of agricultural wastes.

Second, about 20 percent of the U.S. maize crop goes to ethanol plants, and federal regulations mandate gasoline producers mix ethanol with it. Multiple controversies surround this use of corn, but the policies derived from a farm lobby that persuaded Congress to mandate ethanol in gasoline. The mandate is a trivial contribution to renewable energy and the Fourth Energy Transition, but its supporters understandably like it because it boosts farm incomes.

Third, production of ethanol in the United States essentially turns nonliquid fuels like coal and gas into liquid fuels for transport, because the ethanol plants use heat from fossil fuels to make ethanol from grain. The EROI of corn-ethanol plants, therefore, is low, so it produces little or no new energy, but proponents argue that it produces a transport fuel entirely from American resources. This reasoning does not significantly advance the Fourth Energy Transition, even if proponents consider it legitimate. Box 10.1 discusses the ethanol industry further.

Based on all the above considerations, biomass ranks + / −− on **Profits and Risks**. Biomass merits the + because some companies already earn adequate returns from biomass energy. Biomass energy, however, has problems, some of which are trending worse. The returns on biomass energy depend heavily on incentives created by government policies. All other energy sources also depend on government policy, so this factor by itself only merits a − ranking: it's a weakness but not fatal.

The -- rank stems from the vulnerabilities of biomass production to climate change plus the potential for conflicts with food crops. As climate change progresses, growing conditions and water for biomass crops may become unstable and perhaps unsuitable. As the population grows, the need for increased food supplies becomes ever more likely to compete destructively with biomass energy from agriculture. The negative trends on the horizon therefore seem to be against biomass energy.

Even though severe constraints will probably limit the use of biomass in the Fourth Energy Transition, investments in facilities to harvest biomass energy will be secure, so long as a strong sense of rule of law prevails. Thus this source ranks +++ on Security of Access and Profits.

On Size of Resource, biomass ironically merits a rank of --. Because of the large amounts of energy needed to produce bioenergy (i.e., its low EROI), the abundant vegetation over the face of the earth can't actually yield much new, useful energy. Humanity already harvests a large proportion of the theoretical maximum harvest, and taking more will exacerbate environmental degradation. This is a serious weakness of bioenergy that trends even worse.

Biomass used to produce electricity, gas, or liquid biofuels has high flexibility. Feedstock for electrical production can be stored and burned at the rate needed when needed. Gas and liquid biofuels also store well until needed. On Time of Use, biomass ranks +++.

Biomass also ranks well on Geographic Distribution. While many lands are too dry, too wet, too rocky, or too hot or cold, most inhabited areas of the earth have biomass available. Thus almost all people have access to secure supplies, which merits a rank of +++.

On the criterion Protection of Common Resources, biomass energy merits + / --. In certain conditions, harvest of biomass energy reduces emissions of carbon dioxide and/or methane. In these conditions, the reductions of emissions of greenhouse gases represent a strength, but in some cases the reduction amounts to little.

If used to generate electricity, heat, or light—as a substitute for any fossil fuel—then the regrowth of the biomass will reabsorb the carbon dioxide released on combustion. If manure or municipal waste ferments without oxygen to produce methane that is captured and burned, then the decay of the biomass will release carbon dioxide rather than a mix of methane and carbon dioxide; methane is a more powerful greenhouse gas than carbon dioxide.

If used to make a liquid biofuel in a plant very near the harvested biomass and powered by biofuel, then the amount of carbon dioxide

Box 10.1. Biomass Made into Biofuels

Fuel ethanol burns like gasoline, and Henry Ford famously believed that his Model T Fords would probably use ethanol from maize more often than gasoline, because gasoline supplies in the early 1900s did not look like they would be stable and reliable.[a] Petroleum supplies and gasoline, however, proved to be cheaper than ethanol and reliable enough, eclipsing the early thoughts of ethanol-powered mobility. Gasoline worked well in engines, but as engineers sought higher compression before ignition, the fuel-air mixture exploded prematurely, causing "knock." Ethanol mixed with gasoline solved the knock, but invention of tetra-ethyl lead additive eclipsed ethanol even as an additive.[b]

Interest in maize ethanol did not revive in the United States until the petroleum shortages of the late 1970s, and between then and the early 2000s, small amounts of ethanol supplemented gasoline supplies, driven mostly by policies that phased out tetra-ethyl lead, benefited ethanol use by reduced taxes, and mandated government vehicles capable of using E85 (a mixture of 85 percent ethanol and 15 percent gasoline). Interest in ethanol increased further after 2002, when it began to replace MTBE (methyl tertiary-butyl ether, an additive making gasoline burn more completely and thus reducing carbon monoxide), and after 2005, when the Energy Policy Act of 2005 mandated that gasoline contain specified amounts of ethanol, a renewable fuel standard (RFS).[c]

The Energy Independence and Security Act of 2007 (EISA) mandated increased amounts of renewable fuel—ethanol and other materials—in gasoline, growing to 36 billion gallons per year by 2022. Importantly, EISA sought to limit the amounts of ethanol derived from the starch of corn grains and instead promote the use of lignocellulose to make *cellulosic biofuels*.[d] People can't digest lignocellulose,

[a] Bill Kovarik, "Henry Ford, Charles Kettering, and the Fuel of the Future," *Automotive History Review* 32 (1998): 7–27, www.environmentalhistory.org /billkovarik/about-bk/research/henry-ford-charles-kettering-and-the-fuel-of-the-future/, December 18, 2015.
[b] Michael S. Carolan, "A Sociological Look at Biofuels: Ethanol in the Early Decades of the Twentieth Century and Lessons for Today," *Rural Sociology* 74, no. 1 (2009): 86–112.
[c] Barry D. Solomon, Justin R. Barnes, and Kathleen R. Halvorsen, "Grain and Cellulosic Ethanol: History, Economics, and Energy Policy," *Biomass and Bioenergy* 31 (2007): 416–25.
[d] U.S. Environmental Protection Agency, "Summary of the Energy Independence and Security Act," www.epa.gov/laws-regulations/summary-energy-independence-and-security-act, December 18, 2015.

the major structural material of plants like cornstalks and corncobs, so these materials could make biofuel without directly reducing supplies of human food and livestock feed. In addition, wood and woody waste from forest practices could serve as feedstock for making transport fuels.

EISA thus altered the commercial landscape for making fuels that satisfied the RFS. Before EISA, and for a time afterward, most analysts thought ethanol, from lignocellulose but not starch, would soon become the predominant biofuel. Making ethanol from lignocellulose, the assumption continued, would grow into a large industry based on enzymes that could break cellulose and other inedible plant parts into simple sugars for fermentation to alcohol and processing to 100 percent ethanol. EISA mandated a specific timetable for increasing fuels made from lignocellulose and capping starch-based ethanol.

By 2010, it was clear that the EISA timetable would not be met, because companies could not produce the specified amounts of commercial supplies of cellulosic biofuel. For several years, the U.S. Environmental Protection Agency (EPA) and the oil industry disputed resolution of the impossible RFS mandated by EISA, while starch- and sugarcane-based ethanol continued in E10, 10 percent ethanol in gasoline. By 2014, however, small amounts of cellulosic ethanol and other cellulosic biofuels began to enter commerce, and the EPA produced new targets for use of fuel ethanol and other cellulosic biofuels. The new fuels made from lignocellulose included both ethanol and hydrocarbons that more closely mimicked gasoline and diesel fuels from petroleum.[e]

Despite new companies using technologies developed over the past forty years for turning biomass into biofuels, the question remains, Is this pursuit of biofuels really a good idea? More pointedly, Can biofuels reduce or eliminate the emissions of carbon dioxide from petroleum-powered transport, which in 2014 produced 42 percent of U.S. emissions of carbon dioxide? Natural gas and coal produced 26 percent and 32 percent of U.S. carbon dioxide emissions, respectively.[f]

Reducing U.S. emissions of carbon dioxide from liquid transport fuels poses challenges not for the faint of heart, because the amounts of fuel used in U.S. transport per year are so large. Some simple calculations show that ethanol from maize starch by itself has no potential to replace all or even a significant part of gasoline:

[e] Tristan R. Brown and Robert C. Brown, "A review of Cellulosic Biofuel Commercial-Scale Projects in the United States," *Biofuels, Bioproducts, and Biorefineries* 7 (2013): 235–45; Renewable Fuels Association, "EPA Final Rule for 2014–2016 RVOs," www.ethanolrfa.org/wp-content/uploads/2015/12/RFA-Talking-Points-on-2014–2016-Final-RVOs1.pdf, December 21, 2015.
[f] U.S. Energy Information Administration, "Carbon Dioxide Emissions by Fuel," www.eia.gov/environment/, December 22, 2015; percent calculations by author.

- The United States used approximately 137 billion gallons of gasoline in 2014.
- In 2014, ethanol distillers produced about 14.3 billion gallons from about 5.2 billion bushels of corn, or 37 percent of the U.S. corn crop that year.
- The land area planted to corn in 2014 totaled about 95 million acres.
- Thus the land planted to corn produced about 150 gallons of ethanol per acre in 2014.
- If the United States wanted to produce 137 billion gallons of ethanol in a year, it would require 910 million acres and 100 percent of the corn crop, leaving none for livestock and human food or for other products based on corn.
- Even if maize yields per acre doubled, an unlikely increase, the United States would not have enough land for 100 percent of the maize crop to provide 137 billion gallons of ethanol.
- Unfortunately, the total cropland in the United States amounts to only about 408 million acres, of a total of about 2,264 million acres.
- Even worse, ethanol has only about two-thirds of the energy content of gasoline, so the production of 150 gallons of ethanol per acre is equivalent to about 100 gallons of gasoline; 14.3 billion gallons of ethanol equals the energy in about 9.5 billion gallons of gasoline, this from 37 percent of the U.S. corn harvest.
- If the energy content of 137 billion gallons of gasoline had to come entirely from ethanol, the cropland required would be about 1,368 million acres, even more impossible.[g]
- This conclusion even includes efficiency gains in producing ethanol from starch between 1997 and 2014; to obtain the 2014 yields with 1997 technology would have required 343 million more bushels.[h]

These very simple considerations demonstrate that even the United States, the world's largest producer of corn, cannot produce enough corn to supply enough starch-based ethanol to power the country's gasoline-fired transport, even if all food and feed uses of corn were to cease—an impossibility. At this point the question becomes, How much energy exists in agricultural residues that could serve as feed-

[g] Figures calculated by author from data provided by the U.S. Department of Agriculture and the U.S. Energy Information Administration.
[h] Tony Radich, "Corn Ethanol Yields Continue to Improve," *Today in Energy*, May 13, 2015, www.eia.gov/todayinenergy/detail.cfm?id=21212, December 23, 2015.

stocks for making biofuels (assuming that production of biofuels from lignocellulose is feasible)?

Some controversy surrounds estimates of energy in lignocellulose residues, because the number of assumptions needed to calculate estimates is, to put it mildly, very large. Nevertheless, no study I have found suggests that lignocellulosic residues could easily provide a significant replacement for the energy currently provided by petroleum. Consider just a few simple numbers:

- The Intergovernmental Panel on Climate Change estimates agricultural residues produced each year contain between 15 and 70 EJ; a 2014 study calculated 65 EJ produced each year in such residues.[i]
- In 2014, the world produced about 32.4 billion barrels of oil, which had an energy content of about 185 EJ.[j]
- About 64 percent of the oil produced, or about 118 EJ, went to power transport.[k]
- Thus even if all agricultural residues could be collected for making biofuels (they couldn't) and no other uses of these residues were important (but they are), biofuels could not even replace the oil used in transport, let alone all the other uses of oil.

The simple numbers are no more than back-of-the-envelope rough estimates, but their implication points to a serious conclusion that biofuels have no possibility of substituting for petroleum. Could biofuels play a significant if partial role in the Fourth Energy Transition?

This question brings the issue of EROI to the forefront: how much energy it takes to make biofuels from biomass. Again the estimates vary, but a recent comprehensive review concluded that ethanol from sugarcane had the highest possible EROI, but it was less than 5. Other biofuels had an EROI of < 1.85. In other words, for 1 unit of energy invested, you could make back the 1 and have a surplus of about 0.85 units.[l]

On this measure, the use of ethanol from sugarcane might be practical as a source of energy, but its supplies are distinctly limited. Other biofuels have such a low EROI that for all practical purposes they are not primary energy sources.

[i] Niclas Scott Bentsen, Claus Felby, and Bo Jellesmark Thorsen, "Agricultural Residue Production and Potentials for Energy and Materials Services," *Progress in Energy and Combustion Science* 40 (2014): 59–73.
[j] *BP Statistical Review of World Energy,* June 2015, 8.
[k] International Energy Agency, *2014 Key World Energy Statistics,* 33.
[l] Carlos de Castro, Óscar Carpintero, Fernando Frechoso, Margarita Mediavilla, and Luis J. de Miguel, "A Top-Down Approach to Assess Physical and Ecological Limits of Biofuels," *Energy* 64 (2014): 506–12.

released on burning may be somewhat smaller than that released by a fossil fuel with comparable energy. The above considerations all merit a + for biomass energy on Protection of Common Resources.

If, however, fossil fuel–powered trucks must haul harvested biomass long distances and if the processing plant uses coal or natural gas for heat, then the biofuel may result in as much or higher emissions of greenhouse gases compared to using fossil fuel liquids for transport.

As the climate changes and the population grows, the methods for harvesting and processing biomass for biofuels will likely move toward more involvement of fossil fuels in producing the fuels. A destabilized climate will make biomass production more expensive, and a growing population will speak against higher harvest levels. These conditions will favor the "cheaper" fossil fuels as an adjunct to making biofuels. This set of considerations merits the −− ranking on Protection of Common Resources.

Biomass also merits a mixed ranking of + / −− for Distribution of Benefits, Costs, and Risks. In a few situations, biomass energy returns a genuine if small benefit, and hence the +. The real problems arise when biomass energy competes with supplies of food, feed, and fiber. In these cases, lower-income people will lose. The prospects of climate change and population growth suggest the trends will be toward worse adverse effects on lower-income people, and thus the −− ranking.

The rationale for the ranking on Ethics, + / −−, parallels that of the ranking on Distribution of Benefits, Costs, and Risks. When the benefits occur, for example, when biomass is used for electricity, the sharing of benefits will be widespread. Unfortunately, the disadvantages imposed on lower-income people from the likely worsening costs and risks has no moral justification, which leads to the −− ranking.

Finally, biomass ranks + on Aesthetics. Biomass production looks like farming, a landscape that most people find attractive. Biorefineries, perhaps located every 40 to 60 miles to keep transport costs of raw biomass low, look like industrial factories. Most people don't find this sort of installation attractive, but they also don't object once they become used to it. Of all energy sources, biomass may rank the highest on Aesthetics.

Geothermal

Geothermal energy, the second or third most abundant source of renewable energy after solar (see tables 8.1 and 8.2), is a sleeping giant. In theory it offers a huge supply of heat, only a small amount of which currently enters the human energy economy each year. The question hanging over this sleeping giant is, Can humans awake it to serve their purposes with acceptable costs and risks?

Awakening the resource depends almost entirely on transforming the heat energy into electrical energy. Geothermal energy can serve as a locally important energy source without being transformed into electricity. For example, hot springs resorts and recreation areas, like Glenwood Springs, Colorado, or Yellowstone National Park, Wyoming, have long celebrated natural hot water for bathing and the display of mighty geysers and boiling mud pots.

Direct use of geothermal heat (i.e., not transformed to electricity) totaled 424 PJ/year globally in 2010. China was the largest consumer at 75 PJ/year, followed by the United States (57 PJ/y), Sweden (45 PJ/y), Turkey (37 PJ/y), and Japan (27 PJ/y). A total of seventy-eight countries reported some use of direct geothermal heat in 2009, and the 2010 global utilization had increased by 55 percent over 2005. Ground source heat pumps used 47 percent of this heat, followed by 26 percent for heating water, 15 percent for space heating, 6 percent for greenhouses, and some percentage for a number of other minor uses. These uses have proven quite valuable in specific regions for specific uses, but direct uses of geothermal heat still rank as a minor component of the world's energy economy of 500,000 PJ/year (500 EJ/y).[72]

Local uses will undoubtedly continue far into the future, but only transformation of the heat into electricity allows transfer of the energy to distant places for utilization in more diverse ways than making hot water. Development of geothermal electricity, however, has remained small compared to the big-four fuels. Total global installed capacity for geothermal electricity stood at only 200 MW in 1950, and it rose to about 10,900 MW by 2010. In that year, the United States had the largest capacity (about 3,100 MW) and the largest amount of total generation (about 16,600 GWh/year). The top five countries for installed capacity are the United States, the Philippines, Indonesia, Mexico, and Italy.

Unfortunately, tapping the largest amounts of geothermal heat requires enhanced geothermal systems (EGS) (see chapter 8). The U.S. Geological Survey estimates that vapor- and liquid-dominated systems could yield in the United States a total of 3,700 MW, with a 50 percent chance of yielding 9,000 MW. The most recent estimates for EGS in the United States range from 517,800 to as high as 15,900,000 MW.[73] Only successful development of EGS will propel geothermal energy into a major player in the United States with a capacity ranging from about 50 percent of the current electrical generating capacity, 1.07 million MW, to vastly more than that.[74]

Other analysts have a more pessimistic outlook: no more than 70,000 MW of installed global capacity by 2050 based on exploitation of vapor- and liquid-dominated sites. Successful development of EGS could possibly double the installed capacity to 140,000 MW to produce an estimated 8.3 percent of the world's electricity.[75] This is still a significant amount for the Fourth Energy Transition, but the low estimates don't predict a major role for geothermal energy.

What, then, are the prospects for developing geothermal electricity with EGS, and is it a good investment? At least three factors, all discouraging, influence the development of geothermal generating facilities: induced earthquakes, costs and cost flows during development, and water. If successful development proceeds, then challenges will remain in terms of potential land use conflicts and building new transmission lines. Here we'll focus on just the first three issues.

Geothermal electricity involves two processes that can induce seismicity, or earthquakes: injection and/or withdrawal of water from underground deposits and changing the temperature of the rocks beneath the surface. EGS injects water into hot rock formations to crack them so that subsequently water injected to collect heat can move through the fissures and then be pumped to the surface for generating electricity. This "fracking" of hot dry rocks has a high probability of inducing earthquakes felt on the surface.

The first efforts to create EGS began in New Mexico in the 1970s and subsequently moved to projects in the United Kingdom, France, Germany, Japan, Australia, Sweden, and Switzerland. A project near Basle, Switzerland, attracted great notoriety for the earthquakes it induced, which caused government authorities to end the effort. Experience at The Geysers in California demonstrated that simply putting cold water into the geothermal deposits could also induce seismic events.[76]

Methods to mitigate and reduce seismic events may prove successful enough to raise EGS to a fully acceptable practice, and the U.S. Department of Energy has developed a best-practices protocol for geothermal developers to follow. It includes steps such as preliminary screening of a proposed site, communications with the public, establishing criteria for evaluating induced vibrations, monitoring, quantification of hazards and risks, and a mitigation plan with stop-caution-go signals if seismic events occur.[77] In the meantime, however, costs and the flow of costs have hindered development of EGS and other hydrothermal projects.

Costs of geothermal electricity measured by LCOE don't raise alarm (see table 10.2), but the costs are high in another way. A full project may take five to seven years before electricity sales can offset the expenses, but about half of those expenses occur even before construction of the power plant itself begins. Most of these early expenses come in drilling bore holes into the rock, a process that draws heavily on the experiences and methods of drilling for oil and gas.

With so many upfront expenses over many years, the risks are high, and in recent years cheap natural gas in the United States has diminished the interests of utilities in developing geothermal. The twenty-odd major geothermal firms have dealt with these risks sometimes by vertical integration to include all aspects of project development and power plant operations and sometimes with heavy participation by government. Firms active in the United States tend to rely strictly on private capital and separation of exploration-construction companies from power companies.[78]

In addition, competing fuels like gas, solar PV, and wind can be installed in small increments of new capacity. That is, installation can be in small *modules* with short project times and less risk. Geothermal installations are best when exploiting economies of scale, so they do not lend themselves to modularization. Geothermal energy has strong advantages, noted below, but risks and costs create barriers. Compared to the growth rates of solar and wind installations since 2000, geothermal capacity has definitely lagged.[79]

Federal and state policies have supported geothermal development but in ways less powerful than for solar and wind. Most policies supporting geothermal at the federal level provide financial help only after the project begins to generate electricity. Thus for several years a geothermal developer endures heavy costs with limited government support, even before the developer knows whether the site under exploration will be suitable. Policies available from federal and state sources include the following:

- Research to assist development of geothermal technology
- Loan guarantees and/or grants after 2005
- Production tax credits (PTC) or investment tax credits (ITC); between 2009 and 2011, the PTC could be converted to an ITC, and an ITC could be converted to a cash grant.
- Development of EGS projects by the U.S. Department of Defense

- Over one-half of the states have renewable portfolio standards (RPS) for their utility industries, and geothermal electricity can help satisfy these purchase requirements.[80]

Based on the above considerations, EGS merits a rank of + / – on Profits and Risks. The energy source clearly has demonstrated strengths, and with public policy support the + could change to ++ or even +++. At the same time the fuel has weaknesses, giving it the – score. The costs of project development are high, with high risk at the start. The risks are so high that EGS projects may be like nuclear power: only government can take the risks, and without strong intervention by government EGS won't move beyond a minor source of energy. In addition, induced seismicity is a weakness, perhaps not fatal, but that depends on the success of the best-practices protocol to mitigate seismic events. Without government absorption of financial risks and/or without acceptable mitigation of seismic events, the – score could become –– or even –––, signaling that EGS would likely remain a poor or even disastrous investment.

In jurisdictions with strong rule of law, Security of Access and Profits for EGS will be very strong, meriting a rank of +++.

Geothermal energy based on vapor- and liquid-dominated sources alone means geothermal heat already has demonstrated strength in Size of Resource, meriting a rank of +. With strong public policy support and successful mitigation of seismic events, EGS will carry geothermal energy to a rank of +++. Hence the overall rank for geothermal on Size of Resource is + / +++.

In contrast to solar and wind, which show definite weakness on Time of Use, EGS fully merits a rank of +++. EGS power does not depend on weather, and EGS can provide base-load electricity.

Similarly, on Geographic Distribution, EGS merits a rank of +++. Deposits of geothermal heat are not evenly spread over the earth and not equally easy to harvest, but nearly every location on the planet has some resource, and electrical transmission can send the energy to distant places.

EGS, like the other renewable sources, emits low amounts of carbon dioxide compared to the fossil fuels (see table 10.2). On these grounds, geothermal merits a rank of +++ on Protection of Common Resources. If seismic activity accompanies EGS-derived geothermal energy, however, this ranking could drop and become negative.

EGS to produce electricity means that the benefits of geothermal heat can easily spread far distances, meriting a score of +++. At the same time, development of an EGS project may inevitably induce some earth-quakes, causing local distress and damage around the project area. These risks will be borne by a small group, not the widespread benefici-aries of the project. Thus EGS also has a – ranking, and the overall rank for Distribution of Benefits, Costs, and Risks is +++ / –.

On Ethics, EGS merits a rank of +++, provided that those enduring seismic risks have acceptable compensation and the risks are minimized. If those suffering these risks do not find satisfaction, then negative scores would attach to the Ethics ranking.

On Aesthetics, EGS ranks a – score. A geothermal plant is an indus-trial facility, generally seen as "attractive" only by the engineers build-ing it and the owners making money from it. For the public, it's just another factory, often placed in a beautiful rural area. Still, most people will tolerate this weakness in return for the electricity.

CONCLUSION

Two overarching conclusions are shown clearly in table 10.1: every source of primary energy has strengths and weaknesses, and the profiles of the respective sources differ significantly. Judged by the traditional criteria, the energy economy of the future would remain approximately like it is today: based on fossil fuels, supplemented by uranium and renewable energy when economic rationality or political decisions sup-port them. In other words, the energy economy that matured during the Third Energy Transition suffices until depletion of fossil fuels changes the economics and the supplemental fuels come into play. The idea of a deliberately chosen Fourth Energy Transition has no compelling legiti-macy under the traditional criteria.

Adding the new criteria for sustainability alters the situation pro-foundly. With the new considerations, rankings of ––– in table 10.1 cloud the future of the big-four fuels (coal, oil, gas, uranium). The big-four fuels create serious problems, and renewable energy, accompanied by high levels of energy efficiency, indicate that humanity can build a new energy economy, deliberately chosen. Sustainability is the idea that brings the Fourth Energy Transition to the forefront.

Is 100 percent renewable energy really possible? It doesn't exist any-where today, but, as the above discussion shows, it has a good possibil-ity of becoming real, provided genuine challenges find acceptable solu-tions. Many areas currently are steadily increasing their reliance on renewable energy, and so far no insurmountable problems have appeared. The path of prudence, therefore, lies in continuing and enhancing the changes already underway.

Individuals, companies, and governments should clearly ask engineers and scientists to make a renewable, sustainable energy economy the long-term goal and ask them to find workable solutions to challenges. The goal of the Fourth Energy Transition should be a phase-out of the big-four fuels, as close as possible to 100 percent.

In the final chapter, we turn to the major barriers to the Fourth Energy Transition.

Connections with Everyday Life

People who set ambitious goals and plans know that something might come along to derail their chosen course of action. If you, the reader, want to change energy to renewable sources used efficiently, then this chapter is essential. What can go wrong? What do I have to address in order to maximize the probability of success?

As it turns out, lots of things can go wrong, but they are neither inevitable nor insurmountable. They do, however, demand attention, because they can thwart the best-laid plans. Take heart, work hard, and carry on!

CHAPTER 11

Barriers and Challenges

Renewable energy and energy efficiency have already brought about changes in the sources and amounts of energy used, but multiple reasons mandate pushing them as close as possible to 100 percent replacement of the big-four fuels. Modern societies and states cannot sustain themselves on the fuels that underwrote the Third Energy Transition.

What pathways of technological change will relegate the big-four fuels to the past, supplanted by renewable energy sources used very efficiently? Detailed analysis of pathways must await a different book, and we end with brief discussions of the key barriers and challenges standing in the way of the Fourth Energy Transition. None are insurmountable, but left unaddressed many will slow, stymie, and decrease the success of the transition. Political barriers and challenges have the greatest negative potentials, followed by needs for new technologies and institutions and for new perceptions and behaviors.

POLITICAL BARRIERS AND CHALLENGES
Guiding Investment to Energy Efficiency and Renewable Energy

Investment built the infrastructure of the Third Energy Transition. These investment decisions were made primarily by private individuals and companies, sometimes by governments. Government incentives encouraged some investments, but governments affected all investments through rules, regulations, subsidies, and taxes.

Until after the 1970s, investments went almost exclusively to fossil fuels, uranium, hydropower, and efficiency. Increasingly after 1980, concerns about energy, particularly climate change and energy security, prompted more incentives to invest in renewable energy and energy efficiency to reduce carbon emissions. Humanity now must phase out the big-four fuels (coal, oil, gas, uranium) and push energy infrastructure as far as possible toward renewable energy and energy efficiency (see chapters 6–10).

Pushing new investments in energy as far as possible toward renewable energy and energy efficiency generates the challenge that if not overcome is now and will remain the prime barrier to the Fourth Energy Transition. It's very simple: without large investments in renewable energy and energy efficiency, the Fourth Energy Transition won't proceed very far. Not only is the challenge simple, but it's far more political in nature than economic.

If renewable energy had a clear monetary advantage over the big-four fuels, particularly the fossil fuels, then investors would flock to renewable energy and energy efficiency. The fossil fuels would fade into history. Lots of coal and oil would lie waiting, but no one would dig up worthless minerals. This would mimic the technological obsolescence suffered by, for example, firewood compared to coal, the telegraph compared to email, and the horse-drawn buggy compared to the automobile.

Analysis has shown, however, that—in the market conditions governing energy investments—technological obsolescence is not likely to happen for fossil fuels in the next few years (see table 10.1). Too many externalities continue to benefit producers and consumers of the fossil fuels. The rules for their markets evolved to promote them, not renewable energy, and producers and consumers pay nothing for the externalities (except in a few places). Thus returns on investment are higher than if consumers had to pay for externalities.

The political challenge lies in persuading governments to alter the rules of the investment arena. In 2013, the International Energy Agency estimated that $1,600 billion ($1.6 trillion) dollars flowed into energy infrastructure worldwide (see chapter 5). Over the period to 2035, the agency estimated the investment stream at $48 trillion, of which $40 trillion will go to ensure supply and the rest ($8 trillion) to energy efficiency. The $8 trillion going to energy efficiency matches the goal of reducing need for energy supply, but of the $40 trillion, $23 trillion will go to fossil fuel extraction, refining, and transport. Over half of the funds going to supply will serve to offset declining production from existing oil and gas sources and to replace retiring power plants. Of the rest, the agency estimated $6 trillion will go to renewable energy. By 2035, $2 trillion per year will be going to supply and $0.55 trillion to energy efficiency.[1]

Does the amount projected to go to fossil fuels match the needs of the Fourth Energy Transition? Or does it simply reflect the International Energy Agency's prediction of what private and governmentally owned companies will do given the current rules governing energy investment? Certainly fossil fuel companies have a vested interest in mining and selling as much of their product as possible; coal, oil, and gas companies see renewable energy primarily as a competitor, not as a worthy investment.

Fossil fuel companies also see costs of production and projected revenues as the critical issues for investment decision making, not other issues like protecting common resources and health. Governments with national oil companies see oil revenues as crucial to the fiscal health and security of their respective states. Some governments of oil-producing nations and states rely heavily on sales and/or tax revenues from these companies to fund public services. It is too much to expect that private and public investors will—without guidance—see the importance of preserving common resources such as the atmosphere and climate.

A price on carbon emissions and/or mandatory tradable permits to emit carbon would require producers and consumers to recognize that these fuels already impose costs on societies, and only those who are directly affected by the damage have to pay those costs. Subsidies, tax incentives, and mandatory purchase levels for renewable energy and efficiency will guide investors to continue following the money. And investments will go toward advancing the Fourth Energy Transition.

Policies pushing the Fourth Energy Transition will affect the operations of modern economies, but they will not necessarily make total energy costs significantly more expensive for consumers. Citi GPS, a financial consulting division of the large American bank, Citibank, estimated the differences in cash outlays to procure fuels and build new infrastructure for two different energy investment pathways between 2015 and 2040. One was business as usual, with most expenditures going to fossil fuels; and one emphasized low carbon emission energy and significant expenditures in energy efficiency. The expenditures over this twenty-five-year period totaled $192 trillion for business as usual and $190.2 trillion for the low carbon emissions pathway. In short, the pathway more like the Fourth Energy Transition actually cost $1.8 trillion less than proceeding according to current patterns.

To be sure, significant differences separated the two pathways. For example, substituting expenditures to build and operate solar energy farms compared to gas-fired power plants involves heavy capital expenditures up front for solar followed by no expenditures for fuel. Gas, in contrast, has low capital expenditures to build the plant followed by heavy fuel expenses during operations. Moreover, the costs of gas might be extremely volatile and hard to predict, whereas the cost of sunlight remains free and without volatility.[2]

These large sums stagger the imagination, and they clearly indicate the scale of political actions needed. Only organized citizens insisting that their governments shape investment policies have a chance of pushing huge investments in the right directions. Without citizen insistence, it won't happen. No matter how many individuals put solar panels on their roofs, insulate their buildings, and buy efficient electric cars, the guidance to investment won't be large enough. Collective political action must lead the way.

In addition, no plan to promote the Fourth Energy Transition with appropriate policies will succeed unless it has support across a reasonably broad political spectrum. The tasks of remaking national and global energy economies are too large to rest on one or a narrow spectrum of political opinion. For the United States, this means significant bipartisan agreements must underwrite needed changes in government policies. These difficult tasks will face opposition, but they are both possible and necessary.

Resistance from Established Energy Industries, Supporting Industries, and Political Leaders

Substituting renewable energy and efficiency for fossil fuels raises serious challenges to existing energy industries. Investors, management, and labor each have substantial financial and human capital in these industries. Replacement of the fossil fuels means loss of sales to these companies, which potentially means their obliteration if their product lines are narrow. It would be ludicrous to think that the companies will not defend their respective products against renewable energy.

The energy industries have already fought back. For example, major oil companies created and financed the climate change countermovement in the United States.[3] Companies and trade associations have also launched positive arguments that their product cures an ailment associated with one of the other big-four fuels. For example, the Nuclear Energy Institute and the Natural Gas Council advance uranium and gas, respectively, as fuels vital to mitigating climate change, especially as substitutes for coal.[4] Similarly, the Renewable Fuels Association promotes biofuels in place of petroleum as part of a solution to climate change.[5] Taking yet another tack, the American Petroleum Institute has sought to build a reputation as a good citizen for reducing emissions of methane and carbon dioxide to mitigate climate change while at the same time emphasizing the benefits of oil and gas.[6]

A major battleground will be the fate of energy deposits already proven, based on past investments. Citi GPS estimates that $100 trillion of unusable reserves may have to be left in the ground to avoid exceeding the carbon budget compatible with no more than a 2°C rise in temperature.[7] What will happen to these assets if they become "stranded" in the ground? Widespread ownership of equity shares in these energy companies by individuals, insurance companies, and pension funds means that the fate of stranded assets will have repercussions for many people. As yet, no government has legally stranded any asset, but opposition to that possibility already exists.

In addition, supporting institutions may depend heavily on business with fossil fuel companies. Industries such as transport, education, and finance provide vital support in transporting, training staff, and financing, respectively. For example, American railroads earn 18.8 percent of their revenues from hauling coal, their largest customer.[8] Professionals and labor in support institutions may also be in jeopardy and oppose movement to the Fourth Energy Transition.

Industries will attempt to sway public opinion, but the critical debates will be in legislative, executive, and judicial arenas. As just one example, consider the statement of Sen. John Barrasso (R, WY) about President Barack Obama's 2016 decision to reduce leasing of public lands for more coal mining: "Pres. Obama's administration is in a full-scale war with coal communities and families. . . . A moratorium on federal coal leasing puts thousands of people out of work in #Wyo & across the West who are employed in coal production. . . . I'm prepared to act to protect families who depend on coal for their livelihood against this latest attempt to destroy American energy."[9]

Loss of Sales and Tax Revenues to Companies, Communities, and Nations

Coal, oil, gas, and uranium sell for substantial sums (see chapter 4), but it's easier to forget that companies are not the only institutions surviving on these sales revenues. Wages to workers and dividends and capital gains to investors spread the money received, and these constituencies suffer economically when prices or production drops.

Mines, wells, and power plants based on these fuels are usually located in less densely populated areas, and local and state governments often impose taxes on the primary fuels or electricity generated from them. For some countries, revenues from sales and taxes often provide major support for all public programs, so essential government services also suffer when prices and production decline.

Consider a few examples.

- Entergy Corporation wants to close the FitzPatrick nuclear power station in Oswego County, New York. An undetermined fraction of the 615 workers at the plant would lose jobs that pay exceptionally well for the area.[10] New York's Gov. Andrew Cuomo wants to shutter the Indian Point nuclear power plants near New York City on safety grounds but argues FitzPatrick should be kept open for the economic benefits to the area.[11]
- North Dakota tax revenues have fallen as the price of oil declined in 2015, and in October state officials estimated a shortfall of $112 million. State agencies, including universities, began planning for a reduction, which will affect jobs and opportunities for students.[12]
- The Russian Federation derives about one-half of its annual budget from oil revenues, which dropped below $50 per barrel in 2015; $50 was the oil price used to calculate the 2016 budget, and the Russian economy was in recession in 2015.[13]

All technology change generates winners and losers, but the scale needed in the Fourth Energy Transition makes it imperative to find new arrangements for those who suffer dislocation. It defies logic to think that those facing economic disaster will not resist politically. People who may suffer must participate in the debates to offset losses and come to believe that the Fourth Energy Transition will also improve their lives.

Rejection of Climate Science

Climate science has identified carbon dioxide and methane as the most important greenhouse gases. Projections from models indicate that about half the amount of greenhouse gases needed to raise the global mean surface temperature beyond 2°C has already entered the atmosphere (see chapter 6). To keep temperatures from rising higher than 2°C will require reduction of carbon emissions, which is almost certain to require reductions in uses of fossil fuels. Only successful development of carbon capture and storage (CCS) could prevent these emissions from fossil fuel uses, and CCS shows little promise to date of being successful at the scales needed.

As a result, those who accept climate science—the vast majority of the scientific community—have favored policies to encourage and force reductions in uses of fossil fuels. A small minority of scientists, 10 percent or fewer, have continued to express doubts, even in the face of endorsement of the science by many science academies (see chapter 6). Some political and business leaders, however, have embraced the minority viewpoint and rejected the implications of climate science.

Scientists who deny or are skeptical of climate science have opposed government regulations of carbon emissions. Some have added that opposition to fossil fuels condemns the world's poor to endless energy poverty.[14] Those who argue against climate science may couch their arguments in terms of scientific issues, but careful analysis indicates that in fact the opposition is better understood as a fundamental political objection to strong government regulation. Building policies to promote the Fourth Energy Transition requires recognition of the political nature of climate denial and skepticism.

Acceptance of Military Might to Gain Oil Security

Several factors propelled petroleum to a unique position in the global energy economy. Among the most important are (a) the need of many countries to import the material, (b) its primary role in powering transport, and (c) the willingness of governments with powerful militaries to dominate less powerful countries with huge supplies (see chapters 3 and 7). Imperialism subjected people to violence and conquest long before oil's rise to prominence, but by 1900, capture of oil supplies by the powerful became the prize most often sought.[15]

During World War II, the United States began to replace Britain and France as the most powerful country to project military might abroad, and security of oil supplies from the Middle East drove strategic analysis (see chapter 7). As long as the United States or any other country sees imposition of military might to secure oil supplies as the best way to guarantee the stability of its transportation system, the military "solution" will continue to hinder movement to renewable energy.

NEEDS FOR NEW TECHNOLOGIES, EXPERTISE, AND INSTITUTIONS

Political action heads the list of the most important challenges to the Fourth Energy Transition, and existing technology and expertise can support all or almost all of the changes needed. Nevertheless, innovation and new knowledge, housed in new institutions, will ease the processes of overcoming barriers and challenges. The following developments hold key roles.

Electrifying Transport

Globally, transportation powered mostly by petroleum emitted about 14 percent of the greenhouse gases in 2010. In the United States in 2013, transport accounted for 27 percent of greenhouse gas emissions, second only to 31 percent of the emissions from generating electricity.[16] Replacement of fossil fuels (and uranium) generating electricity with renewable energy poses few problems to electricity consumers, because electrical energy works the same regardless of its source. Nevertheless, replacing fossil fuel–powered transport with renewable energy will occur only if electric motors replace internal combustion engines in automobiles, trucks, buses, trains, ships, and airplanes.

Electrified rail transport has existed for many years, but until recently very few electric vehicles plied the roads. By 2016, commercially available electric cars, buses, and trucks indicated that electrification of transport had entered territory formerly occupied only by gasoline and diesel fuel. Some vehicles had combinations of internal combustion engines and electric motors, but others were purely electric. Electric motors can run on battery power or by electricity generated by fuel cells.[17] Renewable electricity can charge batteries or make hydrogen for fuel cells by electrolysis of water.

Large ships, too, have entered the electric arena. The battery-powered *Ampere*, a car and passenger ferry in Norway, won the Ship of the Year award in 2014. Hydropower recharges the lithium-ion batteries that run two 450-kilowatt engines, and each trip requires 150 kWh.[18] Replacing energy-dense jet fuel from petroleum will undoubtedly raise many technical challenges not yet solved. As I write this, however, *Solar*

Impulse 2, an aircraft powered by photovoltaic electricity, is due to land in California from Hawaii on its round-the-world flight. Shorter solar-powered flights had preceded this accomplishment, which comes just 107 years after the first flight across the English Channel.[19]

How far can electrification of transport proceed? Will it be sufficient to completely "win" over fossil fuel–powered mobility? Analysts from the European Union believe that pathways exist to reduce transport sources of greenhouse gases by 50 percent by 2050. This transformation, however, rests on an estimated $20 trillion invested by 2050, alterations in urban design, and policies to increase the costs of carbon emissions, all of which mean serious changes from current patterns.[20] Technology of electrified mobility will improve in the years ahead as investments in research, development, and deployment alter the transport landscape.

Electric Grids, Energy Storage, and New Institutions

Renewable electricity fueled by light, moving air, falling water, hot rocks, and biomass is an extraordinarily flexible and easily moved form of energy. Some sources of renewable energy, such as solar PV, lend themselves to small-scale, locally used production of electricity, but utility-scale production generally offers the best way to tap most renewable energy sources. As a result, a transition to renewable energy means reengineering the electric grid to accommodate high levels of renewable energy, often distributed in many locations.

The challenges of remaking electric grids to transmit and distribute renewable energy are welcomed by eager engineers who love nothing better than puzzles with multiple variables spread over space and time with many acceptable solutions at different levels of cost and complexity. For example, in the United States, the richest sources of renewable energy don't always lie next door to the places needing the electricity. Wind power exists mostly in the sparsely populated Great Plains, and solar power is strongest in the desert Southwest. Making extensive uses of these energy sources may mean moving the electricity to population centers far removed from the sources.

In addition, the intermittency of wind and solar don't mesh well with the regular rise and fall of electricity use during each day and the seasonal variations in supply and uses of power. Electric motors, electric circuits, lighting fixtures, and computing resources all function best with electricity that arrives at a steady voltage and frequency, and some uses depend on availability of power 24 hours a day, 7 days a week, and 365 days per year. Solutions to intermittency, combined with far-flung geographic distribution, must be found to make the Fourth Energy Transition appealing.

Engineers cannot yet specify the exact solutions that will work in a grid with ample, high-quality electricity, but three considerations give confidence that acceptable answers to the many puzzles will emerge over time. First, Denmark, Germany, California, Iowa, and Texas already have grids that operate with significant and increasing levels of renewable energy (see chapter 8). No place has yet reached 80 to 100 percent electricity from renewable, intermittent sources, but grids have absorbed and functioned well with as much as 50 percent renewable, intermittent energy without engineers saying they can go no further.

Second, extensive studies and modeling have not yet identified a technological factor or consideration that suggests higher penetration levels of renewable, intermittent energy simply won't work. One of the most extensive studies, from the U.S. National Renewable Energy Laboratory, found no reason to doubt the ability of the U.S. electric grid to continue to function well in all parts of the country, at all hours, at all times of the year, with 80 to 90 percent penetration levels. Moreover, these models involved no new construction of nuclear or fossil fuel power plants but instead assumed such facilities would steadily retire at the end of their lifetimes. Various means of energy storage stabilized operations of the grid.[21] Clever engineers, given the right marching orders, have a high probability of successfully pushing more renewable electricity onto the grid.

Third, new technology for generation, storage, and transmission of electricity continues apace. In a sense, the world has just begun to explore the potentials for an electrified energy economy fueled by renewable energy without fossil fuels or uranium. Based on past experiences with technology development, we can assume that the technologies available today will improve over time.

Engineering and innovations in technology alone, however, do not have to hold all the answers for achieving a practical, affordable, high-quality, electrified energy economy. Changes in time-of-day pricing can stimulate customers to curtail use of electricity during times of high demand, easing pressures on grids. Arrangements for lower-cost electricity in return for agreement to reduce or forgo electricity under specified conditions can effectively match customer needs with abilities to generate electricity. These are changes from the current philosophy that electricity should be available in any amount at any time you want it, and perhaps some customers will grumble that they don't like the new institutional arrangements. But most people will quickly adapt, as the "new" becomes "normal."

Perhaps one of the more challenging sets of changes may lie not in engineering, new technology, or altered terms of service. Instead, the existing business models of utilities and their financing may not accommodate the increasing penetration of renewable energy into electricity generation. The problem originates with generation from solar and wind that continues apace when conditions are favorable, because the fuel is free. Utilities still operating plants using fossil fuels and uranium can't compete with energy from free fuel, thus crippling the utilities' ability to earn revenue to repay the capital or support the operating and maintenance costs of their plants.[22]

In brief, the technical ability to utilize renewable energy for electrification appears robust. New technology will be developed, and concerted efforts must address a number of behavioral and institutional challenges. The objection to the Fourth Energy Transition on the grounds that renewable energy won't work grows weaker year after year. Challenges exist, but there is no reason yet to think they cannot be met.

Broadened "Energy Expertise"

The energy economy of the modern world owes its existence to science and engineering, and the Fourth Energy Transition will be every bit as expertise-intensive as was the Third Energy Transition. Nonexperts have essentially zero ability to design, build, maintain, and operate a steam engine, a coal-fired power plant, a nuclear power plant, or a solar PV installation. The theoretical knowledge required, by the 1900s, lay firmly embedded in academic departments of science and engineering. Research occurred in universities, government laboratories, and, sometimes, industrial laboratories. Practical knowledge resided almost entirely in business firms, with some support from academic and government departments.

One of the challenges of the Fourth Energy Transition centers on investment in and development of a skilled energy workforce, without which the energy economy would fall apart. Every year, new students must begin the process of learning so that new workers can replace those who retire or leave the workforce. The rest of society depends on the knowledge and competence of the skilled energy workforce.

That said, energy expertise as it currently exists in people and institutions often lacks breadth beyond science, engineering, and practice of the many disciplines that populate the workforce. A petroleum engineer knows about producing and refining oil and gas but not about how to run a nuclear power plant or obtain coal from a mine. An expert in solar PV will not likely know much about wind turbines or hydraulic fracturing for natural gas. Experts in supply of energy may know very little about managing the demand for energy. An expert in fuel pricing generally knows markets for only one or a few fuels and probably knows very little about the mitigation of climate change, the protection of public health, or the management of geopolitical tensions surrounding energy.

In other words, the expertise that developed to build and support the Third Energy Transition was not really energy expertise. It generally took the form of technology-fuel expertise. An expert knew how to procure a fuel and make it do the things people want. Fuel experts generally knew very little about energy economies as a whole or the multiple reasons that the Third Energy Transition cannot be sustained for the indefinite future.

Michael Webber, an engineer at the University of Texas who has led efforts to broaden energy education stated, "Higher education for students interested in energy lacks the cross-disciplinary curriculum that they critically need." Technical students may learn only a single slice of the energy pie. They often learn next to nothing about the broader context in which their specialization sits.[23] Students not majoring in science or engineering generally learn little about energy and energy choices and how these choices affect the larger society.[24]

Unfortunately, when political and business leaders need advice on energy, the expertise they frequently obtain all too often comes from an expert on the production and use of a particular technology or fuel, who may not be a good authority on the energy economy in general. All too often, such narrow expertise reflects the interests of companies or industries seeking to perpetuate their own products and business models. It is not the same thing as knowledge about multiple ways of obtaining vital energy services, some of which might not require the particular fuel of a narrow "expert."

These considerations indicate that moving expeditiously to the Fourth Energy Transition requires investment in and development of new forms of general knowledge and expertise. The specialists who must make complicated technologies work properly need to know their subject matter deeply. At the same time, they must know how their narrow specialty fits within a larger framework of a *sustainable energy economy*.

Citizens, who all too often know nothing about energy, need to learn more so they understand the imperative for change and the range of options and issues. They will not and cannot be experts, but they can know enough to guide investment and expertise in the right directions.

MISLEADING PERCEPTIONS AND BEHAVIORS

Belief That Renewable Energy Is Too Small and Erratic to Work

Surveys and inventories of the size and quality of renewable energy sources indicate that the gross supplies vastly exceed the energy needs of the global energy economy as it now exists. Refined estimates of the net ability of existing and likely new technology to harvest these energy sources show lesser quantities of energy available for practical use, but even these amounts considerably exceed the 500 EJ per year used by humanity (see chapter 8).

Mark Jacobson (Stanford University) and Mark Delucchi (University of California, Berkeley), both engineers, have concluded in peer-reviewed publications that every state in the United States and every country in the world can build practical, workable energy economies based entirely on wind, hydropower, and solar energy. They see no need after the transition for fossil fuels or uranium.[25] Optimistic? Yes. Are they right? Well, we have to wait and see, as the Fourth Energy Transition is still a work in progress.

It's easy to find negative comments about renewable energy and the impossibility that it can totally replace fossil fuels and uranium. A search on Google of, for example, "renewable energy criticism" locates many commentaries, pro and con, mostly by journalists, generally not by professional scientists and engineers. They cannot compel agreement, because for the most part they are incomplete assessments: perhaps they cite a real challenge but then leap to the conclusion that no solutions could exist.

One exception, however, comes from James E. Hansen, a physical scientist, formerly head of the Goddard Space Science Institute at Columbia University and one of the best known climate scientists. Hansen has spoken widely about the absolute necessity of reducing carbon emissions to provide a livable world for future generations. He does not believe, however, that renewable energy provides workable solutions; in his view, nuclear power has a far better chance. Consider this statement he made in 2011: "Can renewable energies provide all of society's energy needs in the foreseeable future? It is conceivable in a few places, such as New Zealand and Norway. But suggesting that renewables will let us phase rapidly off fossil fuels in the United States, China, India, or the world as a whole is almost the equivalent of believing in the Easter Bunny and Tooth Fairy."[26]

Hansen has near hero status among people deeply concerned about climate change, and his climate science contributions significantly advanced the evidence indicating potentially catastrophic effects from carbon emissions. As a climate scientist, Hansen has great credibility; in contrast, his conclusions about (a) the lack of promise of renewable energy and (b) the need to retain uranium as a fuel and build more nuclear power plants don't compel agreement. No careful analysis of either point by him appears in indexes of the peer-reviewed scientific literature. He has not refuted and explained why he thinks the surveys of renewable energy are mistaken.

Hansen correctly recognizes the challenges posed by moving away from the big-four fuels, but his recommendations to push investment toward nuclear power cannot be supported; uranium cannot provide a sustainable energy economy (see chapter 10). As noted above, the Fourth Energy Transition is still a work in progress, but there are many indications that point to its future success.

Failure to Agree on a Framework for Building Public Policy

No universal, objective standards exist for measuring "success" or "adequacy" of an energy economy. Global opinion may agree that some energy services must be available. Examples of services that may have global or near-global status as "essential" are the following:

- production and delivery of clean drinking water to everyone;
- production of sufficient fertilizer to support agricultural yields that provide everybody with an adequate, nutritious diet;
- mobility to move people, food, and goods as needed;
- production and operation of buildings for safe, comfortable living, work, and entertainment;
- enough light;
- good health care for everyone;
- mechanization and automation sufficient to eliminate excessive physical drudgery;
- and adequate communications.

This list may have the correct essential services, but ferocious debates will engulf efforts to agree on the meaning of *sufficient, as needed, adequate, comfortable, enough, excessive*, and *good*. Those advocating a spartan or modest living standard will find it unnecessary, and maybe immoral, to produce energy for "luxury." Those insisting on a more affluent, material lifestyle will consider those plumping for austerity as miserly and desirous of suffering for everyone.

Moreover, people in different areas and people of different ages within an area will have different energy needs and priorities. People with abundant energy services now may think they worked hard and deserve their affluence.

And then there is military power. Armed forces require enormous energy services to build and operate. They always have. Advocates of mighty armies will consider any effort to reduce energy supply a potential threat to national security, a danger to the modern state, and thus unacceptable.

In short, the simple statement, "Use renewable energy efficiently so as to need as little as possible, and keep the demand for energy in balance with the supply," will never suffice to make public policy on energy.

In 2001, the United States had an excellent example of the ferocity and passions underlying energy. The newly installed President George W. Bush launched an immediate effort to build a new federal policy on energy to replace what he considered a hopelessly inadequate one. President Bush appointed Vice President Dick Cheney to head the National Energy Policy Group, composed of the most senior members of the new administration. The production and release of their report, *Reliable, Affordable, and Environmentally Sound Energy for America's Future*,[27] ignited a firestorm of complaints and protests.

The energy industries, however, welcomed the report, as did those who thought the United States had let its vaunted leadership in energy fall by the wayside. Critics, in contrast, said the report reflected the attitudes of those who supplied the big-four fuels, plus hydropower, and minimized the fundamental needs for efficiency and development of renewable resources. Republicans held the presidency and Congress, and the 2001 report laid the groundwork for the Energy Policy Act of 2005, heavily oriented to supplying as much energy as anyone wanted at any time they wanted it. Efficiency and renewable energy received some attention in the act, but it drew its inspiration from the ideas and technology of the Third Energy Transition, not the Fourth.

Vice President Cheney's report still stands as the last comprehensive effort by the federal government to articulate a systematic and holistic vision of the energy future. New laws after 2005 bolstered federal support of renewable energy, but no comparable, governmental effort to articulate a vision and pathway to the Fourth Energy Transition has ever been attempted.

CONCLUSION

The discussion above by no means stands as a comprehensive inventory of challenges and barriers. Others include, for example:

- eliminating energy poverty requires continued and increased uses of fossil fuels;
- carbon capture and storage will enable continued uses of fossil fuels without climate change;
- nuclear power is the best investment for mitigating climate change; and
- increased use of energy is the only pathway to economic growth.

Nevertheless, the barriers identified in politics, technology, knowledge, and perceptions each have the capacity to derail the Fourth Energy Transition. These are the barriers that must fall by the wayside. Continued debates will identify other barriers needing answers and solutions.

Epilogue

The message of this book is, superficially, very simple: humanity must build a new energy economy based as much as possible on renewable energy used efficiently. Accomplishing this task, however, will not be easy. Over a period of three centuries, modern societies traversed the Third Energy Transition, from firewood to coal, oil, gas, and uranium. Production and use of the mineral fuels are now deeply embedded in commerce, politics, laws, institutions, international relations, and the stability of money. People depend on the energy services provided by the big-four fuels, and the frameworks of civilization created during the Third Energy Transition cannot be easily dismantled.

Unfortunately, civilizations based on the big-four fuels are not sustainable into the indefinite future. Depletion ultimately will render these fuels useless, but long before that time they threaten the survival of civilizations using them. Climate change poses the most immediate, pervasive risks to the entire world, and locally these primary energy sources engender geopolitical tensions plus unacceptable health and environmental effects.

To have an energy economy based on renewable primary energy sources, the world must rebuild energy infrastructure with continued investments over the next fifty to one hundred years. Government policies have always structured energy markets, so appeals to let free markets decide make no sense. Politics and public policy shaped our existing energy economy, and only they can guide its replacement by a new one.

Literally trillions of dollars are at stake in this grand enterprise. The new energy economy will provide many opportunities for individuals and private companies, but as some gain others will decline. Jobs will change as old skills become unneeded and people with new skills find employment. Some individuals and communities will thrive; others will face displacement, which in itself threatens health and well-being.

It is difficult to overstate the magnitude of the challenges lying in the path of the Fourth Energy Transition to renewable energy used with high efficiency. Suffice it to say that no country has yet marshaled the political, economic, and moral will commensurate with the tasks of moving away from the big-four fuels. Even when tiny steps have been started, fierce backlashes have occurred.

Perhaps the United States offers the best example.[28] President Barack Obama forged plans to reduce emissions of carbon dioxide and methane to mitigate the risks of climate change. The presidential elections of 2016, however, installed President Donald Trump, who, at this writing, is vigorously working to dismantle every step taken by his predecessor. Why? Because he doesn't believe the risks predicted by climate science. The president has also claimed that climate change is a "hoax," and his administrator of the Environmental Protection Agency disbelieves the major findings of climate science.

Despite the current partisan bickering about climate change and energy choices, it is important to remember that this situation has not always been the norm. Bipartisan agreement on the ratification of the United Nations Framework Convention on Climate Change in 1992 made the United States a world leader in climate science, but that political harmony did not last. American political parties split over ideological arguments about the proper role of government in regulating commercial transactions. Most Democrats accepted climate science and believed it necessary to regulate fossil fuels, but most Republicans opposed such regulation and rejected climate science.

Some Republicans have taken important steps to restore bipartisanship. In 2017, senior officials from previous Republican administrations called for a carbon tax to fight climate change. A Republican congresswoman from New York introduced a resolution in the House of Representatives calling for conservative stewardship of the environment based on science and quantifiable facts. Those already embracing the need to regulate fossil fuels to mitigate climate change must welcome and build on these steps taken by their political opponents.

Only elected leaders in bipartisan agreement have the political legiti-
macy to face the challenges ahead. Prevention of hardship and injustice
will require the inputs of many, but the job of synthesis and reconcilia-
tion requires strong input from top leadership, accountable at the next
election. Democracies have no other pathway to solving difficult problems in acceptable ways.

Units for Measuring Energy and Power

OVERVIEW

Energy, the ability to do work, occupies center stage in the laws of thermodynamics. The concept of energy united scientific studies of (a) electricity and (b) heat with (c) the commercial uses of electricity (telegraphy) and heat (steam engines). The marriage rested on units for measuring energy that served both science and commerce. British scientists working through the British Association for the Advancement of Science led the international effort in the late 1800s to establish standard units of energy, and the units relied on the mechanical equivalent of heat, that is, work (see chapter 2).

FIRST STEPS TO UNITS

Steam engines first pumped water from mines, and a mine owner assessed the work to be done, by a horse or a steam engine, as the pounds of water lifted, against the attractive force of gravity, a certain number of feet. That is, the unit *foot-pound* captured the relevant commercial work needed: multiply the number of pounds to be lifted by the number of feet of lift. An increase in either pounds or feet of lift increased the work involved. The mine owner also wanted to know how long it took to do the work, because faster was better. James Watt standardized *horsepower* as a unit of work per minute, also called power: 33,000 foot-pounds/minute.

Thus in the 1700s, artisans and entrepreneurs invented practical units of measurement for *work* or *energy* and for *power*, even though these concepts had not yet been developed, and no philosophical or scientific scheme united work and energy. The foot-pound and horsepower utilized mass (measured as pounds), length (distance of lifting), and time (minutes). The steam engine utilized heat to do work, but Watt's measurements of work and power did not include heat. Watt instead focused on the mechanical work done, lifting water. In addition, Watt used traditional English units for weight (pounds) and length (foot).

In the 1820s, a French professor, Nicolas Clément, developed *calorie* as a unit of heat, and set the calorie in metric system units, which had been embraced by France during the French Revolution. A calorie was the heat that raised a kilogram of water by one degree centigrade, from 0°C to 1°C. Clément and Sadi Carnot exchanged information, and both explored the efficiencies of the steam engine.[1] Both also understood that a mechanical equivalent of heat existed, but Mayer and Joule have received credit for establishing the equivalency.

Clément's interest in measuring heat derived from his concerns about steam engines and their efficiencies in turning heat into work, and his unit survives today largely as the measure of heat energy contained in food and feed in the United States.[2] Analysts of energy systems seldom use it, because units based on the International System of Units (SI units), which began to develop in 1875, gained acceptance.

A derivative of Clément's calorie, however, does survive as a unit of heat used frequently in commerce and in system-level energy statistics, the British thermal unit (Btu). John Bourne of England in 1865 simply took Clément's calorie in metric units and converted them to English units.[3] Thus Bourne defined a Btu as the amount of heat that would increase the temperature of a pound of water by 1°F. He referred to the French calorie as the "French thermal unit," and his conversion meant that 3.96832 Btu = 1 French thermal unit. Conversely, 0.251996 French thermal units = 1 Btu.[4]

Utilities sell natural gas to customers in the United States in units of *therms*, 100,000 Btu. Analysis of energy consumption by nations often uses units of *quads*, or quadrillion (10^{15}) Btu.

The International System of Units[5]

The International Bureau of Weights and Measures (in French, the official language, Bureau International des Poids et Mesures [BIPM]), with headquarters in Paris, produces the technical work that establishes the nearly universally agreed on definitions of units for various physical concepts and processes. BIPM originated in the Meter Convention, signed in 1875, and operates under the supervision of International Committee for Weights and Measures (CIPM), which in turn works under the authority of the General Conference on Weights and Measures (CGPM). The Meter Convention itself has been amended from time to time since 1875.

A BIPM publication, *The International System of Units (SI)*, summarizes the definitions of SI units, which include *base units* and *derived units*. For the purposes of energy, four base quantities are of greatest importance: *length, mass, time,* and *electric current*. The base units for these quantities are *meter, kilogram, second,* and *ampere,* abbreviated, respectively, *m, kg, s,* and *I* or *i*.

Base quantities and units. The ways in which the standard meter, kilogram, second, and ampere become units on a measuring device or instrument, like a meter stick or a clock, have evolved since 1875, and their current definitions by BIPM are as follows:

- The metre is the length of the path travelled by light in vacuum during a time interval of 1/299 792 458 of a second.
- The kilogram is the unit of mass; it is equal to the mass of the international prototype of the kilogram (a manufactured artifact of platinum-iridium kept by BIPM in Paris under specified conditions).
- The second is the duration of 9 192 631 770 periods of the radiation corresponding to the transition between the two hyperfine levels of the ground state of the caesium 133 atom, at rest at 0 K.
- The ampere is that constant current which, if maintained in two straight parallel conductors of infinite length, of negligible circular cross section, and placed 1 metre apart in vacuum, would produce between these conductors a force equal to 2×10^{-7} newton per metre of length.

Note that the second is defined by a natural event, radiation emitted by a cesium atom, not a manufactured artifact. Reducing the cesium to 0 K (absolute zero) is "manufactured" and ideal, because no material has actually reached 0 K. The definition of meter also rests on a natural phenomenon, the speed of light in a vacuum, but it requires use of the second. Thus a meter also contains an ideal quantity not fully achievable in reality. The definition of a kilogram is purely manufactured: a hunk of metal stored in Paris, which everyone agrees is "1 kilogram."

The ampere's definition is also based on an ideal that can't be achieved in reality. It envisions two parallel wires of infinite length but as close to 0 diameter as possible, both of which are clearly impossible. A *newton* is a force that will accelerate a mass of 1 kilogram by 1 meter per second each second. Both meter and second rely on ideal measurements that can't be precisely achieved in reality. The ampere also says nothing about the nature of electric current; instead it measures only the electromagnetic attractive force that two currents flowing in two wires creates, and when that force is 2×10^{-7} newton per meter of length the currents are 1 ampere.

Ideal components of base quantities and units do not at all diminish their usefulness. Instead, it simply means that all measurements of the physical world, including energy, are approximate. They are accurate to levels achieved by the laboratory instruments used to establish them, and near-global acceptance of them has proven highly useful in science, technology, and commerce.

Derived units. A number of units derived from base units have particular importance in measuring force, energy, and power. They are *newton, joule,* and *watt.*

The *newton (N)* measures force, which, as noted above, is present when a mass accelerates, that is, increases in speed or changes direction. Physics students learn it as $F_{net} = ma$. The newton in terms of base units has the dimensions *kilogram × meters per second per second,* or *kg m/s² or kg m s⁻².*

The *joule (J)* measures energy or work, and it has the dimensions of mechanical work or a mass accelerating through a distance (mad) because of a force. Its dimensions are *kilogram × meters/second² × meter or kg m² s⁻².*

TABLE A.1.1. COMMON SI PREFIXES AND UNITS FOR LARGE QUANTITIES OF
ENERGY AND POWER (E = ENERGY; P = POWER)

Prefix	Factor Larger	Name of Unit (unit components)	Abbreviation
—	$10^0 = 1$	joule (1 kilogram-meter^2/second^2) (E)	J
peta	10^{15}	petajoule (E)	PJ
exa	10^{18}	exajoules (E)	EJ
—	$10^0 = 1$	watt (1 joule/second) (P)	W
kilo	10^3	kilowatt (P)	kW
		kilowatt hour (E)	kWh
mega	10^6	megawatt (P)	MW
		megawatt hour (E)	MWh
giga	10^9	gigawatt (P)	GW
		gigawatt hour (E)	GWh
tera	10^{12}	terawatt (P)	TW
		terawatt hour (E)	TWh

The joule also measures heat by the mechanical equivalent of heat first determined with accuracy by James Prescott Joule. The accepted value of the equivalency is now 1 calorie = 4.185 joules, or 4.185 J, or better said, 1 *kilocalorie =
4,185 J* by definition. Joules can also be stated in Btu, derived from the calorie
as noted earlier, by the appropriate conversion.

The *watt (W)* measures power, or the time rate of expending energy or doing
work. Thus a watt is 1 Joule per second or 1 J/s. It has the basic units of *kilogram × meter/second^2 × meter × 1/second* or *kg m^2 s^{-3}*.

Prefixes of Si Units

Sometimes the quantities of energy in an experiment or in commerce are smaller
or larger than the quantities in 1 joule or 1 watt. Particularly in commerce, the
amounts of energy sold and bought by individuals, companies, and nations are
much larger than 1 J or 1 W. The SI system has standard prefixes to indicate
much larger units, the most common of which appear in table A.1.1.

The most common use of the watt in practical energy studies is as a measure
of energy for individuals and households, kilowatt-hours, indicating the amount
of electrical energy used or purchased. For nations, the unit is megawatt-hours,
gigawatt-hours, and terawatt-hours. In these units, watt (J/s) is multiplied by
the number of seconds in an hour (3,600 s/hour) to produce joules sold or consumed, a measure of total energy.

UNITS USED IN THE COMMERCE OF ENERGY

Of the derived units, the joule and the watt are the SI units for energy and
power. Commerce in electrical energy is based directly on these two units. Elec-

tric utility bills tell individual customers how many kilowatt-hours (kWh) they used and the price of each kWh. A kWh measures joules. A watt is 1 joule per second, a kilowatt is 1,000 watts, and an hour has 3,600 seconds. So a kWh has 1,000 J/s × 3,600 s = 3,600,000 joules, or 3.6 million joules (3.6 × 10^6 joules).

The average American home uses 10,932 kWh per year or 911kWh per month or about 30 kWh per day. Louisiana homes use the most, 15,497 kWh per year, and homes in Hawaii use the lowest amount, 6077 kWh.[6]

During the past year at my home in the San Francisco Bay Area, California, the lowest month of consumption was August (185 kWh) and the highest was November (343 kWh). In May 2016, most of the kWh we bought cost $0.18212 each, and the rest cost $0.22481.

Commerce in electricity at the utility scale generally uses the measurement megawatt-hours (MWh). At the national and global levels, units can be in MWh, gigawatt-hours (GWh), or terawatt-hours (TWh). In 2015, for example, the United States generated about 4,126 TWh of electricity, but the data tables reported "thousand megawatt-hours."[7]

Gas typically does not use the SI units joule or watt for commerce. Instead the most common units in the United States count either cubic feet or therms. For example, the U.S. Energy Information Administration reports gas production and sales at the national level in *billion cubic feet* (Bcf). For some purposes, however, *million cubic feet* (MMcf) or *thousand cubic feet* (Mcf) are numbers that are easier to handle. The abbreviations for these latter two units can be confusing, but M = 1,000 in Roman numerals, and MM = 1,000 × 1,000 = 1,000,000, or 10^6. In countries using the metric system, volumes of gas may appear in cubic meters rather than cubic feet.

Data reported in cubic feet often contain a measure of heat for the fuel, and 1 cubic foot generally has an average heating value of 1,032 Btu per cubic foot. Analogous to the abbreviations for multiples of cubic feet, the Btu content of gas often appears as *thousand Btu* (MBtu) or *million Btu* (MMBtu).

For household-scale sales, utility companies use *therms* or 100,000 Btu. For example, my house used 30 therms in May 2016. In the past year, we used 83 therms in December, the month of highest consumption. Currently, most therms cost about $1.08, but some cost about $1.57.

Conversions of cubic feet and Btu to joules and/or kilowatt-hours are straightforward, and many websites convert any number and unit inserted into other numbers and units. Approximate conversions can be done in your head by remembering the following:

- 1 Btu = 1055.06 J, or 1 Btu ≈ 1000 J (≈ means "approximately equals")
- 1 kWh = 3412.14 Btu, or 1 kWh ≈ 3400 Btu
- 1 cubic foot of gas = 1,032 Btu, or 1 cf ≈ 1000 Btu
- 1 quad = 10^{15} Btu = 1.055 exajoules (EJ) ≈ 1 EJ

Oil commerce in the United States and other places is often transacted in *barrels* (bbl); 1 bbl = 42 U.S. liquid gallons. In 2015, for example, the United

States consumed about 7.08 billion bbl of petroleum products, or about 19.4 million bbl per day.[8]

Oil, coal, and uranium commerce may also be transacted in tons or metric tons, but both the spelling and the conversion to other units vary by country. For example, in the United States a *ton* is 2,000 pounds; a *tonne*, using the French spelling, is 1,000 kilograms. The unit *barrel of oil equivalent* (BOE) is 1 metric ton that burns to release 1.364 million kilocalories or 5.73 gigajoules or 1.64 MWh. Another unit for oil, however, is the *ton of oil equivalent* (toe) = 7.33 bbl or 10 million kilocalories or 42 gigajoules.[9]

CONCLUSION

At this point a reader new to the world of energy numbers may want to throw up his or her hands in utter confusion. The two easy points to keep in mind, however, are that *any unit of energy can be converted to any other unit* (based on the first law of thermodynamics), and *conversion websites and tables are readily available from many sources.*

Conversion to joules puts the energy available from any source into a common measurement that enables comparison of all sources in common units reflecting the work that can be done. When examining energy data, one of the first requirements is to note the units used by the producers of the data. When needed, one can convert multiple units into a common unit.

Production of Heat by Combustion and Fission

OVERVIEW

A fuel or primary energy source is a substance that can produce heat, or a rapid motion of molecules or atoms (see chapter 2). In turn, the heat may be used directly as an energy service, for example, to heat a space or to heat an ore as part of the process of producing a metal. Alternatively, the heat may be transformed into motion, electrical energy, or light by using practical devices or machines of various types. Some of the energy released by a fuel typically includes, in addition to heat, electromagnetic radiation (i.e., light and wavelengths not visible to the eyes).

Each of the big-four fuels (coal, oil, gas, and uranium) derives its value from practical knowledge of how to make these physical substances release energy under controlled conditions. The fossil fuels (coal, oil, and gas) produce heat through combustion; uranium and plutonium produce heat through a completely different process, fission.

Combustion The heart of combustion processes lies in the collision of two different kinds of chemicals (the reactants) to create new chemicals (the products). Fossil fuels consist largely of carbon and hydrogen, plus varying amounts of other chemicals such as oxygen, sulfur, and other elements. When fossil fuel molecules collide with oxygen molecules, under the right circumstances, these reactants interact—burn, in everyday language—and form two new products, carbon dioxide and water. During the collision and interchange of atoms within the reacting molecules, energy is released in the form of heat and electromagnetic radiation (e.g., light).

Consider the simplest fossil fuel as an example: methane + oxygen → carbon dioxide + water + energy. Methane molecules have one atom of carbon and 4 atoms of hydrogen, and oxygen molecules have 2 atoms of oxygen. Carbon dioxide has one atom of carbon and 2 of oxygen, and water has 2 atoms of hydrogen and one of oxygen. In chemical terms, expression of the reaction is

$$CH_4 + 2\ O_2 \rightarrow CO_2 + 2\ H_2O + energy$$
methane + 2 oxygen → carbon dioxide + 2 water + energy

For the reaction to occur, the methane and oxygen reactants must have enough kinetic energy—they must be moving fast enough (i.e., at a high enough temperature)—to interact so that carbon-hydrogen and the oxygen-oxygen bonds break, and then new bonds form between carbon and oxygen and between hydrogen and oxygen.

Chemical bonds between atoms form by atoms sharing electrons, and combustion involves changes in how atoms share electrons. The identities of the atoms, based on the protons and neutrons of the atoms' nuclei, do not change.

Methane is a good fuel, meaning that at normal temperatures the reaction between it and the oxygen in the atmosphere occurs so slowly as to be virtually undetectable. Thus the fuel can be stored for long periods until needed. An electric spark or a lit match supplies an activation energy to ignite the methane and oxygen; that is, their collisions break carbon-hydrogen bonds in combustion, which releases enough heat to keep the methane burning, such as in a gas stove.

One way of describing a fuel rests on the concept of potential energy: breaking the carbon-hydrogen bonds releases more energy than the energy required to form carbon-oxygen and hydrogen-oxygen bonds. The extra energy leaves the burning materials as heat and light. The products of combustion, carbon dioxide and water, have a lower energy level and are thus stable, meaning that they will not react further with oxygen.

FISSION

During fission the nuclei of some atoms—uranium and plutonium—break apart, and the chemical identity of the atom changes to a different element. Neutrons traveling at the right speed can break apart the nucleus of an atom of ^{235}U or ^{239}Pu—the fission event—to form atoms of new elements, plus release more neutrons and a large amount of energy as heat and electromagnetic radiation.

Each fission event releases 2 to 3 more neutrons, which causes further fission events in other nuclei, thus initiating a chain reaction that continues until the amounts of ^{235}U or ^{239}Pu have declined to levels that no longer provide enough new targets for neutrons from other fission events. In everyday language, fission of uranium or plutonium is sometimes called "burning" of the metallic fuel, but it's important to remember that this "burning" in nuclear fuels differs enormously from the combustion of fossil fuels.

If an atom of ^{235}U fissions, the divisions resulting in two product pairs can consist of many different new elements. For example, fission of ^{235}U can result in ^{144}Ba (barium) and ^{89}Kr (krypton), plus 3 neutrons, or in ^{143}Cs (cesium) and ^{90}Rb (rubidium), plus 3 neutrons, or in many other pairs, a total of eleven of which are relatively common. As a result, the fuel rods of a nuclear reactor, after

TABLE A.2.1. COMPARISON OF ENERGY DENSITIES:
COAL, OIL, AND URANIUM

Fuel	Energy Density (kWh/kg)
Coal	8
Mineral oil	12
235-uranium	24,000,000

fission events have occurred, become filled with a host of new chemicals, many of which are themselves highly radioactive. In chemical terms, an example of a fission event is

$$n + {}^{235}U \rightarrow {}^{236}U \rightarrow {}^{144}Ba + {}^{89}Kr + 3n + energy$$
$$neutron + 235\text{-}uranium \rightarrow 236\text{-}uranium \rightarrow barium + krypton + 3 \ neutrons + energy$$

Uranium and plutonium are good fuels in the sense that chain reactions of fission events occur in large numbers only when induced in a nuclear reactor (or nuclear weapon). That is, these metals can be stored reasonably safely until used, although they are radioactive and break down very slowly over thousands of years.

Fissionable uranium and plutonium have high potential energy, which is released when the metals fission by technological design. The fission products have lower energy states, because they will not fission. They may, however, decay radioactively to elements with even lower energy states.

COMPARISON OF ENERGY FROM COMBUSTION AND FISSION

Assessment of the comparative strengths and weaknesses of different fuels must rest on multiple considerations (see chapter 9), one of which is energy density. Energy density measures the amount of energy that one kilogram of fuel delivers during combustion or fission. The number of kWh provided by one kilogram of fuel is a common unit for measuring energy density, and it discloses the amount of fuel needed to generate a specified amount of electrical energy. It is an advantage to have high energy density, because that means less mass of fuel is needed to generate power.

On the criterion of energy density, uranium far outranks the fossil fuels. For example, consider the following (table A.2.1).[1] On the basis of mass, uranium has 2 million to 3 million times as much energy as the fossil fuels. As valuable as this characteristic is, many other factors play an important role in setting the strengths and weaknesses of primary energy sources.

Notes

CHAPTER I

1. Jill D. Pruetz and Thomas C. LaDuke, "Reaction to Fire by Savanna Chimpanzees (*Pan troglodytes verus*) at Fongoli, Senegal: Conceptualization of 'Fire Behavior' and the Case for a Chimpanzee Model," *American Journal of Physical Anthropology* 141 (2010): 646–50.

2. Richard Wrangham and Rachel Carmody, "Human Adaptation to the Control of Fire," *Evolutionary Anthropology* 19 (2010): 187–99.

3. Francesco Berna, Paul Goldberg, Libra Kolska Horwitz, James Brink, Sharon Holt, Marion Bamford, and Michael Chazan, "Microstratigraphic Evidence of In Situ Fire in the Acheulean Strata of Wonderwerk Cave, Northern Cape Province, South Africa," *Proceedings of the National Academy of Sciences*, www.pnas.org/content/109/20/E1215.abstract, April 2, 2012.

4. Wil Roebroeks and Paola Villa, "On the Earliest Evidence for Habitual Use of Fire in Europe," *Proceedings of the National Academy of Sciences* 108, no. 13 (March 29, 2011): 5209–14.

5. Wrangham and Carmody, "Human Adaptation."

6. Richard G. Roberts and Michael I. Bird, "*Homo* 'incendius,'" *Nature* 485 (May 31, 2012): 586–87.

7. Stephen Shennan, "Evolutionary Demography and the Population History of the European Early Neolithic," *Human Biology* 81 (April–June 2009): 339–55.

8. Melinda A. Zeder, "The Broad Spectrum Revolution at 40: Resource Diversity, Intensification, and an Alternative to Optimal Foraging Explanations," *Journal of Anthropological Archaeology* 31 (2012): 241–64.

9. Alfred W. Crosby, *Children of the Sun: A History of Humanity's Unappeasable Appetite for Energy* (New York: Norton, 2006),

10. E. A. Wrigley, *Energy and the English Industrial Revolution* (Cambridge: Cambridge University Press, 2010), 47–52.

11. Donald Worster, *Nature's Economy: A History of Ecological Ideas* (New York: Cambridge University Press, 1977), 291–15.

12. Wrigley, *Energy*, 9–16.

13. Rolf Peter Sieferle, *The Subterranean Forest: Energy Systems and the Industrial Revolution* (Cambridge: White Horse Press, 2001), 80–86.

14. Wrigley, *Energy*, 9–16.

15. Wrigley, *Energy*, 47–52; Arnold Toynbee, *The Industrial Revolution* (Boston: Beacon Press, 1956).

16. Wrigley, *Energy*, 225–26.

17. Wrigley, *Energy*, 51–52; W. Stanley Jevons, *The Coal Question: An Inquiry Concerning the Progress of the Nation and the Probable Exhaustion of Our Coal-Mines*, 2nd ed. (London: Macmillan, 1866).

18. Joel Mokyr, "Editor's Introduction: The New Economic History and the Industrial Revolution," in *The British Industrial Revolution: An Economic Perspective*, ed. Joel Mokry (Boulder, CO: Westview Press, 1999), 1–127.

19. John Hatcher, *The History of the British Coal Industry*, vol. 1, *Before 1700: Towards the Age of* Coal (Oxford: Clarendon Press, 1993), 17–19.

20. Hatcher, *History of the British Coal Industry*, 17–32.

21. Gerard Turnbull, "Canals, Coal and Regional Growth during the Industrial Revolution," *Economic History Review*, 2nd ser., 40, no. 4 (1987): 537–60.

22. Sieferle, *The Subterranean Forest*, 80–86.

23. Mussimo Livi Bacci, *The Population of Europe* (Oxford: Blackwell, 1999), 5.

24. Hatcher, *History of the British Coal Industry*, 18–19.

25. C. Knick Harley, "Reassessing the Industrial Revolution: A Macro View," in Mokyr, *The British Industrial Revolution*, 160.

26. Bacci, *The Population of Europe*, 8; Hatcher, *History of the British Coal Industry*, 31–32; Wrigley, *Energy*, 155.

27. Wrigley, *Energy*, 94.

28. Hatcher, *History of the British Coal Industry*, 24–25, 55.

29. Thomas Southcliffe Ashton and Joseph Sykes, *The Coal Industry of the Eighteenth Century* (Manchester: Manchester University Press, 1929), 5.

30. Thomas Southcliffe Ashton, *Iron and Steel in the Industrial Revolution*, 2nd ed. (Manchester: Manchester University Press, 1951), 28–38, appendix E.

31. Wrigley, *Energy*, 94; expansion figures calculated from table 4.2.

32. Ashton and Sykes, *The Coal Industry*, 33–35; Eric Clavering, "The Coal Mills of Northeast England: The Use of Waterwheels for Draining Coal Mines, 1600–1750," *Technology and Culture* 36, no. 2 (April 1995): 211–41.

33. Giambattista della Porta, *Natural Magick* (New York: Basic Books, 1957), v–ix, 385–94; L. T. C. Holt and J. S. Allen, *The Steam Engine of Thomas Newcomen* (New York: Science History Publications, 1977), 16–22.

34. Holt and Allen, *The Steam Engine*, 24–30; Milton Kerker, "Science and the Steam Engine," *Technology and Culture* 2 (4) (1961): 381–90.

35. Holt and Allen, *The Steam Engine*, 32–40, 44–48.

36. Holt and Allen, *The Steam Engine*, 56–57.

37. John Farey, *A Treatise on the Steam Engine: Historical, Practical and Descriptive* (Newton Abbott: David & Charles, 1971; orig. London: Longman, Rees, Orme, Brown and Green, 1827), 310.

38. David Philip Miller, "Seeing the Chemical Steam through the Historical Fog: Watt's Steam Engine as Chemistry," *Annals of Science* 65, no. 1 (January 2008): 47–72.

39. Farey, *A Treatise*, 311–17.

40. James Watt, *New Invented Method of Lessening the Consumption of Steam and Fuel in Fire Engines* (London: Eyre and Spottiswoode, 1855), Patent No. 913, 1769, http://upload.wikimedia.org/wikipedia/commons/0/0d/James_Watt_Patent_1769_No_913.pdf, August 29, 2012.

41. Farey, *A Treatise*, 346–438.

42. Samuel Smiles, *Lives of Boulton and Watt* (Philadelphia: J.B. Lippincott, 1865), 197–206.

43. Smiles, *Lives of Boulton and Watt*, 207–15.

44. John Kanefsky and John Robey, "Steam Engines in 18th-Century Britain: A Quantitative Analysis," *Technology and Culture* 21, no. 2 (April 1980): 161–86.

45. Kanefsky and Robey, "Steam Engines."

46. A.E. Musson, "Industrial Motive Power in the United Kingdom, 1800–70," *Economic History Review*, 2nd ser., 29 (August 1976): 415–39.

47. Paul P. Bernard, "How Not to Invent the Steamship," *East European Quarterly* 14, no. 1 (Spring 1980): 1–8; Jack L. Shagena, *Who Really Invented the Steamboat: Fulton's Clermont Coup* (Amherst, NY: Humanity Books), 95–96; Charles Frederick Partington, *An Historical and Descriptive Account of the Steam Engine* (London: J. Taylor, 1822), 36.

48. Thomas A. Croal, *A Book about Traveling: Past and Present* (London: William P. Nimmo, 1877), 147, 150.

49. E.S. Bates, *Touring in 1600: A Study in the Development of Travel as a Means of Education* (Boston: Houghton Mifflin, 1911), 285–87.

50. John Bell Rae, *The Road and the Car in American Life* (Cambridge, MA: MIT Press, 1971); Robert Makinnon, "Roads, Cart Tracks, and Bridle Paths: Land Transportation and the Domestic Economy of Mid-Nineteenth Century Eastern British North America, *Canadian Historical Review* 84, no. 2 (June 2003): 177–216.

51. Susan E. Alcock, John Bodel, and Richard J.A. Talbert, eds., *Highways, Byways, and Road Systems in the Pre-Modern World* (New York: Wiley-Black-well, 2012).

52. Bates, *Touring in 1600*, 60.

53. Partington, *Historical and Descriptive Account of the Steam Engine*, 38.

54. Shagena, *Who Really Invented the Steamboat*, 224–27.

55. H.W. Dickinson and Arthur Titley, *Richard Trevithick: The Engineer and the Man* (Cambridge: Cambridge University Press, 1934), 46–47.

56. Shagena, *Who Really Invented the Steamboat*, 328–30, 413–15.

57. Waterways of England and Wales, London Canal Museum, www.canalmuseum.org.uk/history/1750/index1750.htm, December 20, 2012.

58. Gregory Clark, "The Price History of English Agriculture, 1209–1914," *Research in Economic History* 22 (2004): 41–123.

59. Samuel Smiles, *The Life of George Stephenson, Railway Engineer* (Columbus, OH: Follett, Foster and Co., 1860), 71–72.

60. Smiles, *The Life of George Stephenson*, 81–90, 128–29; M. C. Duffy, "George Stephenson and the Introduction of Rolled Railway Rail," *Journal of Mechanical Working Technology* 5 (1981): 309–42;

61. M. W. Kirby, "Stephenson, George (1781–1848)," in *Oxford Dictionary of National Biography* (Oxford: Oxford University Press, 2004; online edition, January 2008), www.oxforddnb.com/view/article/26397, December 22, 2012.

62. James D. Dilts, *The Great Road: The Building of the Baltimore and Ohio, the Nation's First Railroad, 1828–1853* (Stanford: Stanford University Press, 1993), xv–xix.

CHAPTER 2

1. Jennifer Coopersmith, *Energy, the Subtle Concept: The Discovery of Feynman's Blocks from Leibniz to Einstein* (New York: Oxford University Press, 2010), 1–4.

2. Joe Sachs, "Aristotle: Motion and Its Place in Nature," *Internet Encyclopedia of Philosophy*, www.iep.utm.edu/aris-mot/#H7, January 16, 2015.

3. Coopersmith, *Energy*, 14–21, 33; D. W. MacDougal, *Newton's Gravity: An Introductory Guide to the Mechanics of the Universe* (New York: Springer, 2012), 17–36; Ron Naylor, "Galileo, Copernicanism, and the Origins of the New Science of Motion," *British Journal of the History of Science* 56, no. 2 (2003): 151–81.

4. *Oxford English Dictionary*, "gravity, *n.*"

5. Richard Wolfson, *Energy, Environment, and Climate*, 2nd ed. (New York: Norton, 2012), 1, 35–37, 62–90.

6. Carlo Rovelli, *Seven Brief Lessons on Physics* (New York: Riverhead Books, 2016).

7. Bruce Director, "Toppling the Tyranny of the 2nd Law of Thermodynamics," *EIRScience*, March 16, 2012, 33–44, www.larouchepub.com/eiw/public/2012/eirv39n11–20120316/33–44_3911.pdf, April 14, 2015.

8. Edmund Russell, James Allison, Thomas Finger, John K. Brown, Brian Balogh, and W. Bernard Carlson, "The Nature of Power: Synthesizing the History of Technology and Environmental History," *Technology and Culture* 52, no. 2 (April 2011): 246–59.

9. D. S. L. Cardwell, *From Watt to Clausius: The Rise of Thermodynamics in the Early Industrial Age* (Ithaca, NY: Cornell University Press, 1971); Crosbie Smith, *The Science of Energy: A Cultural History of Energy Physics in Victorian Britain* (Chicago: University of Chicago Press, 1998); Coopersmith, *Energy*.

10. J. L. Heilbron, *Electricity in the 17th and 18th Centuries: A Study of Early Modern Physics* (Berkeley: University of California Press, 1979), 169–79; Brian Baigrie, *Electricity and Magnetism: A Historical Perspective* (Westport, CT: Greenwood Press, 2007).

11. Heilbron, *Electricity*, 229–32, 252–55; Michael Ben-Chaim, "Social Mobility and Scientific Change: Stephen Gray's Contribution to Electrical Research," *British Journal for the History of Science* 23, no. 1 (March 1990): 3–24; Roderick W. Home, "Fluids and Forces in Eighteenth-Century Electricity," *Endeavour* 26, no. 2 (2002): 55–59.

12. Heilbron, *Electricity*, 313.

13. Benjamin Franklin, *Experiments and Observations on Electricity Made at Philadelphia in America* (London: E. Cave, 1751), 59.

14. Home, "Fluids and Forces in Eighteenth-Century Electricity."

15. Franklin, *Experiments*, 59.

16. L. Pearce Williams, *Michael Faraday: A Biography* (London: Chapman and Hall, 1965), 284.

17. Heilbron, *Electricity*, 344–500. Gerald Holton and Duane H. D. Roller, *Foundations of Modern Physical Science* (Reading: Addison-Wesley, 1958), 477–83; Baigrie, *Electricity*, 42–47; Peter J. Koehler, Stanley Finger, and Marco Piccollino, "The 'Eels' of South America: Mid-18th-Century Dutch Contributions to the Theory of Animal Electricity," *Journal of the History of Biology* 42, no. 4 (Winter 2009): 715–63.

18. Heilbron, *Electricity*, 491–92.

19. Alexander Volta, "On the Electricity Excited by the Mere Contact of Conducting Substances of Different Kinds," *Philosophical Magazine* 7 (September 1800): 289–312; Allan A. Mills, "Early Voltaic Batteries: An Evaluation in Modern Units and Application to the Work of Davy and Faraday," *Annals of Science* 60 (October 2003): 373–98.

20. Williams, *Michael Faraday*, 58.

21. Humphry Davy, "On Some Chemical Agencies of Electricity," *Philosophical Transactions* 97 (1807): 1–56; Williams, *Michael Faraday*, 24, 237.

22. Williams, *Michael Faraday*, 1–29.

23. Williams, *Michael Faraday*, 139–40.

24. Baigrie, *Electricity*, 63–68.

25. Kenneth L. Caneva, "Ampère, the Etherians, and the Oersted Connexion," *British Journal for the History of Science* 15, no. 44 (1980): 121–38.

26. Baigrie, *Electricity*, 68–72; Williams, *Michael Faraday*, 144–51.

27. Baigrie, *Electricity*, 68–72.

28. Williams, *Michael Faraday*, 151–60; Brian Gee, "Electromagnetic Engines: Pre-Technology and Development Immediately Following Faraday's Discovery of Electromagnetic Rotations," *History of Technology* 13 (1991): 41–72.

29. Michael Brian Schiffer, *Power Struggles: Scientific Authority and the Creation of Practical Electricity before Edison* (Cambridge, MA: MIT Press, 2008), 27–29, 36–37.

30. Williams, *Michael Faraday*, 182–83.

31. Schiffer, *Power Struggles*, 52–53, 75–87.

32. Williams, *Michael Faraday*, 364–464; David Gooding, "Empiricism in Practice: Teleology, Economy, and Observation in Faraday's Physics," *Isis* 73, no. 1 (March 1982): 46–67; David Gooding, "Experiment and Concept Formation in Electromagnetic Science and Technology in England in the 1820s," *History and Technology* 2 (1985): 151–76.

33. John Tully, "A Victorian Ecological Disaster: Imperialism, the Telegraph, and Gutta-Percha," *Journal of World History* 20, no. 4 (December 2009): 559–79.

34. Bruce J. Hunt," Michael Faraday, Cable Telegraphy and the Rise of Field Theory," *History of Technology* 13 (1991): 1–19.

35. Williams, *Michael Faraday*, 291–93.

36. Bruce J. Hunt, "Electromagnetism: Ether and Field," in *Pursuing Power and Light: Technology and Physics from James Watt to Albert Einstein* (Baltimore: Johns Hopkins University Press, 2010), chapter 5.

37. Hunt, "Electromagnetism."

38. Hunt, "Electromagnetism."

39. "What Are Maxwell's Equations?," www.maxwells-equations.com /summary.php, February 1, 2016.

40. John Locke, *The Works of John Locke, in Ten Volumes*, 11th ed., vol. 3, *Elements of Natural Philosophy* (London: W. Otridge and Son et al., 1812), 302; found at Google Books, February 2, 2016.

41. Sadi Carnot, *Reflections on the Motive Power of Fire*, in *The Second Law of Thermodynamics: Memoirs by Carnot, Clausius, and Thomson*, ed. and trans. W. F. Magie (New York: Harper and Brothers, 1899), 1–60; Charles Coulston Gillispie and Raffaele Pisano, "Lazare and Sadi Carnot: A Scientific and Filial Relationship," *History of Mechanism and Machine Science* 19 (2014): 1–13.

42. Cardwell, *From Watt to Clausius*, 220–29.

43. http://hyperphysics.phy-astr.gsu.edu/hbase/thermo/carnot.html#c1.

44. Carnot, *Reflections;* Gillispie and Pisano, "Lazare and Sadi Carnot," 337–38.

45. Gillispie and Pisano, "Lazare and Sadi Carnot," 227–46.

46. Carnot, *Reflections*, 3–4.

47. Cardwell, *From Watt to Clausius*, 229–31; Smith, *The Science of Energy*, 73–76; Coopersmith, *Energy*, 246–51.

48. James Prescott Joule, *The Scientific Papers* (London: Physical Society of London, 1884), 1–3, 10–14, 46–53; Henry John Steffens, *James Prescott Joule and the Concept of Energy* (Dawson: Science History Publications, 1979), 1–17; Schiffer, *Power Struggles*, 162–63.

49. Smith, *The Science of Energy*, 55–73; Coopersmith, *Energy*, 251–59.

50. Coopersmith, *Energy*, 257–58.

51. Smith, *The Science of Energy*, 273.

52. Coopersmith, *Energy*, 257–58.

53. William Thomson, "On an Absolute Thermometric Scale Founded on Carnot's Theory of the Motive Power of Heat, and Calculated from Regnault's Observations," *Cambridge Philosophical Society Proceedings*, June 5, 1848, in "Kelvin, William Thomson, Baron," *Mathematical and Physical Papers* (Cambridge: Cambridge University Press, 1882), vol. 1, 100–106, at http://archive .org/details/mathematicalando1kelgoog, February 14, 2015; Coopersmith, *Energy*, 285–86.

54. Coopersmith, *Energy*, 289–90.

55. Coopersmith, *Energy*, 292–93.

56. Coopersmith, *Energy*, 295–98.

57. William John Macquorn Rankine, *A Manual of the Steam Engine and Other Prime Movers* (London: Richard Griffin and Co., 1859).

58. William Thomson and Peter Guthrie Tait, *Treatise on Natural Philosophy* (Oxford: Clarendon Press, 1867), vi; William Thomson and Peter Guthrie Tait, *Elements of Natural Philosophy* (Cambridge: Cambridge University Press, 1890), vi.

59. William Thomson and P. G. Tait, "Energy," *Good Works* 3 (1862): 601–7.

60. Coopersmith, *Energy*, 305.

61. Smith, *The Science of Energy*, 290–300.

62. Smith, *The Science of Energy*, 268–87.

63. Nahum S. Kipnis, "The Window of Opportunity: Logic and Chance in Becquerel's Discovery of Radioactivity," *Physics in Perspective* 2 (2000): 63–99.

64. Rovelli, *Seven Brief Lessons*.

CHAPTER 3

1. Alberto Clô, *Oil Economics and Policy* (Boston: Kluwer Academic, 2000), 40; English translation of *Economía e política del petrolio* sponsored by the European Society for Scientific Publications.

2. David Held, *Political Theory and the Modern State* (Cambridge: Polity, 1989), 11–12.

3. Christopher Dyer, *Standards of Living in the Later Middle Ages* (Cambridge: Cambridge University Press, 1989), 6–7.

4. Massimo Livi Bacci, *The Population of Europe* (Oxford: Blackwell, 1999), 36; E. A. Wrigley, *Energy and the English Industrial Revolution* (Cambridge: Cambridge University Press, 2010), 61.

5. Clifford R. Backman, *The Worlds of Medieval Europe* (New York: Oxford University Press, 2003).

6. Stephen Church, *King John: And the Road to Magna Carta* (New York: Basic Books, 2015); Frederic W. Maitland, Frances C. Montague, and James F. Colby, *A Sketch of English Legal History* (New York: G. P. Putnam, 1915), 78–79.

7. Colin McEvedy and Richard Jones, *Atlas of World Population History* (Harmondsworth: Penguin Books, 1978), 25.

8. John Hatcher, *Plague, Population, and the English Economy, 1348–1530* (London: Macmillan, 1977), 68–73.

9. Jeremy Goldberg, Introduction to *The Black Death in England*, ed. W. M. Ormrod and P. G. Lindley (Stamford, CT: Paul Watkins, 1996), 8–10.

10. Jim Bolton, "'The World Upside Down': Plague as an Agent of Economic and Social Change," in Ormrod and Lindley, *The Black Death*, 17–78, quote on 78.

11. W. M. Ormrod, "The Politics of Pestilence: Government in England after the Black Death," in Ormrod and Lindley, *The Black Death*, 147–81.

12. Wrigley, *Energy*, 59.

13. Wrigley, *Energy*, 92, 94.

14. John Blair and Nigel Ramsay, eds., *English Medieval Industries: Craftsmen, Techniques, Products* (London: Hambledon and London, 1991), xxvi.

15. Thomas Southcliffe Ashton and Joseph Sykes, *The Coal Industry of the Eighteenth Century* (Manchester: Manchester University Press, 1929), 5.

16. Wrigley, *Energy*, 94.

17. Wrigley, *Energy*, 142, 155.

18. Carl Wennerlind, *Casualties of Credit: The English Financial Revolution, 1620–1720* (Cambridge, MA: Harvard University Press, 2011), 17–19.

19. Historical Association (Great Britain), *Constitutional Documents* (London: G. Bell and Sons, 1914); Steve Pincus, *1688: The First Modern Revolution* (New Haven, CT: Yale University Press, 2009).

20. Pincus, *1688*, 366–99.

21. Wennerlind, *Casualties of Credit*, 69, 108–14.

22. Peter Dorman, *Macroeconomics: A Fresh Start* (New York: Springer, 2014), 139–51.

23. Carolyn Merchant, *The Death of Nature: Women, Ecology, and the Scientific Revolution* (San Francisco: Harper & Row, 1980).

24. Wennerlind, *Casualties of Credit*, 48–68.

25. John Hatcher, *The History of the British Coal Industry*, vol. 1, *Before 1700: Towards the Age of Coal* (Oxford: Clarendon Press, 1993), 16–30, 187–91.

26. Michael W. Flinn, *The History of the British Coal Industry*, vol. 2, *1700–1830: The Industrial Revolution* (Oxford: Clarendon Press, 1984), 26, 114–19, 297.

27. This section is based on Phyllis Deane, *The First Industrial Revolution*, 2nd ed. (Cambridge: Cambridge University Press, 1979), 103–18.

28. Deane, *The First Industrial Revolution*, 117.

29. This section is based on Deane, *The First Industrial Revolution*, 87–102.

30. Dyer, *Standards of Living*, 78–79, 175–77.

31. Sven Beckert, *Empire of Cotton: A Global History* (New York: Alfred A. Knopf, 2014), 56–67.

32. Beckert, *Empire of Cotton*, 73–75.

33. Graham Rogers, "Custom and Common Right: Waste Land Enclosure and Social Change in West Lancashire," *Agricultural History Review* 41, no. 2 (1993): 137–54.

34. Ronald Bailey, "The Other Side of Slavery: Black Labor, Cotton, and Textile Industrialization in Great Britain and the United States," *Agricultural History* 68, no. 2 (Spring 1994): 35–50.

35. Don H. Doyle, *The Cause of All Nations: An International History of the American Civil War* (New York: Basic Books, 2015), 218–27.

36. Michael Brian Schiffer, *Power Struggles: Scientific Authority and the Creation of Practical Electricity before Edison* (Cambridge, MA: MIT Press, 2008), 75–89, 91–103, 137–54.

37. Thomas P. Hughes, *Networks of Power: Electrification in Western Society, 1880–1930* (Baltimore, MD: Johns Hopkins University Press, 1983), 18–46.

38. Graeme Gooday, *Domesticatng Electricity: Technology, Uncertainty, and Gender, 1880–1914* (London: Pickering & Chatto, 2008), 2.

39. Hughes, *Networks of Power*, 47–78.

40. Schiffer, *Power Struggles*, 65–71.

41. Hughes, *Networks of Power*, 79–139.

42. Jacob Schoellkopf, "Niagara Falls History of Power," www.niagarafrontier.com/power.html#Sch, September 22, 2015.

43. Engineering and Technology History Wiki, "Milestones: Vulcan Street Plant, 1882," http://ethw.org/Milestones:Vulcan_Street_Plant,_1882, September 22, 2015.

44. Robert Belfield, "The Niagara System: The Evolution of an Electric Power Complex at Niagara Falls, 1883–1896," *Proceedings of the IEEE 64*, no. 9 (September 1976): 1334–50.

45. Bureau of the Census, *Historical Statistics of the United States: Colonial Times to 1957* (Washington, DC: Government Printing Office, 1960), S15–S18, 506; percent calculated by author.

46. Shelly C. Dudley, "The First Five: A Brief Overview of the First Reclamation Projects," www.waterhistory.org/histories/reclamation/, December 5, 2015.

47. U.S. Bureau of Reclamation, "Hoover Dam," www.usbr.gov/lc/hooverdam/history/articles/chrono.html, December 5, 2015.

48. Rocío Uría-Martínez, Patrick W. O'Connor, and Megan M. Johnson, *2014 Hydropower Market Report* (Washington, DC: U.S. Department of Energy, 2015), 5–6.

49. Bureau of the Census, *Historical Statistics of the United States*, 506, S15.

50. Energy Information Administration, *Annual Energy Review*, table 7.1, www.eia.gov/totalenergy/data/monthly/pdf/sec7_3.pdf, February 24, 2015.

CHAPTER 4

1. Gail Tverberg, "World Energy Consumption since 1820 in Charts," March 12, 2012, http://ourfiniteworld.com/2012/03/12/world-energy-consumption-since-1820-in-charts/, April 3, 2015.

2. B. J. Arnold, "Coal Formation," in *The Coal Handbook: Towards Cleaner Production*, ed. D. Osborne (Burlington, VT: Elsevier Science, 2013), 31–52.

3. Robert Ehrlich, *Renewable Energy: A First Course* (Boca Raton, FL: CRC Press, 2013), 33.

4. U.S. Energy Information Administration, International Energy Statistics, Coal, Production, www.eia.gov/cfapps/ipdbproject/IEDIndex3.cfm?tid=1&pid=7&aid=1; ranks calculated by author.

5. U.S. Energy Information Administration, International Energy Statistics, Coal, Consumption, www.eia.gov/cfapps/ipdbproject/IEDIndex3.cfm?tid=1&pid=7&aid=6; years of consumption from reserves at current rates of consumption calculated by author.

6. U.S. Energy Information Administration, International Energy Statistics, Coal, Consumption, www.eia.gov/cfapps/ipdbproject/IEDIndex3.cfm?tid=1&pid=1&aid=2; consumption compared to production calculated by author.

7. John Hatcher, *The History of the British Coal Industry*, vol. 1, *Before 1700: Towards the Age of Coal* (Oxford: Clarendon Press, 1993), 65–68, 239–

56, 498–503; E.A. Wrigley, *Energy and the English Industrial Revolution* (Cambridge: Cambridge University Press, 2010), 61.

8. Michael W. Flinn, *The History of the British Coal Industry*, vol. 2, *1700–1830: The Industrial Revolution* (Oxford: Clarendon Press, 1984), 23–29.

9. Christopher F. Jones, "A Landscape of Energy Abundance: Anthracite Coal Canals and the Roots of American Fossil Fuel Dependence," *Environmental History* 15 (July 2010): 449–84; *Historical Statistics of the United States, 1789–1945*, "Fuels—Bituminous and Anthracite Coal Pproduction: 1807 to 1945," ser. G-13 and G-14 (Washington, DC: Government Printing Office, 1949), 142.

10. Flinn, *The History of the British Coal Industry*, 2:366.

11. Roy Church, Alan Hall, and John Kanefsky, *The History of the British Coal Industry*, vol. 3, *1830–1913: Victorian Pre-Eminence* (Oxford: Clarendon Press, 1986), 188–203, 282, 599–600, 611–15.

12. Peter Wardley, "The Emergence of Big Business: Employers of Labour in the United Kingdom, Germany, and the United States c. 1907," *Business History* 41, no. 4 (1999): 88–116.

13. U.K. Department of Energy and Climate Change, Historical Coal Data: Coal Production, 1853–2008, http://webarchive.nationalarchives.gov.uk /20121217150421/http://decc.gov.uk/publications/basket.aspx?filepath=statisti cs%2fpublications%2fenergytrends%2f1_20090731152425_e_%40%40_ coalsince1853.xls&filetype =4, April 10, 2015; *Historical Statistics of the United States, 1789–1945*, Labor Force—Industrial Distribution of Employed (NICB): 1900–1945, Series D-68, 65.

14. U.K. Department of Energy and Climate Change, Historical Coal Data.

15. Timothy Mitchell, *Carbon Democracy: Political Power in the Age of Oil* (New York: Verso, 2011), 12–42.

16. *BP Statistical Review of World Energy, June 2014*, p. 30; U.S. Energy Information Administration, Total Primary Coal Production (Thousand Short Tons), www.eia.gov/cfapps/ipdbproject/IEDIndex3.cfm?tid=1&pid=7&aid=1, converted to metric tons by 1 metric ton = 1.102 short tons. Figures calculated by author using production and prices for 2011 (American and Asian prices).

17. James G. Speight, *The Chemistry and Technology of Petroleum*, 4th ed. (Boca Raton, FL: CRC Press/Taylor & Francis, 2007), chapter 1.

18. Kenneth S. Deffeyes, *Beyond Oil: The View from Hubbert's Peak* (New York: Hill and Wang, 2005), 15, 55; Speight, *The Chemistry and Technology of Petroleum*, chapter 3.

19. Geoffrey P. Glasby, "Abiogenic Origin of Hydrocarbons: An Historical Overview," *Resource Geology* 56, no. 1 (2006): 83–96.

20. U.S. Energy Information Administration, Total Petroleum and Other Liquids Production—2014, www.eia.gov/countries/data.cfm, March 30, 2015. Calculations by author of (a) years of consumption and (b) percent of consumption from country's own supply.

21. Countries in the world by population, www.worldometers.info/world-population/population-by-country/, March 30, 2015; figures calculated by author.

22. James A. Clark, *The Chronological History of the Petroleum and Natural Gas Industries* (Houston, TX: Clark Book Co., 1963).

23. Daniel Yergin, *The Prize* (New York: Simon and Schuster, 1991), 19–34, 57–58.

24. "Seven Sisters" referred to Standard Oil of New Jersey, Socony-Vacuum (Standard Oil of New York), Standard Oil of California, Texaco, Gulf Oil, Royal Dutch/Shell, and British Petroleum (Anglo-Persian), plus CFP (Compagnie Française des Pétroles). Yergin, *The Prize*, 503; Anthony Sampson, *The Seven Sisters: The Great Oil Companies and the World They Shaped* (New York: Viking Press, 1975), 5–8.

25. Silvana Tordo, Brandon S. Tracy, and Noora Arfaa, *National Oil Companies and Value Creation* (Washington, DC: World Bank, 2011), 15–18.

26. "Really Big Oil," *Economist*, August 10, 2006, www.economist.com /node/7276986, April 1, 2015.

27. Statista, "Top 10 Oil And Gas Companies Worldwide Based On Market Value (in billion U.S. dollars)," www.statista.com/statistics/272709/top-10-oil-and-gas-companies-worldwide-based-on-market-value/, April 1, 2015.

28. Tordo, Tracy, and Arfaa, *National Oil Companies*, 20.

29. B. Piriou et al., "Potential Direct Use of Solid Biomass in Internal Combustion Engines," *Progress in Energy and Combustion Science* 39 (2013): 169–88.

30. *Historical Statistics of the United States, Colonial Times to 1957* (Washington, DC: Government Printing Office, 1960), Series K-201, Q-56, Q-58, Q-315, Q-317.

31. *BP Statistical Review of World Energy June 2014*, 10, 15, bp.com /statisticalreview; World Bank, *World Development Indicators* database, http:// databank.worldbank.org/data/download/GDP.pdf, April 2, 2015. Author's calculations.

32. U.S. Energy Information Administration, *Monthly Energy Review, February 2015* (Washington, DC: Energy Information Administration, 2015), table 1.3.

33. Malcolm W. H. Peebles, *Evolution of the Gas Industry* (New York: New York University Press, 1980), 1–8.

34. Xiuli Wang and Michael Economides, *Advanced Natural Gas Engineering* (Houston, TX: Gulf Publishing Co., 2009), 5–9.

35. Peebles, *Evolution*, 5–19, 27.

36. Peebles, *Evolution*, 51–57.

37. Peebles, *Evolution*, 27, 33–38.

38. Energy Information Administration, *Drilling Sideways—A Review of Horizontal Well Technology and Its Domestic Application*, DOE/EIA-TR-0565, Washington, DC, 1993.

39. Qiang Wang, Xi Chen, Awadhesh N. Jha, and Howard Rogers, "Natural Gas from Shale Formation—the Evolution, Evidences and Challenges of Shale Gas Revolution in United States," *Renewable and Sustainable Energy Reviews* 30 (2014): 1–28; Douglas Martin, "George Mitchell, a Pioneer in Hydraulic Fracturing, Dies at 94," *New York Times*, July 26, 2013, www.nytimes.com/2013/07/27 /business/george-mitchell-a-pioneer-in-hydraulic-fracturing-dies-at-94.html, May 7, 2015.

40. Ursula Klein, "Apothecary-Chemists in Eighteenth-Century Germany," in *New Narratives in Eighteenth-Century Chemistry*, ed. Lawrence M. Principe (Dordrecht: Springer, 2007), 97–137.

41. Earle R. Caley, "The Earliest Known Use of a Material Containing Uranium," *Isis* 38, no. 3–4 (February 1948): 190–93.

42. Nahum S. Kipnis, "The Window of Opportunity: Logic and Chance in Becquerel's Discovery of Radioactivity," *Physics in Perspective* 2 (2000): 63–99.

43. World Nuclear Association, "World Uranium Mining Production," www.world-nuclear.org/info/Nuclear-Fuel-Cycle/Mining-of-Uranium/World-Uranium-Mining-Production/, May 4, 2015.

44. Stephanie Cooke, *In Mortal Hands: A Cautionary History of the Nuclear Age* (New York: Bloomsbury, 2009), 14.

45. J. Samuel Walker, *Containing the Atom: Nuclear Regulation in a Changing Environment, 1963–1971* (Berkeley: University of California Press, 1992), 4.

46. George T. Mazuzan and J. Samuel Walker, *Controlling the Atom: The Beginnings of Nuclear Regulation, 1946–1962* (Berkeley: University of California Press, 1984), 3–7; Walker, *Containing the Atom*, 2–4.

47. William Beaver, "Duquesne Light and Shippingport: Nuclear Power Is Born in Western Pennsylvania," *Western Pennsylvania Historical Magazine* 70, no. 4 (October 1987): 339–58.

48. Paul R. Josephson, *Red Atom: Russia's Nuclear Power Program from Stalin to Today* (Pittsburgh: University of Pittsburgh Press, 2000), 2; Lorna Arnold, *Windscale 1957: Anatomy of a Nuclear Accident* (Houndmills: Palgrave Macmillan, 2007), 23.

49. U.S. Nuclear Regulatory Commission, *Information Digest, 2014–2015*, NUREG 1350, vol. 26, appendices A and C, www.nrc.gov/reading-rm/doc-collections/nuregs/staff/sr1350/, May 5, 2015.

50. U.S. Energy Information Administration, *Electric Power Monthly with Data for February 2015*, table 1.1, April 2015; percent calculated by author.

51. James M. Jasper, "Gods, Titans, and Mortals: Patterns of State Involvement in Nuclear Development," *Energy Policy* 20 (July 1992): 653–59.

52. Peter A. Bradford, "How to Close the US Nuclear Industry: Do Nothing," *Bulletin of the Atomic Scientists* 69, no. 2 (2013): 12–21.

53. Jeroen de Beer, Ernst Worrell, and Kornelis Blok, "Future Technologies for Energy-Efficient Iron and Steel Making," *Annual Review of Energy and Environment* 23 (1998): 123–205.

54. De Beer, Worrell, and Blok, "Future Technologies."

55. Rochim Bakki Cahyono, Naoto Yasuda, Takahiro Nomura, and Tomohiro Akiyama, "Utilization of Low Grade Iron Ore (FeOOH) and Biomass through Integrated Pyrolysis-Tar Decomposition (CVI Process) in Iron Making Industry: Energy Analysis and Its Application," *ISIJ International* 55, no. 2 (2015): 428–35.

56. World Steel Association, *World Steel in Figures 2014* (Brussels: World Steel Association, 2014), 18.

57. International Energy Agency, *2014 Key World Energy Statistics* (Paris: International Energy Agency, 2014), 6.

58. John "Skip" Laitner, "Understanding the Size of the Energy Efficiency Resource: Ten Policy Recommendations to Accelerate More Productive Investments," *Policy and Society* 27 (2009): 351–63.

CHAPTER 5

1. U.S. Energy Information Administration, *Energy Perspectives 2011*, September 2012, chart 5, www.eia.gov/todayinenergy/detail.cfm?id=11951, April 13, 2015.

2. A.L. Austin and S.D. Winter, *U.S. Energy Flow Charts for 1950, 1960, 1970, 1980, 1985, and 1990* (Springfield, VA: National Technical Information Service, 1973, UCRL-51487); U.S., Congress, Joint Committee on Atomic Energy, *Understanding the "National Energy Dilemma"* (Washington, DC: Government Printing Office, 1973); John H. Perkins, "Development of Risk Assessment for Nuclear Power: Insights from History," *Journal of Environmental Studies and Science* 4 (2014): 273–87.

3. Lawrence Livermore National Laboratory, Energy Flow Charts, https://flowcharts.llnl.gov/, April 26, 2016.

4. Natalie Kopytko and John Perkins, "Climate Change, Nuclear Power, and the Adaptation-Mitigation Dilemma," *Energy Policy* 39 (2011): 318–33.

5. International Energy Agency, *World Energy Investment Outlook: Special Report* (Paris: IEA, 2014), 11; Statista, Global gross domestic product (GDP) at current prices from 2010 to 2020 (in billion U.S. dollars), found at www.statista.com/statistics/268750/global-gross-domestic-product-gdp/, January 8, 2016.

6. World Bank, Gross capital formation (% of GDP), retrieved as WDI table, found at http://data.worldbank.org/indicator/NE.GDI.TOTL.ZS, January 8, 2016; percent calculated by author.

7. Adapted from Kathleen M. Saul and John H. Perkins, "Nuclear Power: Is It Worth the Risks?," in *Green Energy Economies: The Search for Clean and Renewable Energy*, ed. John Byrne and Young-Doo Wang (New Brunswick, NJ: Transaction Publishers, 2014), 290.

CHAPTER 6

1. Lydia Saad and Jeffrey M. Jones, "U.S. Concern about Global Warming at Eight-Year High," March 16, 2016, www.gallup.com/poll/190010/concern-global-warming-eight-year-high.aspx, July 29, 2016.

2. Lawrence Livermore National Laboratory, carbon flowcharts, https://flowcharts.llnl.gov/commodities/carbon, April 27, 2016.

3. Intergovernmental Panel on Climate Change (IPCC), *Summary for Policy Makers, Working Group I Contribution to the IPCC Fifth Assessment Report Climate Change 2013: The Physical Science Basis* (Cambridge: Cambridge University Press, 2013), 4, 11, 13, 17.

4. This statement is not an actual quote from any known living person. It is easy to find media sources, however, that make claims and raise questions very much like those presented here. See, e.g., Rush Limbaugh, on September 27, 2013, www.rushlimbaugh.com/daily/2013/09/27/last_gasp_of_the_climate_change_cult.

5. Spencer R. Weart, *The Discovery of Global Warming* (Cambridge, MA: Harvard University Press, 2003), 2–3.

6. Weart, *Discovery of Global Warming*, 2–3; John Tyndall, "On Radiation through the Earth's Atmosphere," *Philosophical Magazine* 25, ser. 4 (March

1863): 200–206; John Tyndall, *On Radiation: The "Rede" Lecture* (New York: D. Appleton & Co., 1863), 14–16.

7. G. Frederick Wright, "Agassiz and the Ice Age," *American Naturalist* 32 (March 1898): 165–71.

8. Weart, *Discovery of Global Warming*, 11–19.

9. Weart, *Discovery of Global Warming*, 7–8; Svante Arrhenius, "On the Influence of Carbonic Acid in the Air upon the Temperature of the Ground," *Philosophical Magazine and Journal of Science* 41, ser. 5 (April 1896): 239–76.

10. Weart, *Discovery of Global Warming*, 1, 7–8, 24; Gilbert N. Plass, "Carbon Dioxide and the Climate," *American Scientist* 44 (July 1956): 302–16.

11. Fairfield Osborn, *Our Plundered Planet* (Boston: Little, Brown, 1948); William Vogt, *Road to Survival* (New York: W. Sloane Assoc., 1948).

12. Michael Egan, *Barry Commoner and the Science of Survival: The Remaking of American Environmentalism* (Cambridge, MA: MIT Press, 2007), 47–78; Rachel Carson, *Silent Spring* (Boston: Houghton Mifflin, 1962).

13. Reid A. Bryson, "A Perspective on Climatic Change," *Science* 184 (1974): 753–60; Weart, *Discovery of Global Warming*, 71–72.

14. S.I. Rasool and S.H. Schneider, "Atmospheric Carbon Dioxide and Aerosols: Effects of Large Increases on Global Climate," *Science* 173 (July 9, 1971): 138–41; R.J. Charlson, Halstead Harrison, Georg Witt, S.I. Rasool, and S.H. Schneider, "Aerosol Concentrations: Effects on Planetary Temperatures," *Science* 175 (January 7, 1972): 95–96; National Research Council, *Understanding Climatic Change: A Program for Action* (Washington, DC: National Academy of Sciences, 1975).

15. Weart, *Discovery of Global Warming*, 35–38.

16. Scripps Institution of Oceanography, "The Keeling Curve," https://scripps.ucsd.edu/programs/keelingcurve/wp-content/plugins/sio-bluemoon/graphs/mlo_full_record.pdf, April 27, 2016.

17. A. Neftel, E. Moor, H. Oeshger, and B. Stauffer, "Evidence from Polar Ice Cores for the Increase in Atmospheric CO_2 in the Past Two Centuries," *Nature* 315 (May 2, 1985): 45–47; H. Friedli, H. Lötscher, H. Oeschger, U. Sidlerstrasse, and B. Stauffer, "Ice Core Record of the $^{13}C/^{12}C$ Ratio of Atmospheric CO_2 in the Past Two Centuries," *Nature* 324 (November 20, 1986): 237–38.

18. James Hansen and Sergej Lebedeff, "Global Trends of Measured Surface Air Temperature," *Journal of Geophysical Research* 92 (1987): 13,345–72; P.D. Jones, S.C.B. Raper, R.S. Bradley, H.F. Diaz, P.M. Kelly, and T.M.L. Wigley, "Northern Hemisphere Surface Air Temperature Variations: 1851–1984," *Journal of Climate and Applied Meteorology* 25 (1986): 161–79.

19. Bryson, "A Perspective"; Stephen H. Schneider and Clifford Mass, "Volcanic Dust, Sunspots, and Temperature Trends," *Science* 190 (November 21, 1975): 741–46.

20. Weart, *Discovery of Global Warming*, 71–75.

21. Dieter Lüthi et al., "High-Resolution Carbon Dioxide Concentration Record 650,000–800,000 Years before Present," *Nature* 453 (May 15, 2008): 379–82.

22. J. Jouzel et al., "Vostok Ice Core: A Continuous Isotope Temperature Record over The Last Climatic Cycle (160,000 Years)," *Nature* 329 (October 1,

1987): 403–8; J. M. Barnola et al., "Vostok Ice Core Provides 160,000-Year Record of Atmospheric CO_2," *Nature* 329 (October 1, 1987): 408–14; C. Genthon et al., "Vostok Ice Core: Climatic Response to CO_2 and Orbital Forcing Changes over the Last Climatic Cycle," *Nature* 329 (October 1, 1987): 414–18; Urs Siegenthaler et al., "Stable Carbon Cycle–Climate Relationship during the Late Pleistocene," *Science* 310 (November 25, 2005): 1313–17; Maureen E. Raymo and Peter Huybers, "Unlocking the Mysteries of the Ice Ages," *Nature* 451 (January 17, 2008): 284–85.

23. Jeremy D. Shakun, Peter U. Clark, and Feng He et al., "Global Warming Preceded by Increasing Carbon Dioxide Concentrations during the Last Deglaciation," *Nature* 484 (April 5, 2012): 49–54.

24. P. D. Jones et al., "High-Resolution Palaeoclimatology of the Last Millennium: A Review of Current Status and Future Prospects," *Holocene* 19, no. 1(2009): 3–49; Michael E. Mann, *The Hockey Stick and the Climate Wars: Dispatches from the Front Lines* (New York: Columbia University Press, 2012), 34–36.

25. Shaun A. Marcott et al., "A Reconstruction of Regional and Global Temperature for the Past 11, 300 Years," *Science* 339 (March 8, 2013): 1198–1201.

26. Weart, *Discovery of Global Warming*, 57–59.

27. National Research Council, *Carbon Dioxide and Climate: A Scientific Assessment* (Washington, DC: National Academy of Sciences, 1979), 1–22.

28. Naomi Oreskes, "The Scientific Consensus on Climate Change," *Science* 306 (3 December 3, 2004, corrected January 21, 2005): 1686; William R. L. Anderegg et al., "Expert Credibility in Climate Change," *Proceedings of the National Academy of Sciences*, www.pnas.org/cgi/doi/10.1073/pnas.1003187107, 2010.

29. IPCC, *Summary for Policymakers*, 18.

30. Richard B. Alley, *The Two-Mile Time Machine: Ice Cores, Abrupt Climate Change, and Our Future* (Princeton, NJ: Princeton University Press, 2000), 8.

31. IPCC, *Summary for Policymakers*, 5–12.

32. Detlef P. van Vuuren, Jae Edmonds, and Mikiko Kainuma et al., "The Representative Concentration Pathways: An Overview," *Climatic Change* 109 (2011): 5–31; IPCC, *Summary for Policymakers*, 14.

33. IPCC, *Summary for Policymakers*, 20.

34. IPCC, *Climate Change 2013: The Physical Science Basis. Contribution of Working Group I to the Fifth Assessment Report of the Intergovernmental Panel on Climate Change* (Cambridge: Cambridge University Press, 2013), 1054–55, figure 12.5, table 12.2.

35. IPCC, *Summary for Policymakers*,19–29.

36. James Hansen, Makiko Sato, and Reto Ruedy, "Perception of Climate Change," *Proceedings of the National Academy of Sciences*, published online, August 6, 2012, E2415–E2423. doi10.1073/pnas.1205276109.

37. IPCC, *Climate Change 2007: Impacts, Adaptation and Vulnerability, Contribution of Working Group II to the Fourth Assessment Report of the Intergovernmental Panel on Climate Change* (Cambridge: Cambridge University Press, 2007), 845–46.

38. P.A. Stott, D.A. Stone, and M.R. Allen, "Human Contribution to the European Heatwave of 2003," *Nature* 432 (December 2, 2004): 610–14.

39. C. Schär, P. L. Vidale, D. Lüthi, C. Frei, C. Häberli, M. A. Liniger, and C. Appenzeller, "The Role of Increasing Temperature Variability in European Summer Heatwaves," *Nature* 427 (January 22, 2004): 332–36.

40. IPCC, *Climate Change 2007*, 845–46.

41. Food and Agriculture Organization of the United Nations, *FAOSTAT*, http://faostat3.fao.org/faostat-gateway/go/to/download/Q/QC/E, January 10, 2014; percent calculated by author.

42. P. Ciais et al., "Europe-Wide Reduction in Primary Productivity Caused by the Heat and Drought in 2003," *Nature* 437 (September 22, 2005): 529–33.

43. Dim Coumou and Stefan Rahmstorf, "A Decade of Weather Extremes," *Nature Climate Change* 2 (July 2012): 491–96.

44. David Barriopedro et al., "The Hot Summer of 2010: Redrawing the Temperature Record Map of Europe," *Science* 332 (April 8, 2011): 220–24, figure 1, 221.

45. Coumou and Rahmstorf, "A Decade of Weather Extremes,"; Barriopedro et al., "The Hot Summer of 2010"; Randall Dole et al., "Was There a Basis for Anticipating the 2010 Russian Heat Wave?," *Geophysical Research Letters* 38 (2011): doi:10.1029/2010GL046582.

46. Dole et al., "Was There a Basis?"

47. Barriopedro et al., "The Hot Summer of 2010."

48. Statistics from http://faostat.fao.org/; percent calculations by author.

49. Stephen K. Wegren, "Food Security and Russia's 2010 Drought," *Eurasian Geography and Economics* 52, no. 1 (2011): 140–56, table 1, 147; percent calculations by author.

50. IPCC, *Summary for Policymakers*, 12.

51. IPCC, *Summary for Policymakers*, 27.

52. IPCC, *Summary for Policymakers*, 5.

53. Naomi Oreskes and Erik M. Conway, *Merchants of Doubt* (New York: Bloomsbury Press, 2010), 169–215.

54. Robert J. Brulle, "Institutionalizing Delay: Foundation Funding and the Creation of U.S. Climate Change Counter-Movement Organizations," *Climatic Change*, December 21, 2013, doi 10.1007/s10584-013-1018-7; Oreskes and Conway, *Merchants of Doubt*.

55. Paul C. Stern, John H. Perkins, Richard E. Sparks, and Robert A. Knox, "The Challenge of Climate-Change Neoskepticism," *Science* 353 (August 12, 2016): 653–54.

CHAPTER 7

1. W. Stanley Jevons, *The Coal Question: An Inquiry Concerning the Progress of the Nation and the Probable Exhaustion of Our Coal Mines*, 2nd ed. (London: Macmillan and Co., 1866), 1–14.

2. Erik J. Dahl, "Naval Innovation: From Coal to Oil," *Joint Force Quarterly* (Winter 2000–2001): 50–56.

3. Daniel Yergin, *The Prize* (New York: Simon and Schuster, 1991), 134–64.

4. Yergin, *The Prize*, 168–73.

5. Yergin, *The Prize*, 185.

6. Yergin, *The Prize*, 184–206.

7. Anand Toprani, "Germany's Answer to Standard Oil: The Continental Oil Company and Nazi Grand Strategy, 1940–1942," *Journal of Strategic Studies* 37, no. 6–7 (2014): 949–73.

8. Irving Anderson, *Aramco, the United States, and Saudi Arabia* (Princeton, NJ: Princeton University Press, 1981), 37.

9. Yergin, *The Prize*, 283–302.

10. *Historical Statistics of the United States, Colonial Times to 1957* (Washington, DC: Government Printing Office, 1960), Series M-74, M-81, M-82.

11. U.S. Energy Information Administration, *Monthly Energy Review*, February 2015, table 3.1, www.eia.gov/totalenergy/data/monthly/pdf/mer.pdf, March 16, 2015.

12. Yergin, *The Prize*, 450–70.

13. *BP Statistical Review of World Energy June 2015* (London: BP, 2015), 8–9; calculations by author.

14. Stephanie Cooke, *In Mortal Hands: A Cautionary History of the Nuclear Age* (New York: Bloomsbury, 2009).

15. U.S. Energy Information Administration, *2014 Uranium Marketing Annual Report* (Washington, DC: Energy Information Administration, 2015), figure 5, p. 21, table S1a; percent calculated by author.

16. E. I. Abbakumov, V. A. Bazhenov, and Yu B. Verbin et al., "Development and Industrial Use of Gas Centrifuges for Uranium Enrichment in the Soviet Union," *Soviet Atomic Energy* 67, no. 4 (1989): 739–41.

17. B. Cameron Reed, "Centrifugation during the Manhattan Project," *Physics Perspective* 11 (2009): 426–41.

18. Houston G. Wood, Alexander Glaser, and R. Scott Kemp, "The Gas Centrifuge and Nuclear Weapons Proliferation," *Physics Today*, September 2008, 40–45.

19. A. Gopalakrishnan, "Evolution of the Indian Nuclear Power Program," *Annual Review of Energy and the Environment* 27 (2002): 369–95.

20. World Nuclear Association, "The Nuclear Fuel Cycle," www.world-nuclear.org/information-library/nuclear-fuel-cycle/introduction/nuclear-fuel-cycle-overview.aspx, February 26, 2016.

21. Peter C. Burns, Rodney C. Ewing, and Alexandra Navrotsky, "Nuclear Fuel in a Reactor Accident," *Science* 335 (March 9, 2012): 1184–88.

22. The Coalmining History Resource Center, Coalmining accidents and deaths, www.cmhrc.co.uk/site/disasters/, May 18, 2015.

23. West Virginia Office of Miners' Health, Safety and Training, "Production of Coal and Coke in West Virginia 1863–2012," www.wvminesafety.org/historicprod.htm, May 17, 2015; ratio and percent calculations by author.

24. United States Department of Labor, Coal fatalities for 1900 through 2014, www.msha.gov/stats/centurystats/coalstats.asp, May 18, 2015; United States Department of Labor, History of mine safety and health legislation, www.msha.gov/MSHAINFO/MSHAINF2.HTM, May 18, 2015; ratio and percent calculations by author.

25. United States Energy Information Administration, "Coal Production, 1949–2011," table 7.2, www.eia.gov/totalenergy/data/annual/#coal, May 18, 2015; West Virginia Office of Miners' Health, Safety and Training, *2013 Statistical Report and Directory of Mines* (Office of MHS&T), 36.

26. Coal Workers Pneumoconiosis, www.pneumoconiosis.org.uk/coal-workers-pneumoconiosis, May 19, 2015; American Lung Association, Pneumoconiosis (Black Lung Disease), www.lung.org/lung-disease/pneumoconiosis/, May 19, 2015.

27. Eva Suarthana, A. Scott Laney, Eileen Storey, Janet M. Hale, and Michael D. Attfield, "Coal Workers' Pneumoconiosis in the United States: Regional Differences 40 Years after Implementation of the 1969 Federal Coal Mine Health and Safety Act," *Occupational and Environmental Medicine* 68 (2011): 908–13.

28. David J. Blackley and Cara N. Halldin, "Resurgence of a Debilitating and Entirely Preventable Respiratory Disease among Working Coal Miners," *American Journal of Respiratory and Critical Care Medicine* 190, no. 6 (September 15, 2014): 708–9; David J. Blackley, Cara N. Halldin, Mei Lin Wang, and A. Scott Laney, "Small Mine Size Is Associated with Lung Function Abnormality and Pneumoconiosis among Underground Coal Miners in Kentucky, Virginia, and West Virginia," *Occupational and Environmental Medicine* 71 (2014): 690–94.

29. A. Scott Laney, Anita L. Wolfe, Edward L. Petsonk, and Cara N. Halldin, "Pneumoconiosis and Advanced Occupational Lung Disease among Surface Coal Miners–16 States, 2010–2011," *Morbidity and Mortality Weekly Report* 61 (June 15, 2012): 431–34.

30. National Research Council, *Hidden Costs of Energy: Unpriced Consequences of Energy Production and Use* (Washington, DC: National Academies Press, 2009), 75.

31. Michael Hendryx, Kathryn O'Donnell, and Kimberly Horn, "Lung Cancer Mortality Is Elevated in Coal-Mining Areas of Appalachia," *Lung Cancer* 62 (2008): 1–7.

32. Michael Hendryx, "Mortality from Heart, Respiratory, and Kidney Disease in Coal Mining Areas of Appalachia," *International Archives of Occupational and Environmental Health* 82 (2009): 243–49.

33. Michael Hendryx, Evan Fedorko, Andrew Anesetti-Rothermel, "A Geographical Information System–Based Analysis of Cancer Mortality and Population Exposure To Coal Mining Activities in West Virginia, United States of America," *Geospatial Health* 4, no. 2 (2010): 243–56.

34. Melissa M. Ahern, Michael Hendryx, et al., "The Association between Mountain Top Mining and Birth Defects among Live Births in Central Appalachia, 1996–2003," *Environmental Research* 111 (2011): 838–46.

35. Melissa Ahern, Martha Mullett, Katherine MacKay, and Candice Hamilton, "Residence in Coal-Mining Areas and Low-Birth-Weight Outcomes," *Maternal and Child Health Journal* 15 (2011): 974–79.

36. Michael Hendryx and Juhua Luo, "An Examination of the Effects of Mountaintop Removal Coal Mining on Respiratory Symptoms and COPD Using Propensity Scores," *International Journal of Environmental Health Research* 25, no. 3 (2015): 265–76.

37. Jonathan Borak, Catherine Salipante-Zaidel, Martin D. Slade, and Cheryl A. Fields, "Mortality Disparities in Appalachia: Reassessment of Major Risk Factors," *Journal of Occupational and Environmental Medicine* 54, no. 2 (February 2012): 146–56; Jeanine M.Buchanich, Lauren C. Balmert, Ada O. Youk, Shannon M. Woolley, and Evelyn O. Talbott, "General Mortality Patterns in Appalachian Coal-Mining and Non-Coal-Mining Counties," *Journal of Occupational and Environmental Medicine* 56, no. 11 (2014): 1169–78.

38. Michael Hendryx and Melissa Ahern, Reply to Borak et al., "Mortality Disparities in Appalachia: Reassessment of Major Risk Factors"; Borak et al., "Ecological Bias and Data Entry Errors: A Reply to Hendryx and Ahern"; both in *Journal of Occupational and Environmental Medicine* 54, no. 7 (July 2014): 768–69 and 770–73.

39. Robert Godby, Roger Coupal, David Taylor, and Tim Considine, *The Impact of the Coal Economy on Wyoming* (Laramie, WY: Center for Energy Economics and Public Policy, 2015), 12–15.

40. Union of Concerned Scientists, "How Coal Works," found at www.ucsusa.org/clean_energy/coalvswind/brief_coal.html#.VWDLMoaUJsk, May 23, 2015.

41. U.S. Energy Information Administration, *Annual Energy Outlook 2014* (Washington, DC: U.S. Energy Information Administration, 2014), table A-15, p. A-29; Chris Carroll, *Wyoming's Coal Resource, Summary Report* (Laramie: Wyoming State Geological Survey, 2014), 2; percent calculated by author.

42. US Energy Information Administration, Two Wyoming mines accounted for 20% of U.S. coal production by tons in 2012, www.eia.gov/todayinenergy/detail.cfm?id = 10591, May 23, 2015.

43. Natural Resources Defense Council and Western Organization of Resource Councils, *Undermined Promise: Reclamation and Enforcement of the Surface Mining Control and Reclamation Act, 1977–2007* (Washington, DC: Natural Resources Defense Council, 2007), 17.

44. M.A. Palmer et al., "Mountaintop Mining Consequences," *Science* 327 (January 8, 2010): 148–49.

45. Paul R. Epstein et al., "Full Cost Accounting for the Life Cycle of Coal," *Annals of the New York Academy of Sciences* 1219 (2011): 73–98.

46. U.S. Department of State, Beijing Mission, Beijing—$PM_{2.5}$, www.stateair.net/web/post/1/1.html, 22 May 2015; U.S. Environmental Protection Agency, *Guidelines for the Reporting of Daily Air Quality—The Air Quality Index* (Washington, DC: USEPA, 2006), 11.

47. Epstein et al., "Full Cost Accounting," 85.

48. Jevons, *The Coal Question*.

49. Tyler Priest, "Hubbert's Peak: The Great Debate over the End of Oil," *Historical Studies in the Natural Sciences* 44, no. 1 (February 2014): 37–79; Yergin, *The Prize*, 248–50.

50. Priest, "Hubbert's Peak."

51. M. King Hubbert, "Energy from Fossil Fuels," *Science* 109 (February 4, 1949): 103–9.

52. M. King Hubbert, "Nuclear Energy and the Fossil Fuels," Pub. No. 95, Shell Development Company, Houston, 1956, 22; M. King Hubbert,

Notes to Pages 159–170

"Techniques of Prediction with Applications to the Oil Industry," Preprint version, March 1959.

53. M. King Hubbert, *Energy Resources* (Washington, DC: National Academy of Sciences—National Research Council, 1962), 73.

54. Tyler Priest, "Hubbert's Peak."

55. United States Energy Information Administration, U.S. Field Production of Crude Oil, www.eia.gov/dnav/pet/hist/LeafHandler.ashx?n=PET&s=MCRFPUS1&f=A, June 11, 2015.

56. Emma Hemmingsen, "At the Base of Hubbert's Peak: Grounding the Debate on Petroleum Scarcity," *Geoforum* 41 (2010): 531–40.

57. Hemmingsen, "At the Base of Hubbert's Peak."

58. David J. Murphy, "The Implications of the Declining Energy Return on Investment of Oil Production," *Philosophical Transactions A* 372 (2014), (http://dx.doi.org/10.1098/rsta.2013.0126; Charles A.S. Hall and Cutler J. Cleveland, "Petroleum Drilling and Production in the United States: Yield per Effort and Net Energy Analysis," *Science* 211 (February 6, 1981): 576–79.

59. Charles A.S. Hall, Jessica G. Lambert, and Stephen B. Balogh, "EROI of Different Fuels and the Implications for Society," *Energy Policy* 64 (2014): 141–52; David J. Murphy, Charles A.S. Hall, Michael Dale, and Cutler Cleveland, "Order from Chaos: A Preliminary Protocol for Determining the EROI of Fuels," *Sustainability* 3 (2011): 1988–1907 (doi:10.3390/su3101888).

60. Hall, Lambert, and Balogh, "EROI of Different Fuels."

61. Jessica G. Lambert, Charles A.S. Hall, Stephen Balogh, Ajay Gupta, and Michelle Arnold, "Energy, EROI and Quality of Life," *Energy Policy* 64 (2014): 153–67.

62. David J. Murphy and Charles A.S. Hall, "Year in Review—EROI or Energy Return on (Energy) Invested," *Annals of the New York Academy of Sciences* 1185 (2010): 102–18.

CHAPTER 8

1. John H. Perkins, "Development of Risk Assessment for Nuclear Power: Insights from History," *Journal of Environmental Studies and Sciences* 4 (2014): 273–87.

2. Michael L. Ross, "How the 1973 Oil Embargo Saved the Planet: OPEC Gave The Rest of the World a Head Start against Climate Change," *Foreign Affairs*, Snapshot, October 13, 2013, www.foreignaffairs.com/articles/north-america/2013-10-15/how-1973-oil-embargo-saved-planet, January 19, 2016.

3. Energy Policy Project of the Ford Foundation, *A Time to Choose: America's Energy Future* (Cambridge, MA: Ballinger, 1974).

4. Robert H. Socolow, "Reflections on the 1974 APS Energy Study," *Physics Today*, January 1986, 2–9.

5. Arthur H. Rosenfeld and Deborah Poskanzer, "A Graph Is Worth a Thousand Gigawatt-Hours: How California Came to Lead the United States in Energy Efficiency," *innovations* (Fall 2009): 57–79.

6. Amory B. Lovins, "Energy Strategy: The Road Not Taken?," *Foreign Affairs* 65 (1976–77): 65–96.

7. Mikel González-Eguino, "Energy Poverty: An Overview," *Renewable and Sustainable Energy Reviews* 47 (2015): 377–85; Benjamin K. Sovacool, "The Political Economy of Energy Poverty: A Review of Key Challenges," *Energy for Sustainable Development* 16 (2012): 272–82; Benjamin K. Sovacool, Christopher Cooper, and Morgan Bazilian et al., "What Moves and Works: Broadening the Consideration of Energy Poverty," *Energy Policy* 42 (2012): 715–19.

8. Yvonne Y. Deng et al., "Quantifying a Realistic, Worldwide Wind and Solar Electricity Supply," *Global Environmental Change* 31 (2015): 239–52; data from table 9.

9. This section is drawn from Robert Ehrlich, *Renewable Energy: A First Course* (Boca Raton, FL: CRC Press, 2013), 249–342.

10. Christiana Honsberg and Stuart Bowden, *PVCDROM* (www.PVEducation.org/pvcdrom), September 4, 2015.

11. M. Gibbons and C. Johnson, "Relationship between Science and Technology," *Nature* 227 (1 July 1970): 125–27.

12. Florida Solar Energy Center, Cells, modules, & arrays, www.fsec.ucf.edu/en/consumer/solar_electricity/basics/cells_modules_arrays.htm., July 30, 2016.

13. Steven Lacey, "America's Utility-Scale Solar Generation Is 31 Times Higher Today than a Decade Ago," *greentechsolar*, August 28, 2015, found at www.greentechmedia.com/articles/read/us-utility-scale-solar-generation-is-31-times-higher-today-than-a-decade-ag, September 8, 2015.

14. U.S. Energy Information Administration, *Electric Power Monthly for December, 2014*, February, 2015; *Electric Power Monthly for June, 2015*, August 2015; California ISO, *Renewables Watch*, Monday, September 7, 2015, http://content.caiso.com/green/renewrpt/DailyRenewablesWatch.pdf, 8 September 2015. Calculations by author.

15. Harry Wirth, "Recent Facts about Photovoltaics in Germany," Freiburg: Fraunhofer Institute for Solar Energy Systems, Freiburg, 2015, 5.

16. M. J. M. Pathak, P. G. Sanders, and J. M. Pearce, "Optimizing Limited Solar Roof Access by Energy Analysis of Solar Thermal, Photovoltaic, and Hybrid Photovoltaic Thermal Systems," *Applied Energy* 120 (2014): 115–24.

17. This section is drawn from Ehrlich, *Renewable Energy*, 183–218.

18. Pacific Northwest National Laboratory, *Wind Energy Resource Atlas of the United States* (Golden, CO: National Renewable Energy Laboratory, 1986), http://rredc.nrel.gov/wind/pubs/atlas/atlas_index.html, September 13, 2015.

19. Ehrlich, *Renewable Energy*, 183; California Energy Commission, Electricity from wind energy statistics and data, http://energyalmanac.ca.gov/renewables/wind/index.php, July 28, 2015; John K. Kaldellis and D. Zafirakis, "The Wind Energy (R)evolution: A Short Review of a Long History," *Renewable Energy* 36 (2011): 1887–1901; U.S. Energy Information Administration, "Wind Generates More than 10% of Texas Electricity in 2014," http://www.eia.gov/todayinenergy/detail.cfm?id=20051, July 29, 2015.

20. ERCOT, Wind Integration reports, issued daily, www.ercot.com/gridinfo/generation/windintegration/index.html, September 19, 2015.

21. This section is based on Ehrlich, *Renewable Energy*, 219–47.

22. International Energy Agency, *2014 Key World Energy Statistics* (Paris: International Energy Agency, 2014), 24.

23. Electropaedia, "Hydroelectric Power," www.mpoweruk.com/hydro_power.htm, September 23, 2015.

24. International Energy Agency, 2014 *Key World Energy Statistics*, 24.

25. Ottmar Edenhofer et al., eds., *Renewable Energy Sources and Climate Change Mitigation, Special Report of the Intergovernmental Panel on Climate Change* (New York: Cambridge University Press, 2011), 441 (cited hereafter as *IPCC Special Report*).

26. International Energy Agency, 2014 *Key World Energy Statistics*, 19.

27. U.S. Energy Information Administration, "Electric Power Industry Generation by Primary Energy Source Back to 1990," www.eia.gov/state/search/#?1=108&5=124&2=228, September 11, 2015; percent calculated by author.

28. U.S. Energy Information Administration, "Washington State Energy Profile," www.eia.gov/state/print.cfm?sid =WA#73, December 1, 2015.

29. Edenhofer et al., *IPCC Special Report*, 446.

30. This section is based on Edenhofer et al., *IPCC Special Report*, 497–533.

31. Edenhofer et al., *IPCC Special Report*, 183, 503.

32. Power-Technology, "Tidal Giants—The World's Five Biggest Tidal Power Plants," April 11, 2014, www.power-technology.com/features/featuretidal-giants-the-worlds-five-biggest-tidal-power-plants-4211218/, September 24, 2015.

33. Edenhofer et al., *IPCC Special Report*, 504.

34. Edenhofer et al., *IPCC Special Report*, 503, 511–12.

35. E. A. Mason, "From Pig Bladders and Cracked Jars to Polysulfones: An Historical Perspective on Membrane Transport," *Journal of Membrane Science* 60 (1991): 125–45; Karl Wilhelm Böddeker, "Tracing Membrane Science," *Journal of Membrane Science* 100 (1995): 65–68.

36. R. E. Pattle, "Production of Electric Power by Mixing Fresh and Salt Water in the Hydroelectric Pile," *Nature* 174 (October 2, 1954): 660.

37. Edenhofer et al., *IPCC Special Report*, 507, 511–13; Jin Gi Hong et al., "Potential Ion Exchange Membranes and System Performance in Reverse Electrodialysis for Power Generation: A Review," *Journal of Membrane Science* 486 (2015): 71–88; Dutch Water Sector, "Dutch King Opens World's First RED Power Plant Driven on Fresh-Salt Water Mixing," November 26, 2014, www.dutchwatersector.com/news-events/news/12388-dutch-king-opens-world-s-first-red-power-plant-driven-on-fresh-salt-water-mixing.html, September 27, 2015; IRENA, *Salinity Gradient Energy: Technology Brief* (IRENA: Bonn, 2014), 3–4, 9–10.

38. Ehrlich, *Renewable Energy*, 133–54.

39. Ehrlich, *Renewable Energy*, 133.

40. Veronica Dornburg, Detlef van Vuuren, and Gerrie van de Ven et al., "Bioenergy Revisited: Key Factors in Global Potentials of Bioenergy," *Energy and Environmental Science* 3 (2010): 258–67.

41. Mikel González-Eguino, "Energy Poverty: An Overview," *Renewable and Sustainable Energy Reviews* 47 (2015): 377–85; Benjamin K. Sovacool, "The Political Economy of Energy Poverty: A Review of Key Challenges," *Energy for Sustainable Development* 16 (2012): 272–82; Benjamin K. Sovacool, Christopher Cooper, and Morgan Bazilian et al., "What Moves and

Works: Broadening the Consideration of Energy Poverty," *Energy Policy* 42 (2012): 715–19.

42. S. Irene Virbila, "Critic's Choice: Restaurants with Wood-Burning Ovens," *Los Angeles Times*, January 19, 2012, http://articles.latimes.com/2012/jan/19/food/la-fo-0119-critics-choice-20120119, October 3, 2015.

43. U.S. Energy Information Administration, *Electric Power Monthly*, September 2015, tables 1.1, 1.1A, and 1.18B; percent values calculated by author.

44. Edenhofer et al., *IPCC Special Report*, 214; Dornburg, van Vuuren, van de Ven et al., "Bioenergy Revisited."

45. Roger Real Drouin, "Wood Pellets: Green Energy or New Source of CO_2 Emissions?" *environment 360*, January 22, 2015, http://e360.yale.edu/feature/wood_pellets_green_energy_or_new_source_of_co2_emissions/2840/, October 4, 2015.

46. This section is based on Ehrlich, *Renewable Energy*, 155–81. See also National Research Council, *Induced Seismicity Potential in Energy Technologies* (Washington, DC: National Academies Press, 2013), 59–75.

47. Alan R. Carroll, *Geofuels: Energy and the Earth* (Cambridge: Cambridge University Press, 2015).

48. National Research Council, *Induced Seismicity*, 61–62.

49. Edenhofer et al., *IPCC Special Report*, 409.

50. National Research Council, *Induced Seismicity*, 60–61.

51. Calpine, "Geysers by the Numbers," www.geysers.com/numbers.aspx, October 7, 2015.

52. U.S. Department of Energy, Geothermal heat pumps, http://energy.gov/energysaver/geothermal-heat-pumps, October 7, 2015.

53. Mark Z. Jacobson and Mark A. Delucchi, "Providing All Global Energy with Wind, Water, and Solar Power, Part I: Technologies, Energy Resources, Quantities and Areas of Infrastructure, and Materials," *Energy Policy* 39 (2011): 1154–69; Mark A. Delucchi and Mark Z. Jacobson, "Providing All Global Energy with Wind, Water, and Solar Power, Part II: Reliability, System and Transmission Costs, and Policies," *Energy Policy* 39 (2011): 1170–90; Mark Z. Jacobson, Mark A. Delucchi, and Guillaume Bazouin et al., "100% Clean and Renewable Wind, Water, and Sunlight (WWS) All-Sector Energy Roadmaps for the 50 United States," *Energy & Environmental Science*, 2015, doi:10.1039/c5ee01283j; Arjun Makhijani, *Carbon-Free and Nuclear-Free: A Roadmap for U.S. Energy Policy* (Takoma Park, WA: IEER Press, 2007); Walt Patterson, *Electricity vs. Fire: The Fight for Our Future* (Amersham: Walt Patterson, 2015).

CHAPTER 9

1. David Coady, Ian Parry, Louis Sears, and Baoping Shang, *How Large Are Global Energy Subsidies?* ([Washington, DC]: International Monetary Fund, 2015), 29.

2. Jonas Meckling and Llewelyn Hughes, "Globalizing Solar Industry Specialization and Firm Demands for Trade Protection," Berkeley Roundtable on the International Economy, 2015.

3. This section is based on David E. Blockstein and Robert I. Long, eds., *Report from the Summit,* 2nd National Energy Education Summit, June 7, 2016, Session 5 (Washington, DC: National Council for Science and the Environment, 2016), 13–15.

4. International Institute for Sustainable Development, "What Is Sustainable Development?," www.iisd.org/sd/#one, July 15, 2015.

5. Blockstein and Long, *Report.*

6. James Fieser, "Ethics," in *Internet Encyclopedia of Philosophy,* www.iep.utm.edu/ethics/, July 24, 2015.

7. Dean Apostol, James Palmer, Martin Pasqualetti, Richard Smardon, and Robert Sullivan, *The Renewable Energy Landscape: Preserving Scenic Values in Our Sustainable Future* (New York: Routledge, 2017).

8. Robert G. Hunt and William E. Franklin, "LCA, How It Came About: Personal Reflections on the Origin and Development of LCA in the USA," *International Journal of Life Cycle Analysis* 1, no. 1 (1996): 4–7.

CHAPTER 10

1. Ottmar Edenhofer et al., *Renewable Energy Sources and Climate Change Mitigation Special Report of the Intergovernmental Panel on Climate Change* (New York: Cambridge University Press, 2012), 728–33.

2. Theodore M. Porter, *Trust in Numbers: The Pursuit of Objectivity in Science and Public Life* (Princeton, NJ: Princeton University Press, 1996).

3. International Energy Agency, *2014 Key World Energy Statistics* (Paris: International Energy Agency, 2014), 6.

4. International Energy Agency, *Key Coal Trends, 2015* (Paris: International Energy Agency, 2015), 3–5.

5. Yeonbae Kim and Ernst Worrell, "International Comparison of CO_2 Emission Trends in the Iron and Steel Industry," *Energy Policy* 30 (2002): 827–38; Rochim Bakki Cahyono, Naoto Yasuda, Takahiro Nomura, and Tomohiro Akiyama, "Utilization of Low Grade Iron Ore (FeOOH) and Biomass through Integrated Pyrolysis-Tar Decomposition (CVI Process) in Iron Making Industry: Exergy Analysis and Its Application," *ISIJ International* 55, no. 2 (2015): 428–35.

6. Energy Information Administration, *Electric Power Monthly with Data for July 2015* (Washington, DC: Energy Information Administration, 2015), table 1.1; percent calculated by author.

7. International Energy Agency, *Key Coal Trends, 2015,* 7.

8. Michael W. Flinn and David Stoker, *The History of the British Coal Industry,* vol. 2, *1700–1830: The Industrial Revolution* (Oxford: Clarendon Press, 1984), 395–411.

9. Antonia Juhasz, "Shell Is Reeling after Pulling Out of the Arctic," *Newsweek,* October 15, 2015, http://royaldutchshellplc.com/2015/10/13/shell-is-reeling-after-pulling-out-of-the-arctic/, October 20, 2015.

10. Spencer Dale, *The New Economics of Oil* (Oxford: Oxford Institute for Energy Studies, 2015), 2–9; International Energy Agency, *Medium-Term Oil Market Report 2015* (Paris: International Energy Agency, 2015), 10–16; Federal Reserve Bank of Minneapolis, "The Bakken Oil Boom," August 5, 2015,

www.minneapolisfed.org/publications/special-studies/bakken/oil-production, October 22, 2015.

11. U.S. Energy Information Administration, "Crude Oil Proved Reserves (Billion Barrels, 2014)," www.eia.gov/cfapps/ipdbproject/IEDIndex3 .cfm?tid=5&pid=57&aid=6, October 21, 2015.

12. Energy Information Administration, Natural Gas Policy Act of 1978, www.eia.gov/oil_gas/natural_gas/analysis_publications/ngmajorleg/ngact1978 .html, October 26, 2015.

13. Energy Information Administration, "Major Legislative and Regulatory Actions (1935–2008)," www.eia.gov/oil_gas/natural_gas/analysis_publications /ngmajorleg/ngmajorleg.html, October 26, 2015.

14. Benjamin K. Sovacool, "Cornucopia or Curse? Reviewing the Costs and Benefits of Shale Gas Hydraulic Fracturing (Fracking)," *Renewable and Sustainable Energy Reviews* 37 (2014): 249–64.

15. Jeff Combs, "Uranium Markets," *Bulletin of the Atomic Scientists* 64, no. 4 (2008): 48–51; Energy Information Administration, *2014 Uranium Marketing Annual Report* (Washington, DC: Energy Information Administration, 2015), table 51.b.

16. Mycle Schneider and Antony Froggatt, *The World Nuclear Industry Status Report 2014* (Paris: Mycle Schneider Consulting Project, 2015), 7–8.

17. Peter A. Bradford, "How to Close the US Nuclear Industry: Do Nothing," *Bulletin of the Atomic Scientists* 69, no. 2 (2013): 12–21; James M. Jasper, "Patterns of State Involvement in Nuclear Development," *Energy Policy* 20 (July 1992): 653–59.

18. World Nuclear Association, "Supply of Uranium," www.world-nuclear .org/info/Nuclear-Fuel-Cycle/Uranium-Resources/Supply-of-Uranium/, October 29, 2015.

19. Intergovernmental Panel on Climate Change, *Renewable Energy Sources and Climate Change Mitigation, Special Report of the Intergovernmental Panel on Climate Change* (Cambridge: Cambridge University Press, 2012), 732, 980–82.

20. World Nuclear Association, "Climate Change and Nuclear Energy," www.world-nuclear.org/Features/Climate-Change/Climate-Change-and-Nuclear-Energy/, November 7, 2015.

21. Kristin Shrader-Frechette, *What Will Work: Fighting Climate Change with Renewable Energy, Not Nuclear Power* (Oxford: Oxford University Press, 2011), 45–50.

22. Susan E. Dawson and Gary E. Madsen, "Psychosocial and Health Impacts of Uranium Mining and Milling on Navajo Lands," *Health Physics* 101, no. 5 (November 2011): 618–25; Marley Shebala, "Poison in the Earth," *Navajo Times*, July 23, 2009, http://navajotimes.com/2009/0709/072309uranium .php#.VkDsyr-UJsk, November 9, 2015; Joshua Lott, "Once upon a Mine: The Legacy of Uranium on the Navajo Nation," *Environmental Health Perspectives* 122, no. 2 (February 2014): A44–A49; Anna Stanley, "Wasted Life: Labour, Liveliness, and the Production of Value," *Antipode* 47, no. 3 (2015): 792–811; John R. Trabalka, L. Dean Eyman, and Stanley I. Auerbach, "Analysis of the 1957–1958 Soviet Nuclear Accident," *Science* 209 (July 18, 1980): 345–53.

23. Kirk R. Smith, Howard Frumkin, and Kalpana Balakrishnan et al., "Energy and Human Health," *Annual Review of Public Health* 34 (2013): 159–88; Ian Fairlie, "A Hypothesis to Explain Childhood Cancers Near Nuclear Power Plants," *Journal of Environmental Radioactivity* 133 (2014): 10–17.

24. The Chernobyl Forum, *Chernobyl's Legacy: Health, Environmental and Socio-Economic Impacts* (Vienna: International Atomic Energy Agency, 2006), 16.

25. E. Cardis, "Cancer Effects of the Chernobyl Accident," in International Atomic Energy Agency, *Chernobyl: Looking Back to Go Forward* (Vienna: International Atomic Energy Agency, 2008), 96.

26. Ian Fairlie and David Sumner, *The Other Report on Chernobyl (TORCH): An Independent Scientific Evaluation of Health and Environmental Effects 20 Years after the Nuclear Disaster Providing Critical Analysis of a Recent Report by the International Atomic Energy Agency (IAEA) and the World Health Organisation (WHO*, (Berlin: The Greens/EFA in the European Parliament, 2006).

27. Alexey V. Yablokov, Vassily B. Nesterenko, and Alexey V. Nesterenko, eds., "Chernobyl: Consequences of the Catastrophe for People and the Environment," *Annals of the New York Academy of Sciences* 1181 (2009): 210.

28. Maggie Molina, *The Best Value for America's Energy Dollar: A National Review of the Cost of Utility Energy Efficiency Programs* (Washington, DC: American Council for an Energy-Efficient Economy, 2014, Report U1402), iii.

29. Hannah Choi Granade, Jon Creyts, Anton Derkach et al., *Unlocking Energy Efficiency in the U.S. Economy* (New York: McKinsey & Company, 2009), iii.

30. Govinda R. Timilsina, Lado Kurdgelashvili, and Patrick A. Narbel, "Solar Energy: Markets, Economics and Policies," *Renewable and Sustainable Energy Reviews* 16 (2012): 449–65.

31. REC Solar, "Solar Finance 101: What's an SREC and How Much Is It Worth?," November 21, 2013, http://blog.recsolar.com/2013/11/solar-finance-101-whats-an-srec-and-how-much-is-it-worth/, August 1, 2016.

32. U.S. Energy Information Administration, *Electric Power Monthly with Data for August 2015* (Washington, DC: U.S. Energy Information Administration, 2015), tables 1.1 and 1.1A; percent calculations by author.

33. Jason Kaminsky and Justin Baca, "US solar electricity production 50 % higher than previously thought," www.greentechmedia.com/articles/read/us-solar-electricity-production-50-higher-than-previously-thought, September 7, 2015.

34. Mark Bolinger and Joachim Seel, *Utility-Scale Solar 2014* (Berkeley, CA: Lawrence Berkeley National Laborator, 2015).

35. Energy Information Administration, *Electric Power Monthly with Data for August 2015*, table 6.1; percent calculations by author.

36. Mark Bolinger and Joachim Seel, *Utility-Scale Solar 2014*; Herman K. Trabish, "NV Energy Buys Utility-Scale Solar at Record Low Price under 4 Cents/kWh," July 9, 2015, www.utilitydive.com/news/nv-energy-buys-utility-scale-solar-at-record-low-price-under-4-centskwh/401989/, November 17, 2015; Herman K. Trabish, "Austin Energy, First Solar Ink Deal for 119 MW of

Solar," October 30, 2015,www.utilitydive.com/news/austin-energy-first-solar-ink-deal-for-119-mw-of-solar/408275/, November 17, 2015.

37. Svetlana V. Obydenkova and Joshua M. Pearce, "Technical Viability of Mobile Solar Photovoltaic Systems for Indigenous Nomadic Communities in Northern Latitudes," *Renewable Energy* 89 (2016): 253–67.

38. Bouchra Bakhiyi, "The Photovoltaic Industry on the Path to a Sustainable Future—Environmental and Occupational Health Issues," *Environment International* 73 (2014): 224–34.

39. Carlos de Castro, Margarita Mediavilla, Luis Javier Miguel, and Fernando Frechoso, "Global Solar Electric Potential: A Review of Their Technical and Sustainable Limits," *Renewable and Sustainable Energy Reviews* 28 (2013): 824–35.

40. Guido Pleßmann, Matthias Erdmann, Markus Hlusiak, and Christian Breyer, "Global Energy Storage Demand for a 100% Renewable Electricity Supply," *Energy Procedia* 46 (2014): 22–31.

41. Michael Kanellos, "Yingli in Trouble: The 'Number One' Curse in Solar Strikes Again," *Forbes*, May 18, 2015; Nicole Litvak, "Here Are the Top 5 Residential Solar Installers of 2014," *greentech media*, March 10, 2015; Jonas Meckling and Llewelyn Hughes, *Globalizing Solar Industry Specialization and Firm Demands for Trade Protection* (Berkeley, CA: Berkeley Roundtable on the International Economy, 2015).

42. Ryan Wiser and Mark Bolinger, *2014 Wind Technologies Market Report* (Washington, DC: U.S. Department of Energy, 2015), 15.

43. Energy Digital, "Top 10 Wind Turbine Suppliers," April 10, 2015, www.energydigital.com/top10/3705/Top-10-Wind-Turbine-Suppliers, November 23, 2015.

44. power-technology, "Top 10 Biggest Wind Farms," September 30, 2013, http://www.power-technology.com/features/feature-biggest-wind-farms-in-the-world-texas/, November 23, 2015.

45. Terra-gen, Wind Projects, 2015, www.terra-gen.com/Projects/Projects_Wind.aspx, November 23, 2015.

46. Bloomberg Business, "Company Overview of ArcLight Capital Partners, LLC," November 24, 2015, www.bloomberg.com/research/stocks/private/snapshot.asp?privcapId=3207406, November 24, 2015; Bloomberg Business, "Company Overview of Global Infrastructure Partners," November 24, 2015, www.bloomberg.com/research/stocks/private/snapshot.asp?privcapId=30373318, November 24, 2015.

47. power-technology, Alta Wind Energy Center (AWEC), California, United States of America, September 30, 2013, www.power-technology.com/projects/alta-wind-energy-center-awec-california/, November 23, 2015.

48. Wiser and Bolinger, *2014 Wind Technologies Market Report*, 29–34, 46–50.

49. Wiser and Bolinger, *2014 Wind Technologies Market Report*, 62–64.

50. Wiser and Bolinger, *2014 Wind Technologies Market Report*, 10.

51. Energy Information Administration, *Electric Power Monthly with Data for August 2015*, tables 1.1 and 1.1A; percent calculations by author.

52. Alice C. Orrell and Nikolas F. Foster, *2014 Distributed Wind Market Report* (Richland: Pacific Northwest National Laboratory, 2015), 3.

53. Wiser and Bolinger, 2014 *Wind Technologies Market Report*, 3, 7, 10.

54. Wiser and Bolinger, 2014 *Wind Technologies Market Report*, 72–73.

55. Robin Brabant, Nicolas Vanerman, Eric W. M. Stienen, and Steven Degraer, "Towards a Cumulative Collision Risk Assessment of Local and Migrating Birds in North Sea Offshore Wind Farms," *Hydrobiologia* 756 (2015): 63–74, doi 10.1007/s10750–015–2224–2; Laura Ellison, *Bats and Wind Energy—A Literature Synthesis and Annotated Bibliography* (Reston, VA: U.S. Geological Survey, 2012).

56. Rocío Uría-Martínez, Patrick W. O'Connor, and Megan M. Johnson, 2014 *Hydropower Market Report* (Washington, DC: U.S. Department of Energy, 2015), 12.

57. Uría-Martínez, O'Connor, and Johnson, 2014 *Hydropower Market Report*, 22.

58. Uría-Martínez, O'Connor, and Johnson, 2014 *Hydropower Market Report*, 3.

59. Uría-Martínez, O'Connor, and Johnson, 2014 *Hydropower Market Report*, 15–16.

60. Uría-Martínez, O'Connor, and Johnson, 2014 *Hydropower Market Report*, 19.

61. Uría-Martínez, O'Connor, and Johnson, 2014 *Hydropower Market Report*, 15–16; Boualem Hadjeriousa, Yaxing Wei, and Shih-Chieh Kao, *An Assessment of Energy Potential at Non-Powered Dams in the United States* (Washington, DC: U.S. Department of Energy, 2012), vii–viii.

62. Uría-Martínez, O'Connor, and Johnson, 2014 *Hydropower Market Report*, 29.

63. U.S. Department of Energy, "Federal Incentives for Water Power," April 2014.

64. Uría-Martínez, O'Connor, and Johnson, 2014 *Hydropower Market Report*, 17.

65. Cory S. Silva, Warren D. Seider, and Noam Lior, "Exergy Efficiency of Plant Photosynthesis," *Chemical Engineering Science* 130 (2015): 151–71.

66. Clifford Krauss, "Dual Turning Point for Biofuels," *New York Times*, April 14, 2014, www.nytimes.com/2014/04/15/business/energy-environment /dual-turning-point-for-biofuels.html?_r =0, December 18, 2015.

67. Intergovernmental Panel on Climate Change, *Climate Change 2013*, chapter 2, p. 18; calculations by author.

68. Silva, Seider, and Lior, "Exergy Efficiency."

69. Haberl, Helmut, Karl-Heinz Erb, and Fridolin Krausmann, "Human Appropriation of Net Primary Production: Patterns, Trends, and Planetary Boundaries," *Annual Review of Environment and Resources* 39 (2014): 363–91.

70. Gabriel E. Lade and C.-Y. Cynthia Lin Lawell, "The Design and Economics of Low Carbon Fuel Standards," *Research in Transportation Economics* 52 (2015): 91–99; California Air Resources Board (December 2012), "Carbon Intensity Lookup Table for Gasoline and Fuels That Substitute for Gasoline," www.arb.ca.gov/fuels/lcfs/lu_tables_11282012.pdf, August 1, 2016.

71. Bruce Springsteen, Thomas Christofk, Robert A. York, Tad Mason, Stephen Baker, Emily Lincoln, Bruce Hartsough, and Takuyuki Yoshioka, "Forest Biomass Diversion in the Sierra Nevada: Energy, Economics, and Emissions," *California Agriculture* 69, no. 3 (July–September 2015): 142–49.

72. John W. Lund, Derek H. Freeston, and Tonya L. Boyd, "Direct Utilization of Geothermal Energy 2010 Worldwide Review," *Geothermics* 40 (2011): 159–80.

73. National Research Council, *Induced Seismicity Potential in Energy Technologies* (Washington, DC: National Academies Press, 2013), 60–61; Adam H. Goldstein and Ralph Braccio, *2013 Market Trends Report: Geothermal Technologies Office* (Washington, DC: U.S. Department of Energy, 2014), iv.

74. Energy Information Administration, *Electric Power Monthly with Data for August 2015*, table 6.1; percent calculations by author.

75. Ruggero Bertani, "Geothermal Power Generation in the World: 2005–2010 Update Report," *Geothermics* 41 (2012): 1–29.

76. National Research Council, *Induced Seismicity Potential in Energy Technologies* (Washington, DC: National Academies Press, 2013), 8, 60–61, 75.

77. National Research Council, *Induced Seismicity*, 156; Peter M. Meier, Andrés Alcolea Rodríguez, and Falko Bethmann, "Lessons Learned from Basle: New EGS Projects in Switzerland Using Multistage Stimulation and a Probabilistic Traffic Light System for the Reduction of Seismic Risk," *Proceedings World Geothermal Conference*, Melbourne, Australia, April 19–25, 2015, https://pangea.stanford.edu/ERE/db/WGC/papers/WGC/2015/31023.pdf, December 30, 2015.

78. Goldstein and Braccio, *2013 Market Trends Report: Geothermal Technologies Office*, 1–4.

79. Kewen Li, Huiyuan Bian, Changwei Liu, Danfeng Zhang, and Yanan Yang, "Comparison of Geothermal with Solar and Wind Power Generation Systems," *Renewable and Sustainable Energy Reviews* 42 (2015): 1464–74.

80. Goldstein and Braccio, *2013 Market Trends Report: Geothermal Technologies Office*, 29–36.

CHAPTER 11

1. International Energy Agency, *World Energy Investment Outlook, Special Report* (Paris: International Energy Agency, 2014), 11.

2. Citi GPS, *Energy Darwinism II: Why a Low Carbon Future Doesn't Have to Cost the Earth* ([London?]: Citi GPS, 2015), 38, 45.

3. Robert J. Brulle, "Institutionalizing Delay: Foundation Funding and the Creation of U.S. Climate Change Counter-Movement Organizations," *Climatic Change* 122, no. 4 (February 2014): 681–94.

4. Nuclear Energy Institute, "French Lessons: What the Paris Climate Agreement Means for Nuclear," December 17, 2015, www.nei.org/News-Media/News/News-Archives/French-Lessons-What-the-Paris-Climate-Agreement-Me, January 9, 2016; Natural Gas Council, "Solving the Climate Change Puzzle: Natural Gas—Clean, Abundant, Efficient, American," no date, www.ingaa.org/File.aspx?id=8948, January 9, 2016.

5. Renewable Fuels Association, "RFA: EPA's Final RFS Rule Puts Future of Biofuels & Climate Policy in Hands of Oil Industry," November 30, 2015, www.ethanolrfa.org/2015/11/rfa-epas-final-rfs-rule-puts-future-of-biofuels-climate-policy-in-hands-of-oil-industry/, January 9, 2016.

6. American Petroleum Institute, *Energy and Climate Change* (Washington, DC: American Petroleum Institute, 2015).

7. Citi GPS, *Energy Darwinism II*, 82.

8. Association of American Railroads, "Railroads and Coal," July 2015, www.aar.org/BackgroundPapers/Railroads%20and%20Coal.pdf, January 9, 2016.

9. John Barrasso, https://twitter.com/SenJohnBarrasso, January 16, 2016.

10. Tim Knauss, "Closing FitzPatrick Nuke Plant Could Cost CNY Estimated $500 Million a Year," *syracuse.com*, October 12, 2015, www.syracuse.com/news/index.ssf/2015/10/closing_fitzpatrick_nuke_plant_could_cost_cny_economy_500m_a_year.html, January 9, 2016.

11. Robert Bryce, "Nuclear Hypocrisy: Cuomo, Schumer's Odd Outrage over a Closing Plant," *New York Post*, November 16, 2015, http://nypost.com/2015/11/16/nuclear-hypocrisy-cuomo-schumers-odd-outrage-over-a-closing-plant/, January 9, 2016.

12. Rob Port, "With State Hit Hard by Low Oil Prices, North Dakota Universities Brace for Budget Cuts," *Watchdog*, November 30, 2015, http://watchdog.org/249275/state-hit-hard-low-oil-prices-north-dakota-universities-brace-budget-cuts/, January 9, 2016.

13. James Marson and Andrey Ostroukh, "Vladimir Putin Says Russia's Economic Crisis Has Peaked," *Wall Street Journal*, December 17, 2015, www.wsj.com/articles/vladimir-putin-warns-government-may-adjust-budget-over-oil-price-fall-1450346139, January 9, 2016.

14. Alex Epstein, *The Moral Case for Fossil Fuels* (New York: Portfolio, 2014).

15. Daniel Yergin, *The Prize* (New York: Simon & Schuster, 1991).

16. U.S. Environmental Protection Agency, "Global Greenhouse Gas Emissions Data," www3.epa.gov/climatechange/ghgemissions/global.html, January 11, 2016; U.S. Environmental Protection Agency, "Sources of Greenhouse Gas Emissions," www3.epa.gov/climatechange/ghgemissions/sources/transportation.html, January 11, 2016.

17. Union of Concerned Scientists, *Truck Electrification: Cutting Oil Consumption & Reducing Pollution* (Cambridge: Union of Concerned Scientists, 2012).

18. Clean Technia, "World's First All-Electric Battery-Powered Ferry," June 13, 2015, http://cleantechnica.com/2015/06/13/worlds-first-electric-battery-powered-ferry/, January 11, 2016.

19. Maggie Donaldson, "Two Battery-Powered Planes Have Crossed the English Channel," *Business Insider*, July 12, 2015, www.businessinsider.com/a-battery-powered-airplane-has-crossed-the-english-channel-2015-7, January 11, 2016.

20. Felix Creutzig, Patrick Jochem, and Orreanne Y. Edelenbosch et al., "Transport: A Roadblock to Climate Change Mitigation?," *Science* 350 (November 20, 2015): 911–12.

21. U.S. National Renewable Energy Laboratory, *Renewable Electricity Futures Study*, 4 vols. (Golden, CO: National Renewable Energy Laboratory, 2012).

22. Malcolm Keay, *Electricity Markets Are Broken—Can They Be Fixed?* (Oxford: Oxford Institute for Energy Studies, 2016).

23. Michael E.Webber and Sheril R. Kirshenbaum, "It's Time to Shine the Spotlight on Energy Education," *Chronicle of Higher Education*, January 22, 2012, http://chronicle.com/article/Its-Time-to-Shine-the/130408/, December 16, 2016.

24. John H. Perkins, Catherine Middlecamp, and David Blockstein et al., "Energy Education and the Dilemma of Mitigating Climate Change," *Journal of Environmental Studies and Sciences* 4 (2014): 354–59.

25. Mark Z. Jacobson, Mark A. Delucchi, and Guillaume Bazouin et al., "100% Clean and Renewable Wind, Water, and Sunlight (WWS) All-Sector Energy Roadmaps for the 50 United States," *Energy and Environmental Science* (2015), doi:10.1039/c5ee01283j; Mark A. Delucchi and Mark Z. Jacobson, "Providing All Global Energy with Wind, Water, and Solar Power, Part II: Reliability, System and Transmission Costs, Aad Policies," *Energy Policy* 39 (2011): 1170–90.

26. Andrew Revkin, "Jim Hansen Presses the Climate Case for Nuclear Energy," July 23, 2013, http://dotearth.blogs.nytimes.com/2013/07/23/jim-hansen-presses-the-climate-case-for-nuclear-energy/?_r=2, January 15, 2016.

27. National Energy Policy Development Group, *Reliable, Affordable, and Environmentally Sound Energy for America's Future* (Washington, DC: White House, 2001).

28. These concluding paragraphs are derived from John H. Perkins, Paul C. Stern, Richard E. Sparks, and Robert A. Knox, *Do you really want to do something about climate change?*, manuscript in preparation.

APPENDIX I

1. Philip Lervig, "Sadi Carnot and the Steam Engine: Nicolas Clément's Lectures on Industrial Chemistry, 1823–1828," *British Journal for the History of Science* 18, no. 2 (1985): 147–96.

2. James L. Hargrove, "Does the history of food energy units suggest a solution to 'Calorie confusion'?," *Nutrition Journal* 6, no. 44 (2007), doi:10.1186/1475-2891-6-44.

3. *Oxford English Dictionary*, definition of British thermal unit.

4. John Bourne, *Handbook of the Steam Engine*, 3rd ed. (Philadelphia: J.B. Lippincott and Co., 1870), 234.

5. Bureau International des Poids et Mesures, *The International System of Units (SI)*, 8th ed. (Paris: Bureau International des Poids et Mesures, 2006).

6. U.S. Energy Information Administration, "How Much Electricity Does an American Home Use?," www.eia.gov/tools/faqs/faq.cfm?id=97&t=3, July 7, 2016.

7. U.S. Energy Information Administration, *Electric Power Monthly for April, 2016* (Washington, DC: U.S. Energy Information Administration, June 2016), table 1.1.

8. U.S. Energy Information Administration, "How Much Oil Is Consumed in the United States?," www.eia.gov/tools/faqs/faq.cfm?id=33&t=6, July 7, 2016.

9. Iowa State University, Cooperative Extension, *Liquid Fuel Measurements and Conversions* (Ames: Iowa State University, 2008).

APPENDIX 2

1. European Nuclear Society, Fuel comparison, www.euronuclear.org/info /encyclopedia/f/fuelcomparison.htm, June 25, 2016.

Glossary

ATMOSPHERIC ENGINE: Term used to describe the engine developed by Thomas Newcomen in the early 1700s; sudden condensation of steam in a cylinder produced a partial vacuum, and the weight of the atmosphere above a piston forced the piston to move in the cylinder.

BIG-FOUR FUELS: Term used in this book to designate coal, oil, gas, and uranium, all of which are mineral fuels rather than organic fuels; the big-four fuels currently provide almost all energy services and are currently essential for human comfort, safety, and survival.

BIOMASS: Products derived from plants and animals that can be used as a primary energy source.

CHEMICAL BOND: The sharing of electrons by two atomic nuclei; during combustion, chemical bonds break and reform to make new substances and release heat and electromagnetic energy.

CLIMATE: Patterns of temperature, wind, and precipitation over a long period of time; see also *weather*.

CLIMATE CHANGE: Alteration of the patterns of temperatures, wind, and precipitation over time.

COMBUSTION: A chemical reaction in which reacting molecules rearrange atoms by breaking and re-forming chemical bonds to (a) produce new substances and (b) release energy in the form of heat and electromagnetic radiation.

CREDIT MONEY: Pieces of paper issued by a bank that promise to pay the holder of the paper a certain sum at a future date. These "bank notes" used by the holder as currency at first supplemented and then replaced coins of precious metals as money to enable commerce. The first forms of credit money appeared long before the Bank of England, but this bank significantly increased the use of credit money, which is now a hallmark of modern states based on high uses of energy.

ELECTRICITY: The presence of electric attractions and repulsions; generally pictured as the transmission of electric force in a circuit, during which the force produces energy services, e.g., motion, heat, or light

ELECTRON: A particle outside the nucleus of an atom that has a negative electric charge; electrons shared by two nuclei constitute a chemical bond between the nuclei.

ELECTROMAGNETIC RADIATION: Detected in some experiments as waves and in other experiments as particles; exercises both electrical and magnetic attractions and repulsions.

ENERGY: (a) the ability to do work, or accelerating a mass through a distance, i.e., making a physical object move faster; (b) heat, measured by temperature, which reflects the movement of atoms and molecules in material substances; (c) electromagnetic radiation, such as light. (See appendix 1.)

ENERGY CARRIER. See *Secondary energy source.*

ENERGY CONSERVATION: Actions that avoid the need for energy services supplied by primary energy sources.

ENERGY EFFICIENCY: Provision of the same energy service with less energy, or provision of more energy service with the same amount of energy.

ENERGY, KINETIC. See *kinetic energy.*

ENERGY, POTENTIAL. See *potential energy.*

ENERGY SERVICES: The practical, desired results produced by energy.

ENERGY TRANSITION: A change over time in the kinds and relative amounts of primary energy sources used for energy services; see *First, Second, Third,* and *Fourth Energy Transitions.*

ENERGY UNITS: Conventions enabling the measurement of energy; see appendix 1.

ENTROPY: In any real process, such as the operation of an engine, some energy will transform into another form of energy or work, and some will transform into a form, such as heat, incapable of performing work; entropy is the form of energy that cannot perform work.

FEED: Agricultural products, usually grains and grasses, fed to livestock.

FIRST ENERGY TRANSITION: Incorporation of fire from firewood and other biomass into daily human life for heat, light, protection from predatory animals, and cooking.

FIBER: Agricultural and forestry products, such as cotton and timber, used for clothing, paper, and other products.

FIELD: A volume of space influenced by electrical or magnetic repulsions and attractions or by gravitational attractions of a mass.

FISSION: A physical process in which the nuclei of atoms of uranium and plutonium break apart when struck by a neutron.

FOOD: Agricultural products of many types used to feed people.

FORCE: An attraction or repulsion that makes a mass change its velocity (speed and/or direction).

FOSSIL FUELS: Primary energy sources composed of the geochemically transformed remains of biomass that grew millions of years ago.

FOURTH ENERGY TRANSITION: A process currently under way to replace all or a great deal of coal, oil, gas, and uranium with renewable energy sources used efficiently.

GREENHOUSE GASES: Chemical substances that are or become part of the atmosphere and absorb infrared radiation and thus trap heat at earth's surface.

HEAT. See *energy*.

INDUSTRIAL REVOLUTION: Term applied to the rapid technological, economic, and social changes that began in England in the 1700s and replaced agrarian cultures; powered at first by coal, water power, and wind.

KINETIC ENERGY: The energy of a moving body, often expressed as $KE = \frac{1}{2}mv^2$, or ½ × mass of the object × velocity (speed) of object squared

LAWS OF THERMODYNAMICS: Scientific conclusions formulated in the mid- to late 1800s articulating the concept of energy and its behavior; the first law holds that energy cannot be created or destroyed, but it can transform from one form into another; the second law holds that during an energy transformation some energy can do work but inevitably in any real transformation process some energy will appear as heat without doing work.

MINERAL FUEL: Term coined by E. A. Wrigley to distinguish primary energy sources, such as coal, from energy sources dependent on daily and yearly fluxes of solar radiation, such as biomass, water power, and photovoltaic electricity.

NEUTRON: A particle held within the nucleus of an atom that has no electric charge.

ORGANIC FUEL: Term coined by E. A. Wrigley to distinguish primary energy sources such as solar, wind, oceanic phenomena, falling water, and biomass from the mineral fuels (coal, oil, gas, and uranium).

POTENTIAL ENERGY: If physical material is known to be able to produce energy of motion (kinetic energy), heat, or electromagnetic energy (e.g., light), then the material has potential energy.

POWER: The rate in time of energy use to do work (see appendix 1).

PRIMARY ENERGY SOURCE: A substance or process that supplies net energy to people in the form of light, heat, or physical movement; see also *secondary energy source*.

PROTON: A particle held within the nucleus of an atom that has positive electric charge

SECOND ENERGY TRANSITION: Incorporation into human culture of agriculture to produce food and feed rather than relying on hunting and gathering; also called the Neolithic Transition; began about 40,000 years before the present and now includes almost all humans on earth.

SECONDARY ENERGY SOURCE: A material or process that provides energy services but requires a primary energy source for its formation; examples are electricity and hydrogen gas; also called an "energy carrier."

STEAM ENGINE: Term applied to engines developed by James Watt and others that used expanding steam to move a piston in a cylinder; this term distinguishes these engines from Newcomen's atmospheric engine.

SUSTAINABILITY: Defined in 1986 as "[economic] development that meets the needs of the present without compromising the ability of future generations to meet their own needs"; synthesized concerns about economy, environment, and equity.

THIRD ENERGY TRANSITION: A process begun in the 1500s in England in which coal began increasingly to substitute for firewood as a source of heat;

petroleum, gas (first manufactured, later "natural," frequently found with petroleum deposits), and hydroelectricity added to the energy supplies of coal; uranium joined coal, oil, and gas in the mid-1900s.

WEATHER: The condition of temperature, wind, and precipitation at one time or during a very short period; see also *climate*.

WORK: Acceleration of a mass through a distance by a force.

Index